The Forested Land
A HISTORY OF LUMBERING
IN WESTERN WASHINGTON

Robert E. Ficken

Forest History Society, Inc.
Durham, N.C.

University of Washington Press
Seattle & London

Work on this book and its publication were supported by grants to the Forest History Society, a nonprofit, educational institution established in 1946 and dedicated to the advancement of historical understanding of the interaction between human beings and the North American forest environment. Further support for publication was provided by the Ben B. Cheney Foundation.

Library of Congress Cataloging-in-Publication Data

Ficken, Robert E.
  The forested land.

  Bibliography: p.
  Includes index.
  1. Lumber trade—Washington (State)—History.
2. Lumbermen—Washington (State)—History. 3. Forests
and forestry—Washington (State)—History. I. Title.
  HD9757.W2F53 1987   338.4'7674'09797   87-2214
  ISBN 0-295-96416-2

For Matthew

Contents

Illustrations

Acknowledgments

So many debts were incurred in the research and writing of this book that an attempt to list them all would risk unintentional exclusion of deserving institutions and individuals. Therefore, I have concluded to limit the sweep of these acknowledgments. Most of the work was supported by the Forest History Society, and I am grateful to the employees and officers of that organization. I would also like to thank Woody Maunder for initiating my interest in forest history and William G. Reed for his involvement with my endeavors in the field.

My principal research base was the manuscripts collection of the University of Washington Library. Richard Berner, Karyl Winn, and the members of their staff were unfailingly professional and encouraging. The late Arthur McCourt, Donnie Crespo, and Linda Edgerly of the Weyerhaeuser Company Archives made my many visits to that facility pleasant as well as productive. Sally Maddocks, that effervescent archivist, was of great assistance in my usage of Pope & Talbot and Simpson materials. The bibliographical guides to published and unpublished sources prepared by Ronald J. Fahl and Richard C. Davis of the Forest History Society were of enormous value in locating targets of research.

Robert E. Burke of the University of Washington read the manuscript with attention to detail and offered much sound advice. Harold K. Steen and Ron Fahl of the Forest History Society also made helpful suggestions. The observations of a panel of reviewers assembled by the society were of considerable help in revision of the original draft. Errors of fact or interpretation that may remain, of course, are the sole responsibility of the author.

Portions of the book have appeared in article form in the *Journal of Forest History, Labor History, Pacific Northwest Quarterly,* and *Western Historical Quarterly.* Since completion of the manuscript, my knowledge of the subject has been enhanced by new works from several historians: William G. Robbins on the political role of lumber trade associations, Steven Pyne on forest fires, Elmo Richardson on the work of David T. Mason, Murray Morgan on the St. Paul & Tacoma Lumber Company, and Charles Twining on the life of Phil Weyerhaeuser.

As in past historical efforts, Lorraine was a welcome and essential presence. The book is dedicated to our young son Matthew, who was absent at the creation but present at the publication.

Introduction

In the narrative that follows, this book examines the history of western Washington—the far, northwestern corner of the continental United States—from the perspective of its most important natural resource, timber. Unlike that portion of the state east of the Cascades, where lumbering has always been relatively limited in extent, and Oregon to the south, where other forms of economic activity rivaled lumbering from the beginning of white settlement, the history of western Washington cannot be understood without a thorough knowledge of the theme of forest usage. The presence of sweeping forested land between the Cascades and the Pacific was crucial to the course of human habitation in the region. Indians derived shelter, transportation, and fishing implements from the forest. Maritime fur traders, poking into unknown inlets in search of opportunity, found in the export of ship's timber a profitable supplement to their principal business. Expanding upon this interest and reflecting a change of focus from exploitation of animals to exploitation of timber, the wilderness accountants of the Hudson's Bay Company commenced the production of lumber in the Pacific Northwest in the late 1820s.

The vital linkage between forest usage and human life was most clearly demonstrated in the century between American occupation of Puget Sound and the Second World War. Settlement north of the Columbia River, in a tree-choked land believed unsuited to the type of agrarian development under way in the valley of the Willamette, centered on the exploitation of the forest. Americans at Newmarket, Olympia, Steilacoom, Seattle, and other hovel outposts of civilization derived their support from the surrounding forest. Happily, their arrival on Puget Sound coincided with the great gold rush to California, enabling the production in crude, water-powered sawmills of lumber for sale in that land of scarce building materials.

San Francisco merchants, men of New England attuned to the forest and to the sea, made the gold rush the crucial event in the early history of western Washington. Searching out sites for sawmills, they built in the midst of the forested wilderness of Puget Sound and Hood Canal an industrial empire. In bustling saltwater mill towns—Gamble, Ludlow, Blakely, Madison, Seabeck—the trees of the Northwest were transformed into lumber for the construction of buildings and wharves in San Francisco, forging in the process a crucial economic linkage between the Sound and the city by the bay. The first generation of lumber barons made Puget Sound, with only a few thousand white inhabitants, the pivot of a Pacific Ocean commerce that encompassed California and, with the early development of export trade, the ports of South America, Hawaii, Australia, and Asia.

For three decades, the western reaches of Washington Territory prospered and stagnated according to the performance of the San Francisco-owned mill companies. The territory was cut off from America east of the mountains, and lumbermen operated in isolation from political and economic events in the rest of the nation, an isolation that facilitated the gathering of vast timber holdings from the public domain. This era came to an end in the late 1880s with the completion of transcontinental railroads, the granting of statehood, the rapid expansion of population, and the rise in Seattle and Tacoma of centers of influence to rival and then surpass the mill ports. Lumber now flowed to the east over the tracks of the Northern Pacific and the Great Northern, as well as to the south and west on ocean-going vessels. Wealthy men of the Great Lakes, of whom Frederick Weyerhaeuser was the most famous example, arrived to buy timber, build mills, and divert the industry from its original San Francisco focus.

As a part of the nation by virtue of steel and lumber and constitutional bond, the timber-oriented residents of western Washington faced the challenges of modern America. Confronting the challenge of organized labor, lumbermen increased wages, improved working conditions, and sponsored reform legislation in a long effort to thwart the union movement. Confronting the challenge of conservationists, lumbermen adopted practical forestry measures to protect their timber holdings and to stave off those who would impose federal regulation. Confronting the challenge of an industry in a state of chaos, lumbermen sought to bring order from disorder. Whether undertaken by trade associations or under the auspices of the New Deal's National Recovery Administration, however, efforts to restrict production and control prices fell victim to a fundamental fact. Differences between large and small operators, between mill owners and loggers, and between debt-ridden and cash-healthy firms, compounded by personal rivalries and eccentricities, prevented agreement on general courses of action.

Down to the Second World War, lumbering was far and away the predominant economic activity west of the Cascades. Most of the region's industrial workforce was employed in the camps and mills, and the expenditures of employers and employees filtered outward to directly benefit shipyards, bankers, urban merchants, and suppliers of farm produce. The rise of aerospace, electronics, and chemicals during and after the war, however, greatly eroded the traditional importance of forest exploitation. Moreover, the rapid expansion of pulp and paper manufacturing, an offshoot of the need to efficiently utilize expensive timber resources, erected a major rival to the production of lumber. Due to the development of a more complex and less forest-oriented economic structure, as well as to the obstacles presented by the lack of extensive primary documentation for the postwar period, the detailed treatment in this study concludes with the end of the Great Depression.

As economic or aesthetic resource, as provider of day-to-day sustenance or largescale wealth, the forest has been of vital importance to the history of western Washington. An extensive outpouring of books, articles, dissertations, and theses attests to this elemental fact. Although massive in extent, this literature has always exhibited a major lack. General accounts written by so-called popular historians, while often useful, have tended to be long on anecdote and legend and short on research and interpretation. Scholars, in published and unpublished work, have dealt in admirable fashion with a wide variety of subjects, but have not attempted general histories. In this book, offering a comprehensive treatment based upon extensive research in primary sources, the reader will find the means for understanding the history of the forested land of Washington west of the Cascades.

The Forested Land

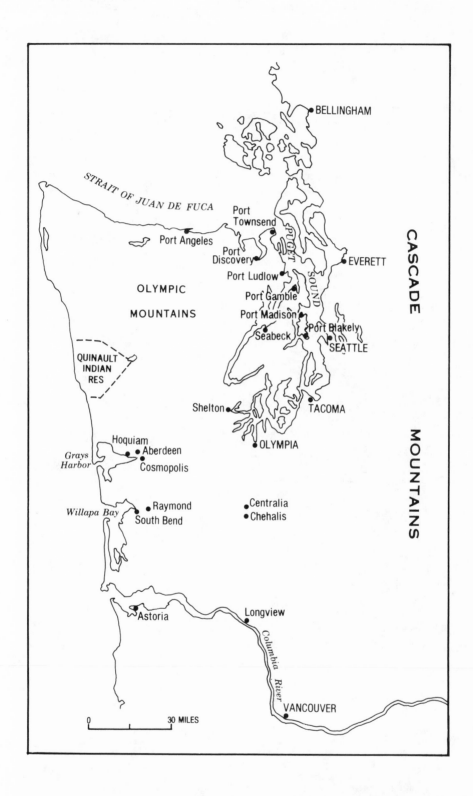

BELLINGHAM

STRAIT OF JUAN DE FUCA

Port Townsend

Port Angeles

Port Discovery

PUGET

Port Ludlow

EVERETT

SOUND

OLYMPIC

Port Gamble

MOUNTAINS

Port Madison

Port Blakely

Seabeck

SEATTLE

QUINAULT INDIAN RES

Shelton

TACOMA

CASCADE

Hoquiam

Aberdeen

OLYMPIA

Grays Harbor

Cosmopolis

Raymond

Centralia

MOUNTAINS

Willapa Bay

South Bend

Chehalis

Astoria

Longview

Columbia River

0   30 MILES

VANCOUVER

1

The Best Ever Planted

The greater part of the whole country as far as the eye can reach is closely covered with pine of several species.—David Douglas[1]

The forests of America, however slighted by man, must have been a great delight to God; for they were the best he ever planted.—John Muir[2]

Deep in the forest of western Washington, there could be found many rugged logging camps: a makeshift shed to shelter the animals, a cookhouse, and a rude bunkhouse for the men. After their evening meal, the loggers gathered in the bunkhouse to relax before turning in. Drying shirts, pants, and socks hung from wires strung in the rafters, the pungent odor mixing with that of tobacco and the occasional unwashed body. Some of the men lay in their bunks reading by candlelight, and others moved to crude tables to play what a visitor called "their everlasting games of cards."[3] And some sat around the stove in the center of the room, their feet stretched out to absorb the warmth, swapping stories. The conversation normally focused on boastings of various sexual exploits, but once in awhile the yarns were drawn from the long-ago age of myth and legend.

There was, for instance, the story of the time when Paul Bunyan, despairing over the mortal illness of his favorite blue ox, began digging a grave amidst the forests of the Pacific Northwest. Happily, Babe the ox recovered after a liberal dosage of an alcoholic concoction prepared by the Big Swede, and the gravesite was abandoned. Filled with water, the empty trench became Puget Sound. The dirt piled up to the east became the Cascade Range, and the great jagged heaps of boulders deposited on the west became the Olympic Mountains. Thus was born western Washington, or so it was claimed by the spinners of tall tales. "And loggers," noted James Stevens, chronicler of the Paul Bunyan stories, "are truthful men."[4]

The Land

Although Paul Bunyan had nothing to do with the actual creation of the region's landscape, the story is no less dramatic for the absence of human intervention. Over millions of years, geological plates shifted and the land surface was uplifted, folded, and eroded, producing a topography of magnificent natural features. The Cascade Mountains, looming like a distant blue wall in summer, and covered with snow

1

in winter, are the dominant environmental feature of Washington state. One hundred miles wide near the Canadian border, the range narrows to half that width near the Columbia River in the south. Towering above the other peaks is the ice-covered bulk of Mount Rainier, described by John Muir as "this grand icy cone."[5] The range as a whole is responsible for the abrupt climatic and vegetational transformation between western and eastern Washington, producing the heavy rainfall and dense timber cover of the former and the arid, relatively treeless environment of the latter.

West of the Cascades, in the great trough running from British Columbia to California, lies Puget Sound, Muir's "arm and many-fingered hand of the sea." During the ice ages, glaciers gouged out the northern portion of the trough in the geological equivalent of Paul Bunyan's gravedigging. As the ice retreated, water filled in the excavation, producing the principal commercial artery of the region and—with bordering forest and towering mountain—an expanse of natural beauty. Visiting Puget Sound in 1879, Muir marveled at the "broad river-like reaches sweeping in beautiful curves around bays and capes and jutting promontories, opening here and there into smooth, blue, lake-like expanses dotted with islands and feathered with tall, spicy evergreens, their beauty doubled on the bright mirror-water." To a nineteenth-century English tourist, a journey on the Sound was "like sailing on a lake in the midst of high woodland."[6]

Between Sound and sea, in the northwest corner of the state, rise the Olympic Mountains, peaks so spectacular that early geographers provided a separate name to distinguish them from the main Coast Range. Viewed from Seattle, the Olympics soar above the Sound, wrote Sinclair Lewis in an early novel, like "frozen surf on a desolate northern shore."[7] Glaciers flowed from the central massif of the Olympics into the surrounding hills, colliding with the Puget Sound glacier in prehistoric times. Rugged terrain and isolation made the Olympics one of the most inaccessible portions of the United States. Accounts of expeditions into the mountains in the closing years of the nineteenth century rival in hardship those of contemporary explorations of Africa and the Amazon Basin, even though the intrepid and often inept explorers labored only a few dozen miles from the urban centers of Puget Sound.

South of the Olympics, the main geographical features are Grays Harbor and Willapa Bay, the latter known as Shoalwater Bay until late in the nineteenth century. Harbor and bay were formed by the sinking of the coastline, submerging the mouths of rivers beneath the sea. Both offered hazards to navigation, with difficult bar crossings, extreme tidal action, and treacherous mudflats and shoals. An officer of the Wilkes expedition, surveying Grays Harbor in 1841, found that because of these obstacles, it possessed "but few facilities for commercial purposes." As for Willapa Bay, an early pilot guide warned mariners that "at low tides about one half of its area is laid bare."[8] The result of such problems was that the coastal region remained isolated for several decades after whites began moving into Washington.

These, with the Columbia River on the south and the Strait of Juan de Fuca—corridor from the ocean to Puget Sound—on the north, dominate the landscape of Washington west of the Cascades. Mountains, water, and forest distinguish the region from the remainder of the state. The joining of two disparate geographical zones into a single political unit in the mid-nineteenth century was a wondrous example of politics ignoring environment. Beginning in the earliest period of territorial government, residents on both sides of the mountains strived to rectify what to them was a gross error. Efforts to divide Washington at the crest of the Cascades, however, always failed, indicating that geographical education is not an important component in the training of politicians and bureaucrats.

The Forest

Streaming off the Pacific, marine air collides with the Cascade barrier, resulting in mild temperatures and heavy rainfall on the westside and markedly different conditions on the east, the mountains making for one of the sharpest climatic divisions in the world. Some 150 inches of rain a year falls on the coast of the Olympic Peninsula. Although lesser amounts bless the area inland, the entire region receives a liberal dampening. "Predictions respecting the weather," observed Grays Harbor lumberman George Emerson in 1906, "are not well adapted to the establishing of a reputation as a prophet, except it be that the man on all occasions predicts rain in [western] Washington."[9] Heavy precipitation combined with suitable soil conditions to produce an extensive forest cover west of the Cascades.

This forest, containing most of the state's commercially exploitable timber, extended from mountainside to water's edge. It was such a claustrophobic feature that one nineteenth-century visitor believed the only way of escaping the enveloping foliage was to book passage on a Puget Sound steamer. According to an amateur calculation, the amount of timber in the region would suffice for construction of a densely-packed city of four-story buildings forty miles long and twelve miles wide. Another estimate concluded that a street between New York and San Francisco could be lined with wooden structures, with enough lumber left over to plank the sidewalks.[10]

To all observers, the grandeur of the forest was its most striking feature. An admirer of the early lumber industry reported that the "whole wooded landscape is of such grand proportions that one looks upon the millions of broad acres covered with tall firs as a vast field of waving shrubbery." A pioneer recorded in his diary that the trees of the Northwest were "so thick tall and straight that it must be seen to be believed." Even lumbermen, usually portrayed as hardened destroyers of the environment, could be humbled by their surroundings. George Emerson, for instance, wondered if trees "talk, with their quiet murmerings" and confessed that he could not "watch an axe or a saw enter their flesh without a shudder."[11]

There was good reason for the humility. The forest was centuries old, individual trees growing and decaying with only modest interference by native inhabitants. Here indeed was evidence for the limited harnessing of the earth's resources by human beings. Two characters in one of Archie Binns's novels, for example, rest on the stump of a tree they have just chopped down and begin counting the rings. Nine inches from the outer edge they have reached back beyond the pilgrims and as the late afternoon light fades they stop counting, still not at the tree's center. The abashed settlers gather up their tools and, writes Binns, find "their way home through the dusk of prehistoric woods."[12]

The greatest of these forest trees and the one most favored by lumbermen was the majestic Douglas-fir. To David Douglas, the pioneer botanist who secured the first scientific samples of the tree and became its namesake, it was "one of the most striking and truly graceful objects in Nature."[13] Concentrated in the tidewater region of western Washington and Oregon, Douglas-fir normally grows to a height of around 200 feet, although individual trees exceeding 300 feet were occasionally found by early white settlers. Because of its sensitivity to shade, the species has branches only on the upper reaches of its straight trunks and usually grows only in extensive, pure stands.

Douglas-fir became the basis for the development of large-scale lumber manufacturing in the Pacific Northwest. The great strength of lumber produced from the tree made it ideally suited for use in heavy construction. The western portions of Washington and Oregon, containing at least eighty percent of the Douglas-fir in the United States, became the principal supplier of lumber for the construction of buildings and wharves in San Francisco and Honolulu and for the support of mining operations in Australia, Peru, and Chile.

Other commercial species became valuable only with the development of new types of forest products and with the increasing value of timberland. Western hemlock, for instance, thrives in the rain-drenched coastal areas of the Northwest, especially on the western slopes of the Olympics. Because of its rapid growth and thick foliage, hemlock often drives out the shade-sensitive Douglas-fir on tracts that have been burned or logged-off. Hemlock was a difficult tree for lumbermen. Popular suspicion of its quality as lumber resulted in a limited market and in numerous efforts to devise a new name for the species so as to disguise its identity. Not until the twentieth-century expansion of the pulp and paper industry, which possessed chemical processes for the breaking down of hemlock fibers and required a cheap supply of wood, would hemlock become commercially desirable.

Like Hemlock, Sitka spruce is concentrated in the coastal region, where rainfall is greatest and the soil is abnormally moist. The trees occasionally exceed 250 feet in height, but are usually much smaller. The commercial importance of spruce derives from a combination of straight grain, lightness, and strength. The traditional market for spruce lumber was limited, focusing on such items as boxes, musical instruments, and racing shells. During the First World War, however, spruce enjoyed a great demand because of its suitability for the construction of airplanes.

Western redcedar, in contrast to hemlock and spruce, has been utilized throughout the history of human habitation in the Pacific Northwest. Growing for the most part in damp, well-shaded areas, cedar is light and can be worked with relative ease. And it is resistant to decay, making it of especial importance for construction purposes in the soggy climate of the region. Cedar was crucial to the Indian way of life, was favored by settlers for the building of cabins, and would become the basis for the Northwest's shingle industry.

These great trees—fir, hemlock, spruce, and cedar—were found almost exclusively west of the Cascades. On the eastside, the principal species, pine, was concentrated in the northeastern corner of the state. The presence of the four major commerical species west of the mountains and the development of a giant industry based upon their exploitation provided another divisive factor within the boundaries of that geographical monster that is the state of Washington.[14]

The Indian and the Forest

One of the most egregious errors in the popular conception of Northwest history is the view that the forest remained unaffected by human intervention until the arrival of white settlers and lumbermen. For centuries, the forests supposedly matured, decayed, and reproduced themselves in splendid isolation. Such a view is totally erroneous, for the native inhabitants of the region engaged in serious exploitation of the forest, violating Alexis de Tocqueville's dictum that the "Indians occupied [the land] without possessing it."[15] Living in river valleys, on Puget Sound, and along the coast, Northwest Indians enjoyed a rich life based on fishing. This lifestyle provided time for contemplation and the devising of means for the utilization of the surrounding forest.

Wood was used for a wide variety of items, ranging from harpoon shafts, clothing, baskets, and mats to such products as gum and tea. The most important usage of wood was in the construction of houses and canoes. Here, cedar, its ease of workmanship of crucial importance in light of the relatively primitive nature of Indian technology, became a key factor in the lives of native peoples. Villages not located near stands of cedar had to rely on the occasional drift log or obtain timber through trade with more fortunate localities. James Swan, a close nineteenth-century observer of Indian life, described the laborious work of felling timber: "Their method was to gather round a tree as many as could work, and these chipped away with their stone hatchets till the tree was literally gnawed down, after the fashion of beavers."[16] The task was eased through early acquisition of axes from white traders.

Floated to the village, the logs were split into planks for house construction. The boards came "off very smooth," according to Swan, and were among the most prized Indian possessions. To assemble a house, the planks were fastened to vertical poles with crude ropes fashioned from cedar twigs. The entire process, from logging to completion of the structure, was difficult and time-consuming and may have accounted for the fact that several families shared each dwelling.[17]

Northwest Indians were quick to adopt the white style of building, but were slow to abandon their canoes, so important in a life oriented toward fishing and travel by water. Canoes were normally fashioned from a single cedar log, the interior hollowed out with wedges and chisel, and the exterior smoothed with a hand adze, all the work of a highly-skilled craftsman. "In constructing their canoes," observed Indian agent George Gibbs, "the Indians use no lines or artificial aid. The whole is modified by the eye."[18] The size of canoes varied, depending on the intended usage: fishing or whaling, peaceful travel or wartime raiding. The larger varieties were usually obtained by Washington tribes through trade with the Indians of Vancouver Island.

The small number of Indians in relation to the extensive forest cover meant that native inhabitants made only a limited encroachment on the environment. Aside from primitive logging, the Indians burned prairies and trees to encourage growth of such natural food crops as camas and to enlarge the habitat of game animals. Burning may also have been designed to enhance natural vistas, at least according to George Vancouver, who had a fine eye for landscaping. Describing a cleared area in May 1792, Vancouver noted that it was situated "on the most pleasant and commanding eminences, protected by the forest on every side, except that which would have precluded a view of the sea." In some areas, burning had a considerable impact on the makeup of the forest. By destroying underbrush and shade, it aided the growth of Douglas-fir, which might otherwise have been smothered by the more rapidly growing hemlock and cedar.[19]

Thus, Indians modified the forest environment through utilization of timber and through burning. They were not the passive occupiers of their natural surroundings, but the active manipulators of those surroundings. Most whites failed to appreciate or even to comprehend Indian usage of the forest. To the explorer David Thompson, the natives were "worthless, idle, impudent Knaves" who could not possibly possess the skill or energy to chop down giant trees. The practice of burning was generally misunderstood as an example of Indian carelessness with fire. Lumbermen took the standard view of native depravity. The potlatch ceremony, in which an Indian demonstrated social status and respect by giving away his possessions, was a source of considerable consternation. Why, exclaimed Josiah Keller, first manager of Pope & Talbot's Port Gamble sawmill, "they give away even to their wearing apparal sometimes."[20]

White explorers, traders, and settlers found the region west of the Cascades ideally suited for the exploitation of the environment. Among other wonderful features, giant trees covered the land down to the water's edge, where ships could anchor and take aboard cargoes of timber. The first whites encountered native inhabitants who had adapted themselves to the environment, albeit with differing technology and differing concepts about the relationship between human beings and nature. How the whites would yoke the environment to their needs remained to be seen.

2

So Delightful a Prospect

The country now before us . . . had the appearance of a continued forest extending as far north as the eye could reach, which made me very solicitous to find a port in the vicinity of a country presenting so delightful a prospect of fertility.—George Vancouver, 1792[1]

[Fort Vancouver] will in Two Years Hence be the finest place in North America, indeed I have rarely seen a Gentleman's Seat in England possessing so many natural advantages and where ornament and use are so agreeably combined.—George Simpson, 1825[2]

On a late Sunday afternoon in March 1778, with a storm gathering, the English vessels stood in toward shore in search of the Northwest Passage. Through his glass, Captain James Cook sighted in on a cape he named Flattery. "There is a round hill of a moderate hieght [sic] over it," Cook recorded in his journal, "and all the land on this part of the coast is of a moderate and pretty equal hieght well covered with wood and had a very pleasant and fertile appearance." Cook was forced to put to sea, it being folly to remain close to a strange coast in foul weather. The expedition again approached land off Nootka Sound on the west coast of Vancouver Island. Thus, although sighting Cape Flattery at the entrance of the Strait of Juan de Fuca, Cook missed the strait itself. "It is in the very latitude we were now in," he wrote, "where geographers have placed the pretended *Strait of Juan de Fuca,* but we saw nothing like it, nor is there the least probability that iver any such thing exhisted."[3] Cook repaired his masts with timber cut at Nootka, spent the summer exploring the northern waters, and sailed for Hawaii, there to meet his violent death. His probe toward the entrance of Fuca's strait and the loading of timber at Nootka were the first stirrings leading to exploration of the Northwest coast and to the exploitation of its forest resources.

The Spar Trade

Captain Cook believed that he had disproved the claim of Greek pilot Apostolos Valerianos—known to history as Juan de Fuca—to have discovered the Northwest Passage in 1592 while in command of a Spanish expedition sailing from Mexico. Despite this alleged discovery, Spain did not express an interest in the region to the north until the 1760s, when the extension of settlement into Alta California created the strategic need for a protective buffer. Following upon Cook's voyage, the Anglo-American involvement, in contrast, was commercial in nature. Linking the Northwest coast with China, maritime traders concentrated on the exploitation of furbearing animals, but were quick to recognize the potential value of timber. Writing from Nootka in 1786, an officer of the East India Company noted that "there is no doubt that the timber with which this coast is covered (and which in its size and fine grain is nowhere to be excelled) would compose a valuable addition to our trading, as this article carries a very advanced price in China and is always in demand there, especially such as is fit for masts and spars."[4]

As they expanded the geographical horizons of the coast region, rediscovering the Strait of Juan de Fuca and finding Shoalwater Bay, British and American merchants were not hesitant to take advantage of this handy supplement to their main business. In late June 1788, for instance, John Meares, the most energetic of the early traders, arrived off Cape Flattery to exploit the trade in timber. Sailing south along the coast, he was overwhelmed by the scenery: "The appearance of the land was wild in the extreme,—immense forests covered the whole of it within our sight down to the very beach . . . the force of Southerly storms was evident to every eye; large and extensive woods being laid flat by their power, the branches forming one long line to the North West, intermingled with roots of innumerable trees, which had been torn from their beds and helped to mark the furious course of their tempests."[5] Unhappily, his plans were soon dashed by the forces of nature and man. On the return voyage to China, Meares was forced to dump his timber cargo overboard in the midst of a storm. And two vessels left behind at Nootka were seized by the Spanish, initiating the diplomatic controversy that ended Spain's imperial venture on the northern coast.

Reacting to the Nootka incident and to the competition of American traders, the British government dispatched the *Discovery* and the *Chatham* under the command of George Vancouver in the spring of 1791. Arriving on the coast in April 1792, Vancouver spent three weeks exploring the southern shore of the Strait of Juan de Fuca. True to the aesthetic disposition of eighteenth-century Englishmen, he and his crew were impressed by the beauty of their forested surroundings. "The country before us," Vancouver wrote in his journal at Port Discovery, "exhibited everything that bounteous nature could be expected to draw into one point of view . . . I could not possibly believe that any uncultivated country had ever been discovered exhibiting so rich a picture." To Archibald Menzies, the expedition's naturalist, the countryside offered "a beauty of prospect equal to the most admired Parks in England."[6]

Vancouver also took note of the worldly value of the forest, a perception heightened when the *Discovery*'s fore-topsail yard gave way
and the spare proved to be defective. "It was a very fortunate circum-
stance," he observed, "that these defects were discovered in a country
abounding with materials to which we could resort; having only to
make our choice from amongst thousands of the finest spars the world
produces." The resources of the country and the need for repairs made
clear to the explorers what the fur traders had already discovered. The
region's trees, Lieutenant Peter Puget pointed out, "would answer for
Masts or Yards for any Frigate in the Service."[7]

From mid-May to mid-June, Vancouver explored Puget Sound, filling his journal with additional admiring references to the park-like natural surroundings. Meanwhile, Robert Gray, master of the Yankee
trading ship *Columbia,* made discoveries that would provide the most
important legal pretext for American claims to the Oregon Country.
On the morning of May 7, 1792, Gray discovered Grays Harbor. Five
days later, he sailed south along the coast and passed over shoals into
another apparent harbor. "When we were over the bar," Gray wrote,
"we found this to be a large river of fresh water, up which we steered."
Thus did Gray stumble upon the entrance to the River of the West.
Gray spent a week in the river, which he named after his vessel, trading with Indian villages and taking on wood and water before returning
to his wanderings along the coast.[8]

Learning of Gray's findings, Vancouver determined to devote the
fall of 1792 to his own investigation. Reaching the Columbia on October 19, the *Chatham* crossed the bar and came to anchor near two
American vessels. Over the next three days, Vancouver failed in repeated efforts to enter the river with the *Discovery.* Frustrated, he sailed
for California, there to regroup his expedition. Vancouver's embarrassment over his inability to pass the bar justified a harsh critique of
the Columbia's commercial purposes. The river could be entered, he
wrote, only by small vessels and then only "in very fine weather."[9]
Publicizing the hazards at the mouth of the Columbia, Vancouver pointed
to factors that would in subsequent years influence the development
of lumbering in the Northwest.

The discoveries of Gray, rather than the explorations of Vancouver,
captured public attention and symbolized the temporary passing of su-
premacy in the maritime fur trade to the Americans. Between 1790
and the end of the War of 1812, the great majority of vessels on the
coast were owned by New England merchants eager to supplement the
traffic in furs with other items that could be sold at a profit. The market
for ship's timber, moreover, expanded with the development of the
Sandwich (or Hawaiian) Islands as a stopping-off point on cross-Pa-
cific voyages. Because of the "secure harbour" at Honolulu, a British

visitor to the islands noted in 1810, and "the facility with which fresh provisions can be procured, almost every vessel that navigates the northern Pacific puts in here to refit."[10] Spars and masts could be secured only with great difficulty in the islands and there was a ready demand for timber cut on the American coast.

To men of the sea, exploitation of the forest came naturally because of their need for timber to repair vessels and because of their exposure to the ports of the Pacific. All that was needed to transform the trade in spars into a trade in lumber was the building of sawmills on the Northwest coast. Those visiting the Pacific Northwest by the overland route, in contrast, had little to say about the commercial possibilities of timber. The difficulty of transportation meant that the forest represented more of an obstacle than an opportunity to men of the land.

When the Lewis and Clark expedition reached the lower Columbia River in November 1805, for instance, its members made only perfunctory journal entries about the timber along the banks. Establishing winter quarters on the coast, the explorers set about cutting timber for construction of the post known as Fort Clatsop. "We are much pleased to find that the timber Splits most butifully," wrote William Clark. Lack of progress, though, forced Clark to appropriate "a load of old boards" from an abandoned Indian village in order to complete the shelters.[11] Over the dreary winter months, Lewis and Clark filled their journals with complaints about the rain and with detailed scientific observations about the surrounding trees. In the spring, the expedition began its long trek back to civilization to report to Thomas Jefferson about the future of the fur trade. Their concern directed toward the land, Lewis and Clark ignored the potential of timber. Plans for a concerted assault on the forest would be left to subsequent American ventures.

John Jacob Astor launched the greatest of these wilderness forays with his plan for a trading empire at the mouth of the Columbia. Astor's scheme suffered numerous setbacks. Soon after its arrival in 1811, the vessel *Tonquin* was destroyed and its crew killed in a clash with Indians on Vancouver Island. The delayed arrival of the overland portion of the expedition increased the gloom at Astoria on the south bank of the Columbia. Adding to the depression was the oppressive nature of the surrounding forest. The country, wrote Robert Stuart, is "so remarkably broken, and heavily timbered, that it is seldom possible to distinguish an object more than 100 yards [away]."[12] Hunting was impossible in such terrain and firearms quickly rusted in the constant rain. Nevertheless, the men had to shoulder guns and axes and march into the forest each morning to cut timber.

The activities of the Astorians offer a fine glimpse of logging in the wilderness. The timber party would select a suitable tree, set aside their guns, and erect a scaffold around the base of the forest giant, upon which four men would commence chopping. "At every other stroke," observed Alexander Ross, "a look was cast round to see that all was safe; but the least rustling among the bushes caused a general stop." When it was determined that Indians were not on the verge of attacking, work resumed. There was even more anxiety when the tree was about to fall. The loggers would study the situation, continued Ross, "fifty different times, and from as many different positions, to ascertain where it was likely to fall, and to warn parties of the danger." As often as not, the tree fell in an unexpected direction and sent other trees crashing to the earth, "keeping us in awful suspense, and giving us double labor to extricate the one from the other."[13]

War between Great Britain and the United States relieved the nervous loggers of their labors. Late in 1813, Astoria was ceded to the English North West Company. Following the war, the two nations negotiated a "joint occupation" agreement, allowing their citizens free access to the Oregon Country west of the Rockies and north of California. The spar trade increased in importance during these postwar years. In 1816, for example, the New England firm Bryant & Sturgis ordered one of its vessels bound for the Columbia River to "take a large number of spars between decks, which will find a good market in China." Other cargoes were carried to South America. All vessels calling at the Columbia, wrote one mariner, take "on board plenty of . . . spars."[14]

Timber was a significant factor in the campaign of Representative John Floyd of Virginia in the early 1820s to disavow joint occupation and annex Oregon to the United States. The arguments of Floyd and his expansionist supporters focused on the fur trade and maritime commerce, but also looked to the future in calling for the exploitation of the forest. "The region of country from the ocean to the head of the tide water [on the Columbia]," Floyd argued in January 1821, "is heavily timbered, with a great variety of wood, well calculated for ship building." Traders, he reflected in December 1822, "have found a profit in cutting lumber on the Columbia river, and shipping it to Chili and Peru. This trade, at no distant day, is destined to employ many individuals, and to contribute largely to the wealth of the Territory of the Oregon."[15]

Debated between 1821 and 1825, Floyd's scheme failed in the light of congressional reluctance to expend the necessary money or to antagonize England. The Pacific Northwest was so far distant, contended one opponent, that it could "never be of any pecuniary advantage to the United States."[16] In the absence of any apparent public advantage, the region would remain the preserve of private commerce. The half century between Captain Cook's arrival at Cape Flattery and the defeat of Floyd's occupation plan witnessed the development of the maritime fur trade and the linking of the Pacific Northwest with the ports of South America, Hawaii, and East Asia. Those who grasped the potential of the coastal timber resources were seafaring traders accustomed to exploiting nature and to seeking out distant markets and sources of supply. The founding of the lumber industry in the Northwest, appropriately, was linked to the further expansion and organization of the trade in furs.

Hudson's Bay Company Lumbering

With the amalgamation of the North West and Hudson's Bay companies in 1821, the latter firm assumed control of the post at the mouth of the Columbia. Late in 1824, George Simpson, resident governor of Hudson's Bay Company operations in North America, arrived on his first visit to the Pacific coast. The following spring, Simpson supervised the transfer of the company's principal depot to a site on the north bank of the Columbia eighty miles above its mouth. The location, he observed, commanded "an extensive view of the River the surrounding Country and the fine plain below which is watered by two very pretty small Lakes and studed as if artificially by clumps of Fine Timber."[17] Simpson named the new post Fort Vancouver and secured the appointment of John McLoughlin as chief factor for the Pacific Northwest.

A prime topic of concern for Simpson and McLoughlin, as activity at the new fort commenced, was how to secure the most efficient operation of the business. Of especial concern was the regrettable fact that vessels and men employed in the gathering of furs were idle during the off season for trading. Returning from a visit to Fort Vancouver in 1828, Simpson reported the solution to the company's governing committee in London: "during the dead Season of the year, . . . when little can be done on the coast, we propose employing the Vessels in carrying Timber, either to the Southern Coast wherever a market can be found for it, or to the Sandwich Islands." Such a venture, Simpson pointed out, would not only cover the company's shipping "Expenses throughout the year, but yield a very handsome profit." Expansion of the firm's business to encompass the lumber trade would also forestall a similar move by the Americans, who were increasing the exportation of spars from the Columbia. "We must avail ourselves of all the resources of this Country," McLoughlin contended, "if we have to Compete for the trade of it with the Americans as we may depend they will turn everything they possibly can to account."[18]

If the Fort Vancouver sawmill, which began operation in 1828, was to succeed in better employing the company's resources, profitable markets for lumber had to be located. "Now that our Saw Mill is in operation," Simpson wrote, "we can supply timber of various kinds in such quantity as to meet all demands either in the Sandwich Islands, or at the Spanish Missions." Information received at the fort indicated that lumber sold for between $60 and $200 per thousand feet at these locations and Simpson was confident of securing a considerable return. "That it [lumber] is an article of Trade in great demand," he stated, "there is no manner of doubt." Careful investigation, though, was required to ascertain the validity of the reports.[19]

Initially, California seemed to offer the most likely outlet. The redwoods of the Santa Cruz Mountains and the northern coast were too immense for manufacturing in the primitive sawpits that constituted the provincial lumber industry. To many observers, moreover, the Mexican residents of California were simply too lazy to exploit their own abundant natural resources and would be ready customers for lumber shipped down the coast. In addition, the export of lumber from Fort Vancouver would help to counter the expansion of Russian trading activities at Bodega Bay and Fort Ross on the California coast. Thus, there were several reasons for believing that the Mexican ports would provide the prime outlet for the Hudson's Bay Company. As James Douglas, McLoughlin's assistant, observed of the practical situation in California during an inspection trip, "timber is in fact inconveniently scarce."[20]

But in the absence of money, California trade was conducted on a barter basis. The only items available for exchange, moreover, were tallow and hides, the latter known on the coast as "California bank notes." There was little demand for these items at Fort Vancouver and trade to the southward was necessarily limited in extent. During the 1830s, the demand for lumber in California was increasingly met by Thomas O. Larkin and other American merchants residing in the Mexican province.[21] Although a Hudson's Bay Company trading station was maintained on San Francisco Bay for several years, some other market for lumber had to be located.

Fortunately, Hawaii—and especially the port of Honolulu—offered possibilities as attractive as those to the south were disappointing. The influx of missionaries and merchants in the 1820s created a demand for lumber that could not be supplied from local resources. White residents of the islands found it all but impossible to erect what they considered proper dwellings and a lucrative demand for prefabricated houses quickly developed. Here, then, was a promising market for Columbia River lumber. Honolulu, Simpson assured McLoughlin, "does afford an outlet for a considerable quantity." The Hudson's Bay Company lost no time dispatching cargoes, the vessels returning to the Northwest coast with sugar, molasses and fresh produce. "We intend to follow up the business," McLoughlin wrote in late 1830, "and supply the Owhyee Market with as many Deals [boards] as they will purchase provided the business pays."[22]

An agency was established in Honolulu and the company became one of the leading participants in the trade between the islands and the mainland. The market for lumber, though, was easily overstocked and Simpson concluded by 1842 that the "Sandwich Island market does not hold out such prospects as I was led to expect." Still, he was convinced that the trade had produced "moderate profits" for the company and had made possible the efficient handling of the firm's shipping.[23] The effort of the Hudson's Bay Company established Hawaii as a major market for Northwest lumber, a position the islands would retain for the remainder of the nineteenth century.

The success of the sawmill affirmed the shrewdness of Simpson's and McLoughlin's original perception. Even when Honolulu lumber prices were depressed, McLoughlin pointed out, they would "still . . . pay the expense of the vessel." The chief factor always had "a cargo ready to be shipped" because lumber had proved to be a more valuable adjunct to furs than such experiments as the sale of wheat to the Russian posts in Alaska. "It is evident when the facility of procuring wood is taken into consideration," he informed the governing committee, "in comparison to the difficulty of raising Grain, that we make more by employing our Vessels . . . in taking wood to Oahu, than taking Grain to Sitka."[24] The mill having proved its utility, plans were developed to improve manufacturing capacity at Fort Vancouver.

In March 1838, James Douglas, in charge of the fort during McLoughlin's advance, informed Simpson that if the mill could be operated on a steady basis, "we could furnish if necessary at least double the quantity of Lumber annually sold in the Sandwich Island market." But the constant "fracture of machinery and other accidents" meant that the saws were frequently stilled. Reconstruction of the mill, the solution to this problem, was completed in the spring of 1838. The new plant, wrote an American traveler, was "a scene of constant toil." Gangs of Hawaiian laborers were constantly at work cutting down trees and dragging logs to the mill, where the saws were manned throughout the day and into the night. Continued mechanical problems, though, gave rise to what Douglas described as "a thousand vexatious interruptions" and forced additional improvements in the mill during the late 1830s and the early 1840s.[25]

Although the output of lumber was puny by later standards, the Columbia River manufacturing center was an impressive sight to overland travelers. Samuel Parker, for instance, was traveling down the river by boat in October 1835 when: "Unexpectedly, about the middle of the day, on the north shore in a thick grove of large firs, I saw two white men with a yoke of oxen drawing logs for sawing." Four years later, an American arriving on a dark night was guided as if by a beacon of civilization by the sawmill's "dim light . . . two miles above on the northern shore." In November 1843, John Charles Frémont, paddling down the Columbia with a small party of men, "heard the noise of a saw-mill at work on the right bank." The men drew in their paddles and, "letting our boat float quietly down, we listened with pleasure to the unusual sounds."[26] The mill was the first sign of civilization encountered by travelers since their departure from the jumping-off points on the Missouri River and invariably triggered an emotional response.

Lumber manufacturing occupied an increasingly important place in the thinking of the Hudson's Bay Company's managers. George Simpson had early dismissed the possibility of expanding the company's non-fur operations into the area north of the Columbia River. That country, he noted in 1826, "is covered with almost impenetrable Forests." But for a variety of reasons—American competition on the Columbia, transportation difficulties at the river's mouth, the decline of the fur trade—Simpson soon became convinced that the firm would eventually have to transfer operations to the north and that it would have to lay more stress on economic activities other than the gathering of furs. As early as 1830, he suggested that the Pacific Northwest headquarters should be moved to Puget Sound, "where there are good harbors, fine timber, and opportunities for agriculture."[27]

In 1833, the company established a trading post at Fort Nisqually on southern Puget Sound. An extensive investigation was made in the same year of Whidbey Island in the northern Sound with an eye to locating suitable mill sites. Three years later, Simpson instructed McLoughlin to carry out a survey of the "coast and islands inside the straits of De Fuca." The survey was to concentrate on discovery of locations possessing "harbour anchorage, timber & c., for the security and convenience of shipping, likewise for trade." Operations north of the river were expanded in 1838 with the founding of the Puget's Sound Agricultural Company. This subsidiary firm established a farming settlement on the Cowlitz River, the standard overland route between the Columbia and the Sound, and assumed management of the Nisqually post. Soon, a thousand acres of land were in cultivation on the Cowlitz, and herds of cattle, sheep and horses grazed on the plain at Nisqually.[28]

The arrival of American merchants and missionaries in the valleys of the Columbia and the Willamette in the late 1830s caused the Hudson's Bay Company to accelerate the transfer of its operations. In October 1839, James Douglas took possession of the falls near the mouth of the Deschutes River—not far from the site of Olympia—on south-

ern Puget Sound. The falls, he reported, would make "a good mill
seat," as they were only a short distance from the Sound "with a clear
open passage into it and abundance of fine timber in the vicinity."
Simpson proposed that the company build a sawmill at this "most conveniently situated" spot. "There is a great deal of fine timber both on
the banks of the Chutes River and of the Sound," he informed the
governing committee, making the area ideally suited for the manufacture of lumber.[29]

Wishing to isolate the company as much as possible from American
inroads, Simpson eventually concluded that the southern tip of Vancouver Island was to be preferred to the mainland around Puget Sound.
This view led to the founding of Fort Victoria, but the company was
still not ready to abandon plans for the mainland. In March 1845,
Douglas informed Simpson that the "long contemplated plan of making
the Straits of De Fuca, the centre of a more extensive plan of operations, has at last been forced upon us." Douglas planned a trip to Puget
Sound, there to "choose a site and get a Saw Mill built, in some convenient situation, between Nisqually and Fort Victoria, from whence
we may ship wood at every season of the year."[30] Despite the abrogation of the joint occupation agreement by the United States, the company still hoped to manufacture lumber on the Sound.

More than jurisdiction over settlers was involved in the final determination of the future of the Oregon Country. The American government was well aware of the valuable resources of Puget Sound and
was determined to secure the land north of the Columbia as part of
the boundary settlement. The authorities in London, in contrast, concluded that those resources were not worth the risk of war. The region
was dismissed by the British foreign secretary as containing nothing
but "miles of pine swamp." Although insisting on retention of Vancouver Island, Great Britain gave up the mainland south of the 49th
parallel, to the disgust of Hudson's Bay Company officials. James
Douglas, miffed at the "yielding mood of the British Ministry," contended that the boundary question had been settled "on a basis more
favourable to the United States than we had occasion to anticipate."[31]

Under the terms of the treaty, the property rights of both the Hudson's Bay Company and the Puget's Sound Agricultural Company were
confirmed. American settlers, though, were quick to encroach on the
holdings of the two firms. The Fort Vancouver sawmill continued to
operate, despite the efforts of a number of Americans to file claims
on the site. One settler even occupied the mill and wrecked the machinery before being evicted. "No people can be more prejudiced . . .
than the Americans in this country," complained Douglas, "a fact so
evident to my mind, that I am more suspicious of their designs, than
of the wild natives of the forest."[32] After years of depredations against

company property, a special commission was created to adjudicate a settlement. In 1869, the commission ruled in favor of a payment of $650,000 to the company as compensation for relinquishment of its claims in Oregon and Washington. Thus ended a half century of business in that portion of the Pacific Northwest now part of the United States.

Symbolizing the end of one era and the beginning of another, the building of a sawmill on the Columbia River was one of the most significant events in the history of the Pacific Northwest. Through the introduction of industrial activity, the mill was the harbinger of immense change in the region's economy. In its selection of markets, the Hudson's Bay Company established trade patterns that would be followed by lumbermen for the remainder of the nineteenth century. And in its removal from the Columbia to the inland sea of Puget Sound, the company recognized the latter's superior navigational facilities and timber resources, a recognition that would be confirmed by subsequent lumber manufacturers. Of most importance, the mill operation demonstrated the profitability of lumbering and dramatized the shift from exploitation of animals to exploitation of timber.[33] For over a decade, the Fort Vancouver sawmill constituted the entire lumber industry of the Pacific Northwest, but it was on the verge of being supplanted by Americans.

3

Gold and Lumber

Vessels of any size . . . run in with goods & pay from 18 to 20 cts. a
foot for hewn timber [and] a man can get pay for clearing his claim.—
Alfred Hall[1]

I have often wondered what we would have done had it not been for this
find [timber], for in the course of seven weeks three of us marketed eight
hundred dollars worth of logs that enabled us to obtain flour, even if we
did pay fifty dollars a barrel, and potatoes at two dollars a bushel, and
sometimes more.—Ezra Meeker[2]

On a rainy mid-November day in 1851, the vessel *Exact* deposited a
small party of settlers at Alki Point, a wind-swept spit of sand on the
southern shore of Elliott Bay. Led by Arthur Denny, the men and women
were in the midst of erecting crude log shelters when the brig *Leonesa*
nosed into the bay. "Seeing that the place was inhabited by whites,"
Denny later recalled, "the captain came on shore seeking a cargo of
piles, and we readily made a contract to load his vessel." The Denny
party thus discovered what all early arrivals on Puget Sound recognized: in a country covered with trees, the only means of support in
the initial period of settlement was the cutting of logs to supply ships
arriving from California. This was clearly demonstrated in February
1852 when most of the Alki residents relocated on the eastern shore
of the bay. Searching for an anchorage sheltered from the wind, Denny
was motivated by the fact that as timber was "the only dependence for
support in the beginning, it was important to look well to the facilities
for the business."[3] Laying out their new village, which they named
Seattle, the settlers demonstrated the crucial importance of forest resources to the pioneers of western Washington.

Lumbering on the Frontier

As obstacles to be removed and resources to be exploited, trees were
a central facet of the frontier experience throughout the course of the
white occupation of North America. Axes and saws were as important
to pioneers advancing the line of settlement as were rifles and plows.
Following upon this pattern, lumber production in the Oregon Country
increased as large numbers of Americans began arriving in the mid-
1840s. The Island Mill Company, founded by missionaries at Oregon
City in 1842, was acquired by George Abernethy in 1846 and began

19

shipping lumber to California and Hawaii. John McLoughlin, who had retired from the Hudson's Bay Company, competed for this trade from his mill at the falls of the Willamette. In the closing years of the decade, Portland surpassed Oregon City as the leading manufacturing center on the river. By the time of the California gold rush, the two dozen sawmills in Oregon were visited on a regular basis by vessels seeking cargoes of lumber.[4]

Mill construction also took place along the Columbia River. In 1843, Henry Hunt and Tallmadge Wood built a sawmill on a bluff overlooking the river thirty miles above Astoria. George Abernethy sold his Oregon City mill in 1849 and transferred his lumbering operations to Oak Point on the lower Columbia. Oak Point was, according to a visitor, "a splendid site for a shipping and lumbering trade" and it was for many years the leading producer of lumber on the river. At Astoria, a steam-driven mill was in business by 1850.[5] These sawmills—at Oregon City, Portland, and on the Columbia River—were erected by pioneer merchants who saw in trees the means for the securing of profit. Americans with similar visions were also moving north into the isolated country around Puget Sound.

At the time of the Oregon boundary dispute, it was alleged that the Hudson's Bay Company had conspired to prevent Americans from entering the region north of the Columbia. Actually, most emigrants were in search of farms and naturally preferred the lush soil of the Willamette Valley to the tree-infested country beyond the river. On Puget Sound, contended a Protestant missionary, "there are but few places, and these are very small spots, where any thing can be raised." Settlement, noted another early Oregonian, would be slow to develop on the Sound because of "the immensely heavy growth of timber."[6] The inclination of a farming people, rather than the greed of the Hudson's Bay Company, diverted attention to the south. Rather than keeping Americans out, moreover, the firm provided crucial assistance to the first settlers on Puget Sound.

Kentucky native Michael Simmons arrived in Oregon in late 1844 and spent the winter splitting shingles at Fort Vancouver. The following spring, he examined Puget Sound and concluded that the falls of the Deschutes River were an ideal location for a settlement. In October 1845, Simmons led his wife and children and six other families to the Deschutes, where they founded the village of Newmarket. The party carried a letter from John McLoughlin to William F. Tolmie, agent at nearby Nisqually, requesting assistance. Adopting the practice followed by the company at Vancouver, Tolmie agreed to accept shingles in payment for sugar, coffee, produce, and other supplies. "They allowed us anything we wanted out of the store," remembered a Newmarket resident. The shingles were shipped to Victoria and then exported to Honolulu, a policy that was continued even when the Americans produced more than could be sold at a profit.[7]

The Hudson's Bay Company also supported the sawmill venture begun by Simmons and his neighbors in late 1847 with formation of the Puget Sound Milling Company. Simmons acquired the machinery from the company, making payment with the initial output of lumber. When production commenced in January 1848, local Indians watched in amazement as the first machine-driven business enterprise on Puget Sound got underway. Most of the lumber, taken in rafts to Nisqually, was sold to the Hudson's Bay Company. Quantities also went to settlers in the Newmarket area and to the U.S. military post established at Steilacoom.[8]

Shortly after the mill began turning out lumber, one of the original partners, Edmund Sylvester, moved a short distance to the shore of the Sound and founded the town of Olympia. "I was satisfied," Sylvester later recalled, "these Falls would be improved and would bring shipping in after lumber and they could not go above here; so that I would be right handy to a market for everything I could raise." Sylvester was joined at Olympia by Simmons when the latter sold his share in the Newmarket mill. In January 1850, the brig *Orbit* arrived in search of piling for the San Francisco gold rush market.[9] Acquired by Simmons, the *Orbit* began regular service to San Francisco, establishing a commercial linkage between the Sound and California.

California Gold

At work on the millrace of John Sutter's American River sawmill in early 1848, James Marshall discovered flecks of gold in the water. The news of Marshall's discovery, carried to all parts of the world, revolutionized conditions on the Pacific coast. Tens of thousands of eager gold seekers flocked to the Sierras with the "expectation," observed James Fenimore Cooper, "of getting rich in a month."[10] Almost overnight, the sleepy village of Yerba Buena was transformed into the bustling emporium of San Francisco, greatest of all boom towns. The discovery of gold in California was also the most important event in the history of western Washington lumbering, leading directly to the building of vast sawmills and the founding of industrial enterprise in the forested wilderness. That such a crucial event took place hundreds of miles from Washington was highly symbolic, because the resultant demand for lumber in San Francisco and the heavy northern investment of the city's merchants established Puget Sound as an economic dependency of California.

By early 1849, reports of gold in California had reached the eastern United States and thousands were on their way to the new El Dorado.

Those in the greatest hurry to reach the gold fields chose the route across the isthmus of Panama. Among the early arrivals by way of Panama were Andrew J. Pope and Frederic Talbot of Maine, who soon began operation of a lumber yard under the name Pope & Talbot. The most profitable way of reaching California, though, was the passage around the southern tip of South America. Merchants choosing this route carried a wide variety of items for which sale could be anticipated in San Francisco, with lumber, prefabricated houses, and other building materials making up the bulk of the shipments.[11] William C. Talbot, Asa Mead Simpson, and William Renton, men who would become important figures in the Pacific Northwest, brought cargoes of lumber around the Horn in this early period. Considering the demand in California, the lengthy voyage was a worthwhile investment of time and money.

All travelers arriving by sea completed their journey at San Francisco, the point of transfer for supplies and passengers from ocean-going vessels to river steamers. The most striking feature of the city, to the discomfort of new arrivals, was the shortage of housing. In a futile effort to keep pace with demand, construction went on as rapidly as lumber and laborers could be secured. Thousands of people took up residence in San Francisco each month, and houses, hotels and stores were thrown up without regard to the price of building materials. The transformation of San Francisco was, to say the least, a pleasant development for merchants in lumber.

Equally as pleasant was the need for lumber in Sacramento, Stockton, and other new cities on the inland waters. "The growth of towns in California," noted an 1850 account of life in the state, "is so rapid, that before you can sketch the last, a new one has sprung into existence." Under the pressure of this growth, the price of lumber climbed, reaching the phenomenal level of $600 per thousand feet in mid-1849. At these prices, lumber was one of the most valuable commodities in California, perhaps second only to gold itself. "Wherever the place of a board, a shingle or a brick can be filled by anything else," wrote one inhabitant of San Francisco, "no matter what, it is done."[12] Demand for most items was continually swamped by heavy importations from the Atlantic coast, but not so with lumber.

By the early 1850s, the focus of the trade in lumber began to shift in a highly significant manner. Most of the lumber sold in California had been imported from New England. But the development of a more accessible source of supply on the Columbia River and on Puget Sound foretold the rapid demise of the Yankee connection. "Oregon will supply this State," Andrew J. Pope pointed out in December 1850, "with all but pine Bds at as low rate as they can be got here from the States."[13] Henceforth, merchants in lumber like Pope & Talbot, Renton and

Simpson looked to the north, rather than to their eastern associates, to supply their yards.

In their search for heavy stands of timber and sawmill sites, the prospective lumber barons expanded their horizons far beyond the bay of San Francisco. The Santa Cruz Mountains to the south were rejected as too rugged and inaccessible, and the little-known bays of the northern California and Oregon coast were passed over as too dangerous, although some mills were erected at Mendocino and Humboldt. As for the Columbia River, the hazards of its bar crossing were so well known as to discourage all but the most reckless investors. Sea captains were known to fortify themselves with strong drink before attempting to cross the bar, while passengers often fell to their knees in prayerful thanksgiving once safely across.[14] Similar considerations caused rejection of Grays Harbor and Shoalwater Bay, the latter a name not calculated to stimulate economic development.

Then, at last, the lumber merchants came to the inland sea of Puget Sound. Here, there were safe anchorages and superb stands of timber, trees that were for all practical purposes free for the taking. Here, wrote an early booster of Washington Territory, was "one vast harbor, safe and secure." And here was the perfect location for the building of a great industry to meet the demand for lumber in California. To inexperienced observers, it seemed strange for this industry to be located so far from the principal market, considering the extensive tracts of timber along the hundreds of miles of intervening seacoast. Andrew Pope's brother, visiting from Maine, discovered that "the reason is almost imperative [as] there is not a safe harbour between S.F. and the Sound."[15] With good weather, vessels could carry lumber to San Francisco in a few days' time. Any delays were more than compensated for by the ease of navigation on the Sound and the ready availability of timber.

Pioneer Lumbermen

The location of mill sites on Puget Sound, the organization of men and capital, the building of manufacturing plants, and the acquisition of shipping required the work of several years. In the meantime, California was serviced by underfinanced pioneer lumbermen. Everyone in the Oregon Country seemed determined to ship lumber to San Francisco in the aftermath of the gold strike. Even John McLoughlin, stirring in his retirement years, planned to build a sawmill on the Strait of Juan de Fuca. The founders of Port Townsend at the entrance to Puget Sound supported themselves by cutting spars for ships bound for the Bay Area. Nicholas Delin, a Swedish immigrant, built a sawmill on the site of Tacoma in 1852. Delin's mill was typical of the early manufacturing facilities, as a contemporary description noted that it "had a fault, apparently beyond cure, of turning out boards thicker at one end than the other and sometimes thicker in the middle than at either end."[16]

Of the early lumbering centers, the most thriving was Steilacoom, located just north of Nisqually. Steilacoom was small and primitive and the residents gathered around huge bonfires to ward off the chilly gloom of the Puget Sound winter, but it was still the most impressive settlement in the region. Visiting in early 1853, Ezra Meeker found seven ships at anchor loading lumber and piles, several logging camps, and the pervasive clamor of industrial activity. "The descent of timber on the roll-ways," he remembered, "sounded like distant thunder, and could be heard almost all hours of the day, even where no camps were in sight, but lay hidden up some secluded bay or inlet." Here, Meeker believed, was the focal point for the future of Puget Sound.[17]

Although Steilacoom thrived, the best-known of the early milling ventures was located to the north in the village of Seattle. Arriving at that point in October 1852 with the intent of building the first steam-powered sawmill on the Sound, Henry Yesler found the waterfront land around Elliott Bay already claimed. Two of Seattle's founders, David Maynard and Carson Boren, however, recognized the value of a mill and in a moment of enlightened self-interest surrendered portions of their claims to Yesler. Building on pilings thrust into bayside tidal flats, he was ready to produce lumber by the following March. "It would be folly to suppose," observed the Olympia *Columbian*, "that the mill will not prove as good as a gold mine to Mr. Yesler."[18]

Yesler's mill did in fact prove an immediate bonanza for the residents of Seattle. Settlers eagerly supplied logs for the plant from their claims, those of excessive enthusiasm occasionally felling trees on top of their cabins. Most of the male inhabitants worked in the mill at one time or another. Seattle prospered and began to see itself, with notable lack of objectivity, as the leading settlement on Puget Sound. At least a dozen cargoes of lumber were shipped to San Francisco in 1854, and timber was also sent to Hawaii and Australia. "The mill runs night and day," wrote one Seattleite, "and cannot supply the demand for lumber." But prices fell the following year and the brief period of good fortune came to an end. Yesler lacked financial backing, shipping connections and a dependable sales outlet in San Francisco and could not compete with the heavily-capitalized firms established in the middle years of the decade.[19]

Seattle's pioneer lumberman at least avoided the sort of problems confronting manufacturers dependent on water power, problems that were nicely demonstrated in the case of the first sawmill built on Bellingham Bay on the northern Sound. Guided by Indians, Henry Roder and R. W. Peabody located what seemed to be an ideal site for a water-driven mill in late 1852. Peabody constructed the buildings, Roder secured machinery and a crew and everything was in readiness by the middle of 1853. But, recalled one of the employees, "as the summer advanced the water in the Creek decreased, and when the mill was ready to commence cutting, the Creek that in the middle of winter discharged nearly a million gallons a minute, did not have sufficient water to saw through a log without stopping." This rather startling development, as might be expected, "greatly dampened the ardor of the owners." Roder and Peabody lacked the capital to adequately improve the stream, lumber prices fell and the partners, their friendship put to the test, had a falling out.[20] The mill finally burned, ending one of the more futile and embarrassing lumbering ventures in the history of western Washington.

Trees were as important to those who came to the Northwest to secure farms as to those who came to build sawmills and sell lumber. Settlers arrived short of supplies and cash and faced the need to acquire at least the former while clearing their claims. Many worked for a time in sawmills or logging camps to support their families. With labor a scarce commodity, wages were attractive. Unskilled workers could earn $4 a day in the camps and a man with a yoke of oxen could expect up to $30.[21] New arrivals usually worked for a month or two before moving to their claims.

Once on their farms, settlers were "kept busy," as one recalled, "hewing down the monarchs of the forest." Those fortunate enough to live near sawmills or on tidewater could earn a considerable amount of money selling logs and piling. While significant in terms of providing support, these ventures were normally undertaken in casual fashion. Ezra Meeker, for example, set out from his claim near Kalama on the Columbia River with a raft of logs for George Abernethy's Oak Point mill. The current, though, carried Meeker past Oak Point and on to Astoria, where he sold his logs for two dollars a thousand feet above the price offered by Abernethy.[22] Such were the vagaries of pioneer industry.

Logging was an extremely dangerous occupation for settlers lacking expertise and suitable equipment. Ezra Meeker kept fires burning in holes bored into the interior of trees to fell timber. He ventured from his cabin one morning "to start afresh the fires, if perchance, some had ceased to burn." Unexpectedly, a weakened forest giant "began toppling over toward me. In my confusion I ran across the path where it fell, and while this had scarce reached the ground, a second started to fall almost parallel to the first, scarcely thirty feet apart at the top, leaving me between the two with limbs flying in a good many directions." Only the fortunate presence of heavy underbrush, which cushioned the weight of the falling trees, allowed the flustered Meeker to make "a marvelous escape."[23] The close call was an ever-present hazard.

The continuing linkage between the pioneer and the forest was demonstrated by the activities of Michael Luark, an emigrant from Indiana, who arrived in the Pacific Northwest in 1853. Luark worked for a few days squaring timber at the mouth of the Cowlitz River before crossing to Puget Sound. There, he spent three weeks as a sawyer in a Newmarket mill, earning $40 in spite of the fact that the plant often "got out of working order." With this stake, Luark located a claim near Olympia and began clearing land. From time to time, he left his farm to seek work in the mills at Newmarket or Steilacoom and on one occasion ventured as far as the high-and-dry sawmill on Bellingham Bay. Finally, Luark was employed for several weeks cutting piles for shipment to San Francisco. For his efforts, Luark received enough money to send for his family.[24]

To early Washington pioneers like Michael Luark and Ezra Meeker, forest resources were crucial to survival in the difficult initial period of locating and clearing farms. Some settlers became so entranced with the surrounding timber that they abandoned farming to become lumber magnates. Others found the pile business so lucrative that they continued for years to supply the California market. Most, however, worked in the woods and in the mills only to provide a financial basis for their farms and families. Once that was accomplished, they could stop being loggers and mill workers and become farmers.

In 1853, the land north of the Columbia River was separated from Oregon and formed into the new Territory of Washington. The formation of the new territory reflected the changing nature of the trade in lumber. The rush of gold seekers to California made possible the transformation of lumbering in western Washington. The primitive, under-financed mills of the pioneer lumbermen were succeeded by steam-driven, heavily capitalized plants, linked to California by fleets of sailing vessels. The gold rush created a large and expanding market for Northwest lumber. And it brought to the Pacific coast merchants with the knowledge and the money to see to the proper exploitation of that market. For the remainder of the nineteenth century, when residents of Puget Sound referred to "the city," they referred not to Seattle, Olympia, or other local communities, but to San Francisco. The major decisions affecting the development of western Washington would be made in paneled offices overlooking the golden bay, from which ships sailed to extract the timbered wealth of the Northwest for the building of California.

4

The Live Yankee

There are now about 7000 white inhabitants in the Territory, many of whom have commenced the lumbering and fishing business; and I know of no part of the world where there is so good a field for enterprising men, at the present time, as Washington Territory.—Business circular, 1853[1]

The active and enterprising Yankee—whose habitation is every where—is among us, and is making *his mark* in the great progress of affairs. Here he (the live Yankee) finds full scope for the exercise of his prolific genius. . . . . We learn that some 12 saw-mills have been erected on the sound by the Bostonians alone within the past 18 months.—Oregon City *Oregon Spectator*[2]

In mid-summer 1853, the schooner *Julius Pringle,* commanded by William C. Talbot of Maine by way of San Francisco, passed Cape Flattery and entered the Strait of Juan de Fuca. With his partners Andrew J. Pope, Josiah Keller, and Charles Foster, Talbot sought a harbor that was secure, surrounded by timber, and far from existing settlements, a location where a great empire of lumber could be erected without interference from outsiders. Anchoring the *Pringle* at Port Discovery, Talbot set out in a small sailboat to inspect Hood Canal and Puget Sound, accompanied in a canoe by Cyrus Walker, a former Maine schoolteacher and surveyor. On the east side of the canal, near its entrance, they found a two-mile long bay named Teekalet, entered on the charts of explorer Charles Wilkes as Port Gamble. The bay was lined with timber and featured a jutting spit of grass-covered sand upon which a mill and a town could be constructed. Arranging to load the *Pringle* with lumber at Henry Yesler's Seattle sawmill, Talbot began work on the buildings and awaited the arrival of Keller, sailing around Cape Horn with boilers and machinery.

Captain Talbot was convinced that he had made the correct decision in settling on Teekalet. "We have got the finest location on the whole of Puget Sound," he wrote Foster, "the best Harbor & best place for vessels to lay & come to." Timber was readily available on the bay or from the shores of Hood Canal. And the mill under construction would easily dominate the fast-growing lumber industry of the Puget Sound country. "I have been up the Sound & seen some of those manufacturing lumber," Talbot reported, "& if they can live doing it the way they manage I think we can make money fast enough." When the mill began production, he concluded, "we shall make more lumber

27

than all the other Mills put together on the Sound."³ Erection of the
mill commenced the era in which the industry would be managed—
and the economy of western Washington dominated—by capitalists
based in San Francisco. The mill at Teekalet, later augmented by purchase of plants at Port Ludlow and Utsalady, would be the leading
force in that California-dominated industry for the remainder of the
nineteenth century.

Merchants of Lumber

Puget Sound experienced an outbreak of sawmill fever in the early
1850s. Over two dozen mills were at work by middecade, several of
them large steam-driven plants erected by San Francisco investors. In
the spring of 1853, William Sayward built the first mill at Port Ludlow
on the western shore of Hood Canal. Captain William Renton, one of
the great figures of nineteenth-century Washington lumbering, also arrived in 1853 to build a sawmill at Alki Point on Elliott Bay. The
following year, Renton moved his mill across the Sound to Port Orchard, where he produced 10,000 feet of lumber a day. By 1863, the
timber around Port Orchard had been logged off and the captain had
lost the sight in his right eye in a mill accident. At Port Blakely, a
deserted bay opposite Seattle on the southern tip of Bainbridge Island,
Renton in that year oversaw yet another transfer of his lumbering operations, expending $80,000 on construction of a modern sawmill.⁴

Among other important operators, George Meigs opened a mill at
Port Madison, a deep-water Bainbridge Island harbor, in 1853. At Seabeck on the eastern shore of Hood Canal, the Washington Mill Company was founded in 1857 by the San Francisco merchant house of
Adams, Blinn & Company. By then, the connection between Puget
Sound and California had been firmly established. Disenchanted visitors like J. Ross Browne might disparage the Sound as "a good country for coarse lumber, and nothing more," but lumber was making
western Washington into a commonwealth of frontier industry. The
transfer of the federal customs house from the southern Sound to Port
Townsend symbolized the transformation of the region's economy,
Olympia and Steilacoom giving way to the northern timber-oriented
communities.⁵

Although the firms operated by Renton, Meigs, and the others were
large for their time, they were dwarfed by the Teekalet mill. In their
form of organization, in their methods of operation, and in their success, the owners of that mill set the standard for the rest of the industry. On December 20, 1852, an agreement was formalized in East
Machias, Maine, organizing the Puget Mill Company "for the purpose
of manufacturing lumber in the Territory of Oregon at Puget Sound."

Two-thirds of the company's capital of $30,000 was supplied by the San Francisco firm of Pope & Talbot. The remaining third was contributed by Foster & Keller of East Machias in the form of the schooner *L. P. Foster*. Under the agreement, Andrew Pope would handle sales from San Francisco and William Talbot would "select a suitable site" on Puget Sound for construction of a sawmill. Josiah Keller would acquire boilers and machinery in Boston and then "proceed with all possible despatch" in the *Foster* to the Pacific Northwest. There, he would become manager of the mill at a salary of $187.50 per month.[6]

Keller arrived at Teekalet in September 1853 to find Talbot's buildings near completion. The long voyage had been uneventful, although stowage of the unwieldy boilers on the *Foster*'s deck had caused the vessel to take on a great amount of water. "We come on a *raft of Steam Boilers*," Keller informed Foster. As work on the mill progressed, Keller and his family adjusted to the conditions of life on the Northwest frontier, especially the constant rain, which rendered everything a depressing shade of gray and flooded the structures thrown up by Talbot. "Soon our house top will begin to leak," Keller wrote on one melancholy winter day, "& under the doors it comes with a rush."[7]

Getting the mill into working order and looking after the supply of logs kept Keller's mind off such distressing aspects of life. The mill was started in late November and production thereafter expanded at a rapid pace. "We shell it out pretty fast when in our glory," Keller reported. Four million feet of lumber was produced in 1854 and in the following year fifty-two vessels sailed from Port Gamble, which gradually replaced Teekalet as the name of the town. Within months of its founding, the Puget Mill Company was the leading operation on Puget Sound. "I think we do about as much business as all the rest of the Sound," Andrew Pope boasted. The company's lumber, he contended, "stands so much better in all the Markets than any lumber made on the Coast." Only the risk of fire, he believed, stood in the way of fabulous earnings.[8]

Much of this success, in the long run as well as in the short-term, was due to the policy of investing earnings in plant improvements. The company, Pope pointed out, must "keep up with the demand, and as others are to build Mills to cut about what we now cut we want to cut more so that we can maintain the preference we have always had." Pope did not mind "using up all the earnings" because the firm "must keep moving." The most significant improvement was construction of a second sawmill at Gamble in 1858. Outfitted with two circular saws, it produced lumber more rapidly than the older mill, although the circulars turned a good deal of raw material into sawdust. Increased capacity more than compensated for this defect, according to Pope, as a circular mill would "cut Lumber cheaper than any Mill of the same cost" and would allow the company to "cut all the cargoes we can sell."[9]

Gamble's dominant position also resulted in part from the relative harmony among the partners. Like all of the firms founded on Puget Sound, the Puget Mill organization was owned by a San Francisco–based company. One associate supervised sales from California, another ran the mill, and others handled assorted tasks. For most of the companies, disagreement among these individuals was a constant source of trouble. William Renton's first partners, for instance, convinced him to build at Alki, a dangerous location because of high winds and tides. A later associate was "more sharp & tricky than capable," wrote an R. G. Dun & Company credit reporter, "& all contracts with him should be well defined." George Meigs was placed in serious financial difficulty by the mismanagement of his partner and San Francisco agent, William Gawley. According to a private report, Gawley was "a hard drinker" who had adopted "the habit of signing & endorsing notes in blank by the dozen & putting them into the hands of his Bookkeeper & others to fill up & negociate."[10]

There were, as well, some disputes between the Pope & Talbot and the Foster & Keller interests in the early years at Port Gamble. When Keller died in 1862, however, Andrew Pope and William C. Talbot acquired full control of the mill company. Believing that the mill manager should be a partner in the firm, Pope and Talbot turned to Cyrus Walker, the only employee remaining from the days of 1853. Offered a loan for purchase of a ten percent interest in the Puget Mill Company, Walker was at first reluctant to become a partner. The stereotypical Maine Yankee, he worried about going into debt, a concern confided in old age to Edwin G. Ames. "On account," Ames recorded, "of his conservative way of doing business, . . . and his objection to borrowing capital, he had to be coaxed into this enterprise." Finally giving in, Walker accepted the offer and commenced a period of four decades in which he would be the leading lumberman in the Pacific Northwest.[11] Because of the mutual respect and friendship among Pope, Talbot and Walker—a relationship ultimately furthered by Walker's marriage to Talbot's daughter—the Puget Mill Company's business was carried on without the internal quarreling that characterized most of the other lumbering concerns.

The founders of the Puget Mill Company, a contemporary observer later recalled, "belonged to that class of men who do not idly wait for something to turn up, but were full of energy and push." Through careful planning, conservative financing and good fortune, the partners built the most powerful of the early firms, the only one to endure into the modern era. By the late 1860s, Pope & Talbot was worth at least $2 million and was, attested a credit report, "the soundest & heaviest

firm on this Coast."[12] Pope, Talbot, and Walker stood at the forefront
on the San Francisco–directed transformation of Puget Sound.

Early Mill Towns

Gamble, Ludlow, Seabeck, and the other lumbering ports were, remarked an early resident of Puget Sound, "like little kingdoms, a law
unto themselves." Their splendid isolation precluded outside interference, by government agents, business competitors, or early-day labor
organizers. The lumber companies managed affairs as they saw fit,
expelling workers who caused trouble and, through the company store,
monopolizing trade with employees and local farmers. Timber, in the
absence of federal authority and large numbers of settlers, was free
for the taking. Visiting the mill towns, wrote another knowledgeable
observer, was like returning to "feudal times, these great mill-owners
have such authority in the settlements."[13]

A massive cloud of smoke trailing over the treetops informed travelers on Puget Sound that they were nearing one of the ports. The
sawmill sat next to the wharf on the waterfront. Straggling up the muddy
bluff were the buildings of the town: store, manager's house, cottages
for married workers, and hotel for single employees and visitors. Most
towns also had a schoolhouse, a church, a saloon, and a field for base-
ball, that sport having become the rage on the Sound by 1870. The
buildings were of white-painted clapboard and were often surrounded
by picket fences, presenting an ambience of New England in the drizzly
wilderness. Visiting the tumbledown ruins at Port Ludlow, one of the
characters in Archie Binns's *The Timber Beast* observes that "any one
who's been in Maine could tell where the first big-time lumber men
came from."[14]

Isolation marked the lives of residents, whether partners in the firm
or unskilled workers. Most of the ports lacked road connections with
other communities, and steamboats provided the only link with the
outside world. There were no newspapers, and the only entertainment
was provided by occasional touring shows. In all of Kitsap County,
wrote a resident of Seabeck, "the only professional men . . . were
two or three doctors, and a half dozen gamblers." Small wonder, then,
that leisure time was taken up in drinking, fighting, and gambling and
that the towns had problems with crime. When a party of men and
women hiked overland from Puget Sound to Port Gamble in 1865, they
were met, recorded one of the hikers, "with almost as much excitement
on the part of the mill people, . . . as if we had risen from the water,
or floated down from the sky, among them."[15] All diversions, however
modest, were a welcome break in routine.

Resident managers, men who were usually partners in the mill com-

pany, ruled over life in the lumber ports. Affairs were "conducted
along simple lines," Edwin G. Ames of the Puget Mill Company recalled, "one man was boss, his will was law, and the policy was to
do as he said." Still, the manager had to rule by example as much as
by fiat, and his practical authority was somewhat limited. At Seabeck,
the Washington Mill Company's Marshall Blinn worked with longshoremen to load vessels. Josiah Keller spent his days in the mill, and
then sat up much of the night working on accounts and correspondence. Blinn and Keller, the blood of puritanical Yankees coursing
through their veins, struggled without success against the preference
of employees for alcohol and amorous encounters with local Indian
women.[16] Such burdens occasionally produced outbursts of resentment
toward partners working in well-appointed San Francisco offices.

In the early years, the companies acquired only their mill and town
sites, one of the partners filing a claim in his own name under the
preemption laws. Josiah Keller, for instance, was the original claimant
at Port Gamble, which he facetiously referred to—government land
policy being designed to advance the interests of genuine settlers—as
"our farm." The lack of relevant laws for the sale of federal forest
land inhibited purchase of timber, but there was actually no need to
buy raw material. "No one can find fault with them but *uncle sam,*"
one visitor noted of those stealing public timber, "and he is *far distant.*" The mills usually contracted logging out to independent operators, agreeing to purchase logs at a specified rate and to supply towage, boom chains, and oxen.[17] Such arrangements allowed the mills
to concentrate on the manufacture of lumber and were the norm in the
industry until the end of the nineteenth century.

Concern over the stability of the workforce also marked the industry.
Initially, lumbering firms recruited young single men in the lumbering
and shipbuilding communities of Maine. Attracted by better jobs in
other mills or by opportunities in the urban areas along the eastern
shore of the Sound, these men tended to remain on the job only a short
time. Many expedients were adopted in the effort to retain the hard-
working sons of New England. Wages were set at a relatively high
rate for the time, care was devoted to serving the most palatable food,
and provision was made for amusement. In early 1859, for instance,
a dance hall was erected at Port Gamble, Andrew Pope observing that
"we have really got some very valuable men here, and if we can make
them contented by laying out a few hundred doll[ar]s I think it a good
investment." The labor shortage remained a major problem until the
Fraser River gold rush of the late 1850s brought to the Sound a large
and diverse group of men, many of whom gave up on the mines for
jobs in the mills.[18]

The lumber produced by these workers was loaded, green and wet, directly onto vessels waiting at the tiny mill wharves. "The logs are taken right out of the water and sawed, thence to the Ship," reported a captain loading at Port Gamble. Ships usually tied up with their sterns against the wharf and the lumber was loaded by Indian laborers through special ports. Lumber was piled on deck to complete loading and the entire process consumed a good deal of time. At Gamble in 1859, for example, the *Mayflower* required over two weeks to take aboard 330,000 feet of lumber.[19]

Early on, it became evident that efficient operations required the ownership of vessels. "We make so much lumber we must have vessels continually loading," Pope pointed out. With shipping constantly available, costs could be reduced to a minimum and profits obtained from cargoes on the return voyages to the mills. "By owning the Vessels," Pope observed, "we make the thing pay all round." By 1861, the Puget Mill Company controlled a fleet of eleven vessels and the other mills operated a smaller number of ships. With the coming of the great lumber firms, the Strait of Juan de Fuca was filled with the sails of ships departing for and arriving from San Francisco. Each of the lumber vessels averaged seven voyages a year to the California metropolis, providing the weather was good and the demand attractive. Because high winds made entering and leaving the strait a problem, the mills resorted to the use of tugboats for towage of lumber carriers between the ports and the open Pacific.[20]

Linking mills and markets, the fleet of schooners, brigs, and barks carried away the forested wealth of Puget Sound, leaving only the wages paid to employees and the small amount expended locally for supplies. In the early days of the industry, many territorial residents complained about the colonial status of their region. "Are the people of the Territory, generally," questioned the Olympia *Pioneer* in December 1853, "advised as to the MANNER in which our lumbering interest is controlled? Are they aware that, to a very undue extent, it is controlled by irresponsible SHARPERS and SPECULATORS, resident in Sacramento, San Francisco, and elsewhere along the coast?"[21] Reflecting a belief that the future of the region depended on the development of agriculture, this outlook ignored a number of practical considerations. The land was covered with giant trees, local capital was insufficient to develop the industry, the smallness of the territory's population necessitated the export of lumber, and most investors preferred life in the splendor of San Francisco to residence on the drizzly Puget Sound frontier. Given the circumstances, western Washington could have become nothing but a supplier of lumber of California. The evident desire of early industry critics to make the region a self-sufficient commonwealth of farmers and locally owned industry was but an exercise in wishful thinking.

The managers of the lumbering concerns were, in effect, the representatives of colonial powers, administering economic decisions made in distant San Francisco. This status was reflected in the political influence wielded by the mill companies. The lumber ports were owned by the mills and county government was manipulated with ease. Local taxes were thereby kept to a minimum and vital roads and bridges secured at the expense of the public. Soon after arriving on Puget Sound, for instance, Josiah Keller saw the necessity of becoming a commissioner of Kitsap County. "We must see that we are not required to do too much," he observed, because "we are to be the tax payers in our County." Other lumbermen occupied important positions in local government and were active in the territorial Republican party.[22] Aspiring politicians, sensitive to the power and needs of mill owners, declined to join in the early attacks on the industry.

Most of the lumber ports—Gamble, Madison, Blakely, Seabeck—were located in Kitsap County on the jutting peninsular thumb between Hood Canal and Puget Sound. This concentration enabled the mills to dominate local affairs and to occupy a leading position in the territory, for Kitsap was the richest county in western Washington. Thickly covered with timber and surrounded by saltwater, it was the focus of early industrialization. In 1870, despite a population of less than a thousand, Kitsap had the third highest property valuation in the territory. Its towns, small as they were, ranked among the leading urban areas, Port Gamble being the fifth largest community on Puget Sound.[23]

With its steam-driven mills, bustling ports, and fleets of ships, the lumber industry fastened itself upon western Washington in the years following the rush for gold in California. "A larger percentage of the inhabitants of this Territory," noted a government official, "is engaged in commercial and manufacturing pursuits than of any new State or Territory in the Union." The 1870 federal census reported that $1.3 million of the $1.9 million invested in Washington manufacturing was in lumbering, that two-thirds of the manufacturing wages in the territory were paid to sawmill workers, and that thirty-one of the thirty-eight steam engines in use in Washington powered the saws producing lumber.[24] Founded and directed from San Francisco and serving the demand for lumber in California, the industry had become the economic mainstay of the region west of the Cascades.

Foreign Trade Patterns

At best an unpredictable mechanism, the economy often deflects the plans of investors. An industry built to supply California, for instance, soon found itself unable to depend on that unstable market. The erratic nature of the gold rush economy—speculative mania followed by financial panic—caused wild fluctuations in demand. "Business is now very dull here," wrote a San Francisco merchant in mid-1853, "and goods are coming tumbling down every day in price." By the following year, the city's economic life was in a state of collapse, symbolized by the dramatic flight of the over-extended and ironically nicknamed speculator, "Honest Harry" Meiggs. The expansion of sawmilling on Puget Sound, moreover, compounded the problem of a declining market with that of overproduction. "We have shipped rather too much lumber to San Francisco," observed Josiah Keller in a fine rendition of the traditional business lament, "but could not well avoid it as we cannot always see into futurity." He calculated that, at best, the Puget Mill Company met expenses on its 1854 shipments to California.[25]

In order to stay in business, the San Francisco–based firms had to locate alternative markets in foreign ports. Almost from their founding, the new Puget Sound mills shipped lumber to Hawaii, a market that had been ignored by manufacturers during the early years of the gold rush. Lumber exports to the islands in 1855 increased four times over the level of the previous year. The Puget Mill Company moved aggressively to control the Hawaiian market, establishing connections with a Honolulu merchant house and dispatching the *L. P. Foster* on four voyages to the islands in 1855. These cargoes of lumber became an important facet of Pacific trade, mill company vessels returning to San Francisco or to the Sound with sugar and produce. Although Hawaii, like California, was troubled by economic instability, the growing population of Honolulu and the expansion of sugar plantations established the islands as the most dependable foreign outlet for Northwest lumber.[26]

The discovery of gold in Australia in 1851 initiated a wild scramble to the southern continent and produced additional overseas demand for lumber. Even before construction of the Teekalet mill, Andrew Pope had recognized that there was "quite a feeling" in San Francisco for the sale of lumber in Australia. Once the mill was in operation, cargoes were dispatched on the long voyage—two months at sea was considered very good time—to the southwest Pacific. Lumber was, in most instances, shipped unsold or on shares with Australian merchants and the initial profits were substantial. But the unstable nature of the gold rush economy combined with the shipment of excess lumber to disappoint exporters. "The news from Australia," reported William C. Talbot in early 1859, was "awfull." In addition, a limited demand in the United States for Australian products necessitated unprofitable return voyages in ballast and caused most lumbermen to avoid the market after the late-1850s.[27]

Although there was some initial enthusiasm for trade with China, the Asian mainland failed to become a significant outlet for the Puget

Sound mills. As for the west coast of South America, the mining regions of Peru and Chile required large shipments of lumber. Fabulous profits could be earned in the markets of Callao and Valparaiso. But the volatile state of demand in those ports gave pause to many mill owners. Conservative firms like the Puget Mill Company preferred to concentrate on Hawaii, where business was more reliable, and left the Latin trade to speculation-oriented competitors.[28]

Through vigorous exploitation of markets and control of shipping, the Port Gamble sawmill quickly matched its leading position in San Francisco with dominance of the foreign trade. The only cargoes lost to other firms, claimed Andrew Pope, were those which could not be loaded at Gamble. This achievement, according to some in the industry, was also the result of a ruthless approach to competitors. William J. Adams of the Washington Mill Company, for instance, accused Pope of bribing prospective customers not to send their vessels to Seabeck. When a particularly attractive order was lost to Port Gamble, Adams fumed: "Well God damn them they are welcome to her, but we will get even with them some how or other as sure as there is a God in Israel!" For all his blustering, though, Adams failed to establish a foothold in the foreign trade and, with other lumbermen, had to rely for the most part on California sales.[29] The ability of the Puget Mill Company to sell at home or abroad gave to it a considerable business advantage.

The trade patterns established in these years remained unchanged for three decades. Most of the lumber produced on Puget Sound was sold in California, while Hawaii provided the main foreign market. Large quantities of lumber were sold in some years on the west coast of South America and an occasional vessel sailed to Australia or China. Prior to the completion of transcontinental railroads to Puget Sound, the economic well-being of western Washington was more closely linked to that of Honolulu, Callao, Sydney, and Melbourne—more closely related to Hawaiian sugar and Latin American and Australian minerals—than to developments in the domestic American economy east of the Cascades. Northwest lumbermen were among the leading traders in the international emporium that was the Pacific Ocean.

The Indian Danger

As the operators of bustling sawmills set down in the wilderness, Puget Sound lumbermen perceived in the Indian inhabitants of western Washington a great danger. At Port Gamble, wrote Josiah Keller, the mill was "surrounded by these miserable though considered friendly, Clalms." A major conflict began in the fall of 1855 when fighting east of the Cascades between Indians and prospectors spread across the mountains. Seattle inhabitants sought protection in a fort built from Henry Yesler's lumber and similar fortifications were thrown up at the other sawmills. The blockhouse at Port Gamble was the "safest of any on the Sound" according to Andrew Pope. The arms available at the mill—pistols, rifles, muskets, and swivel cannon—were augmented with purchases in Victoria. "If they should show any fight in day time," Pope wrote of the local Indians, "they will meet a warm reception, but we fear they may burn us in the night."[30]

Commercial activity was endangered even more than the lives of settlers. Over half the adult white males in the territory joined the volunteers summoned to fight Indians, forcing a sharp curtailment of lumber production. "The mills above I learn have all stopped," reported Keller in November 1855, "& we may have to stop ours by & by to assist our friends above." If the natives were not promptly and soundly defeated, Governor Isaac Stevens informed the territorial legislature, "the whole industrial community [of Puget Sound] would be ruined." With so much at stake, lumbermen became vigorous supporters of the war effort. The Puget Mill Company was the leading supplier of provisions for the volunteers and the regular army contingents, accepting government scrip in payment. "If we could have these Debts guarrunteed," Pope observed, "we should not want more than one Year of such trade to make a pile." By the spring of 1856, the company had made at least $30,000 from the war.[31]

Profits, however, were only a secondary consideration. More important was the need to destroy the Indians and end the threat to the orderly conduct of production. "We are obliged to furnish supplies liberally," Keller explained, "in order to secure ourselves as much as possible in carrying on our business." The Port Gamble mill could be kept running, he contended, only if the war was brought to a rapid and successful conclusion. "If we cannot run the mill," he wrote, "it is all up for the present." Supplies must be provided, even when payment was uncertain, because the troops "should not now leave the field until matters are satisfactorily arranged."[32]

The war on Puget Sound petered out by the end of 1856, with the final battle taking place across the bay from Port Gamble. All that remained was to devise a permanent solution to the Indian problem, a solution that was both clear and bloody to Josiah Keller. "I had hoped that the necessity of destroying these miserable creatures with powder & ball would never occur here," he had written at the beginning of the war. By the end of the fighting, though, he was sure of what must be done: "My doctrine & I think it would be my example . . . would be to *shoot & hang* without regard to treaty or treaties." Moreover, the "bad white men in the Territory," defined as those who had befriended Indians, "should be killed with them."[33] Amounting to a campaign of genocide in the service of commerce, Keller's proposal was too extreme for most Indian-haters, who were satisfied with the surrender and execution of native leaders and the confinement of the tribes on reservations.

With the war at an end, Governor Stevens announced that once again "the Sound is whitening with the sails of commerce." The conflict, in the opinion of settlers, set the territory back a decade and rumors of Indian attack continued to frighten residents of Puget Sound. Although of modest dimensions compared to conflicts in other parts of the nation, the war demonstrated the precarious nature of industrialization on the frontier. It revealed, as well, the pervasive influence of lumbermen and the importance of commercial considerations in early Washington history. "I shall ever feel," a satisfied Josiah Keller wrote, "that we, 'the P. Mill Co.' have done but a duty they owed to themselves & their business, & the inhabitants of the Territory generally in furnishing supplies & necessaries for the Vol. force."[34]

Puget Sound Focus

The lumber industry established on Puget Sound in the decade prior to the Civil War differed considerably from that to the south in Oregon and to the north in British Columbia. In 1860, there were four times as many sawmills in Oregon as in Washington. But the amount of capital invested in Washington was three times that south of the Columbia River and the territory produced a much larger amount of lumber. The mills of Puget Sound, obviously, were bigger and more heavily capitalized. Although Oregon mills occasionally mounted stiff competition in San Francisco, they usually focused on local business.[35] By 1860, the Northwest lumber industry was centered on western Washington.

Lumbering north of the international boundary also failed to keep pace with the Sound. On Vancouver Island, the Vancouver's Island Steam Saw Mill Company was the first important firm, founded in 1851 by employees of the Hudson's Bay Company. A decade later, Captain Edward Stamp, agent for a London syndicate, began production of lumber at Port Alberni. At Burrard Inlet on the mainland, a sizable steam-powered mill was built in 1863. Crossing from Vancouver Island, Stamp founded the Hastings Mill on the south shore of the inlet in 1865. By that time, there were seven steam mills north of the line, but only those at Burrard Inlet were equipped for competition with the American plants.[36]

Like their counterparts in Oregon, the British operators were hampered by a lack of capital. Moreover, American tariff laws restricted shipments to California and prevented the development of export business. The value of Washington Territory's lumber production in 1869, for instance, was double the value of British Columbia exports for the entire decade of the 1860s. Despite the indignant protests of Vancouver Island and Burrard Inlet lumbermen, furthermore, American mills sold a considerable amount of lumber north of the line. British Columbia lumbering did not begin to expand until completion of the Canadian Pacific Railroad in the 1880s.[37]

Puget Sound mills dominated lumbering in the Pacific Northwest. San Francisco lumber merchants avoided Oregon because of the hazards of the Columbia River and Vancouver Island and Burrard Inlet because they were in British territory. They preferred to invest on Puget Sound, where the obstacles of nature and politics were few and the chances of profitable returns the greatest. This concentration of capital may have made the Sound a colony of San Francisco, but it also made the Sound the leading economic region in the Pacific Northwest.

A great industry had been put in place on Puget Sound. That industry made possible the economic advancement of the Sound, providing employment, markets for the produce of farmers, and trade for urban merchants. Twenty-nine of the thirty-two sawmills in Washington Territory in 1860 were located west of the Cascades. Of these, the nineteen on the Sound turned out seventy-one million of the seventy-seven million feet of lumber produced in western Washington in that year.[38] Linked to San Francisco and the ports of the Pacific by vessels laden with lumber, the industry made the Sound a vital part of the economy of the Pacific Rim. Lumbermen had built up organizations featuring sawmills, vessels, and lumber yards on the bay of San Francisco. What remained was to complement these facilities with holdings of timber.

5

The Loose Laws

Was there ever national blunder so great—ever national crime so tremendous as ours in dealing with our land?—Henry George[1]

Whatever was done was done under the loose laws and the government was to blame for the loose laws. But back of all this is the very outstanding fact that for a hundred years or more, Uncle Sam has said to the world, as the old ditty has it, "Then come along, come along, make no delay, for Uncle Sam is reach enough to give you all a farm."—George S. Long[2]

Hazard Stevens, attorney for the Northern Pacific Railroad, visited Hood Canal in the summer of 1871. His small steamer coursed through water turning from bluish green to whitecapped gray as a stiff breeze drove over its surface. Logging camps lined the banks of the canal and rafts were being assembled for delivery to sawmills. At one camp, a half-mile long chute extended down a hillside, ending at the edge of a bluff two hundred feet above a small bay. "The logs come down with terrific velocity," wrote Stevens, "leap out of the chute, clear over the beach, and strike the water with a report like a gun." Stevens landed at each of the camps, determined that the logs had been cut on public land, and ordered them seized on behalf of the government.[3] Although federal authorities had been unable to prevent timber stealing in western Washington, the Northern Pacific was determined to end the practice. Why trespassing on public land was of concern to a private corporation revealed much about the status of timber in the early decades of the Pacific Northwest lumber industry.

Timber Fraud

The sawmills founded on Puget Sound in the years following the California gold rush made no effort to acquire timber. Early on, however, the more perceptive mill owners foresaw the approaching need to support their operations with holdings of land. "I do not think the timber inexaustible," Andrew J. Pope observed in 1858. "I think in a very few years the handy timber will be hauled or burnt unless there is a large tract on the English side up towards Rusia." Loggers had made only a modest dent in the forests around the Sound, but the removal of timber from the shoreline immediately adjacent to the mills increased the cost of logging and raised the question of the future availability of raw material.[4] The problem was to find a viable method for the purchase of timber.

Public land had to be surveyed before it could be sold, but small appropriations, rugged terrain, and inclement weather greatly delayed the work of surveyors in the Pacific Northwest. Slowness aside, there was no federal legislation expressly providing for the sale of timberland until 1878. This was somewhat surprising in that one of the traditional aims of government land policy was to raise revenue. But the other historical purpose was to encourage settlement in the west. The latter objective, combined with the nation's agrarian ethos, meant that laws were designed to accommodate settlers, rather than lumbering and other business enterprises. The land of western Washington, complained Andrew Pope, was "good for nothing only for timber." Farmers, he asserted, "would starve on such land." Yet government policy was directed toward those attempting to hack farms out of the forested wilderness, a misdirection that would ultimately force widespread fraud on Northwest lumbermen.[5]

Thanks to local officials, there was in the early 1860s a brief opportunity for mill owners to purchase timber. The federal government had reserved land in Washington for the support of a territorial university. Such lands were normally sold only after the achievement of statehood, but a special commission headed by the Reverend Daniel Bagley began selling the land set aside for the University of Washington in 1861. This action, Josiah Keller reported from Port Gamble, represented "a good opportunity for us." Purchasers were allowed an unlimited amount of land and were apparently permitted to make their own selections from government holdings at a price of $1.50 an acre. Lumbermen were the principal beneficiaries of the Bagley program. Acquiring its first timberland, the Puget Mill Company purchased over 7,000 acres, one-seventh of the total sold. George Meigs, a member of the university board of regents, bought 3,300 acres, Marshall Blinn acquired 2,300 acres for the Washington Mill Company, and smaller amounts of land went to other firms.[6]

To many observers, the close connection between lumbermen and public officials and the heavy sales to the mills indicated the existence of a conspiracy to defraud the government. The university land sales, a Seattle newspaper charged, were "characterized by gross extravagance if not downright fraud." A federal investigation, based on allegations that Bagley had exceeded his authority, appeared likely, especially as his records were in sorry shape and gave no accurate picture of the sales made. The political strength of lumbermen, however, was apparently sufficient to prevent invalidation of the transactions. "I can see no reason," wrote the commissioner of the General Land Office, "for subjecting to the penalties of the law, honest purchasers . . .

because of error committed by others legally authorized to sell as Agents of the Territory." Bagley and the mill owners, the commissioner concluded, had "no doubt, acted in good faith."[7] This judgment, from the available evidence, seems incorrect. The friendly relations between Bagley and lumbermen, the role of Meigs in determining university policy, the departure from standard practice with regard to territorial sale of public land, and the heavy purchases made by the mills suggest the existence of an agreement to provide funds for the university and timber for the industry.

There were additional means—some legal and some not—of acquiring land during this period. Responding to demands from Puget Sound residents and eager to secure revenue for prosecution of the Civil War, the government offered for sale nearly three million surveyed acres in the summer of 1863. Land not sold at the initial auction was available for purchase in unlimited amounts for a minimum price of $1.25 an acre. Over 50,000 acres—"an endless amt. of timber land" according to a credit report—were acquired by the Puget Mill Company and the Port Madison and Seabeck mills were also heavy purchasers. Cash sales declined rapidly in importance after 1870, as the best surveyed land was sold and much of the remainder was included in the Northern Pacific land grant.[8]

Fraudulent use of the Homestead Act, the government's principal means of encouraging settlement in the west, was also resorted to by lumbermen. Although the mills purchased some claims from genuine settlers, most were acquired through the use of dummy entrymen. The number of such purchases is impossible to determine, but it would seem that progressive-minded historians have exaggerated the extent of fraud. Procedures under the Homestead Act, for one thing, were cumbersome. Moreover, the alternatives of legal purchase under the cash sale system and, after 1878, of the easily-manipulated Timber and Stone Act reduced the importance of the Homestead Act.[9] Besides, prior to the 1870s, there was little immediate need to acquire timber.

Federal laws prohibiting the removal of timber from public land were simply ignored by loggers. Arriving on Puget Sound in the summer of 1861, the new United States attorney, John McGilvra, discovered that "there is now and has been for years past great and continued trespass upon the government lands, in the cutting and removing of timber by Mill Companies and those who supply said Companies with logs." The Sound mills, he reported, "are almost wholly supplied with logs from the government lands." McGilvra quickly secured the indictment of Josiah Keller, Marshall Blinn, and several other lumbermen for theft of timber. Returning from court, Keller informed his partners in the Puget Mill Company that he had been "Indighted for $10,000, which if we are eventually convicted for, the fine will be three times the amt. . . . . [If] in the end we are obliged to pay *$30,000 or $40,000* & costs, you will probably find it out."[10]

A man of considerable ambition for political office and personal wealth, McGilvra soon came to the realistic conclusion that achievement of these goals in Washington Territory would be hampered, if not prevented, by a campaign against the lumber industry. A "vigorous prosecution" of lumbermen, he now determined, would force the sawmills to go out of business. The mills, moreover, were "undoubtedly of much advantage and benefit to this sparsely settled region and to stop these mills would injure the prosperity and prospects of the Territory very much." Whether or not the accommodation between McGilvra and the mill owners involved more than his recognition of reality is difficult to determine. In a surviving letter, though, Cyrus Walker notes that "Mr. McGilvery is evidently in the market and I think he puts to high a price on his services," suggesting a corrupt relationship between the federal prosecutor and the industry.[11]

Those indicted for trespass in 1861 received only modest fines. Edwin G. Ames of the Puget Mill Company later described the court proceedings, as recounted to him by Walker. "At a time agreed upon," Ames wrote, "they appeared in court, plead guilty to the charge of stealing Government timber, were fined a nominal fine, and sentenced to one day in jail, and I have heard that after the sentence was passed that with the judge and the sheriff, and a box of cigars, they went into jail, turned the key, lighted their cigars, staid a few minutes, and then went out, having satisfied the law."[12] Such proceedings solved the immediate problem to the satisfaction of all concerned—especially if the cigars were of high quality—but McGilvra was determined to arrange a permanent agreement between the government and the mill owners.

McGilvra concluded that the wisest course was for the government to assess a charge for timber removed from its lands. "It seems to me," he wrote, "that for the present at least a *rate of stumpage* should be established for this section & that it should be such an one as will insure to the General Government something more than a nominal revenue." Supported by land officials in Olympia, McGilvra's recommendation was accepted in the nation's capital. "Where parties will comply with such terms," Land Commissioner J. M. Edmunds agreed, "it would be better to accept them as a basis of compromise, than to go into the Courts." A fee of fifty cents per thousand feet was ordered to be assessed, beginning a period in which the government, in roundabout fashion, went into the business of selling logs to the mill companies.[13]

In the opinion of land officials, McGilvra's supervision of the fee system left much to be desired. He exacted, after meetings with mill owners, a stumpage fee of fifteen cents, ignoring the fifty-cent rate contained in his instructions. Moreover, while McGilvra set the fee and saw to its collection, he relied on figures submitted by the mills to determine the amount of timber cut, submissions that no doubt greatly underestimated actual removals. And he tried to assert complete control over federal land on Puget Sound, despite repeated orders to restrict his activities to cases of trespass before the courts.[14] McGilvra's actions established a pattern in which federal attorneys, normally ambitious politicians, worked to undermine the efforts of the General Land Office to combat fraud in the territory.

When McGilvra left office in 1865—to devote himself to a career in land speculation, often in association with Cyrus Walker—the stumpage assessment procedure was firmly established. Subsequent debate focused on the size of the fee rather than on the wisdom of the policy. In mid-1865, for example, the rate was increased to $2.50 per thousand feet. Under pressure from shocked lumbermen, the land office reduced the fee to $1 in early 1866 and at the end of the year made a further cut to twenty-five cents. A more assertive land commissioner restored the $2.50 assessment in late 1868 and ordered officials in Olympia "hereafter in no case to compromise with any trespasser, unless there are found to exist such extenuating circumstances, as to satisfy you that the party was acting in good faith and with no intent to defraud the government."[15]

Lumbermen objected to the imposition of large fees and outright theft continued to be widespread. "From 1860 to 1870," wrote an informed observer of life on Puget Sound, "men logged wherever they pleased on government land, and this without settlement or purchase or interference from the government." An investigation concluded that $40 million worth of timber was stolen on Puget Sound by the late 1870s and that the Puget Mill Company was the principal culprit. Officials in Washington, D.C. might strive to enforce the law, but their efforts were undermined by unenthusiastic and often corrupt subordinates in the Pacific Northwest, as well as by the sheer magnitude of the task. "The perpetration of innumerable frauds," admitted the land commissioner in 1875, was "among the traditions of this office."[16] This tradition was for a time abandoned when a powerful corporation decided that self-interest required an end to the theft of the public's timber resources.

The Northern Pacific's War on Fraud

In 1864, the federal government agreed to subsidize construction of the Northern Pacific Railroad from the Great Lakes to Puget Sound. For each mile of track laid through states, the railroad would receive the alternate sections of land in a twenty-mile wide strip on either side of the line. In territories, the strips would extend to forty miles. By 1882, Northern Pacific holdings in Washington Territory were reported as 7.7 million acres, an empire of land equal to Massachusetts, New Jersey and Maryland combined. Of this total, two million acres consisted of commercial timberland.[17] Although construction of the railroad encouraged emigration to the Pacific Northwest and made possible the shipment of commodities—wheat, lumber, shingles—to the east, the Northern Pacific's policies antagonized many residents of the region. The operators of the Northern Pacific realized a fundamental truth: the hauling of passengers and freight was not the only means of making money from railroading.

Among other things, the tangential benefits of railroad construction brought the Northern Pacific into the fight against removal of timber from unsurveyed public land. In locating an exact route from the Columbia to Puget Sound, engineers tried to lay track through the most heavily timbered areas, so that valuable timberland would be included in the land grant. Illegal logging on these lands, it was clear to Northern Pacific management, threatened to reduce their ultimate value. "It is reported that there is much saleable Timber being stolen from Government Lands on the Sound," wrote railroad official J. W. Sprague, "and of course the 'odd sections' that will be acquired by this Co. will be stripped with the rest." The land would not be transferred to the Northern Pacific until it was surveyed, Sprague pointed out, "but the loss of the Timber would in the end be our loss."[18] Thus did the railroad become the self-appointed protector of the public domain.

In April 1871, Sprague ordered Hazard Stevens, the Northern Pacific's Puget Sound attorney and the son of Washington Territory's first governor, to "ascertain the amount of stealing that is going on along the Sound where we are interested . . . and take steps as seem to be necessary to stop it." Sprague also requested that Captain B. B. Tuttle, the federal timber agent in western Washington, aid Stevens. Since Tuttle did not receive a sufficient salary from the government, Sprague agreed to pay "a liberal price" for his services. In the past, the agent had exercised his responsibility to prevent timber theft with a peculiar discretion. Noting that he never raided camps cutting logs for the Puget Mill Company, a knowledgeable observer of the industry wrote that "it was supposed he was *bribed*." Among loggers, continued Eldridge Morse, Tuttle's activities were regarded as "simply a thieving arrangement for the benefit of himself and his friends." In 1871, then, Tuttle sold himself to the railroad in time-honored fashion. "His per diem and reasonable expenses from Government," Sprague observed, "together with the large monthly allowance from this Co. ought to make him zealous in discharge of his duties."[19]

The General Land Office was happy to accept the assistance of the railroad. U.S. attorney Leander Holmes, following in the tradition of John McGilvra, had longed worked to hamper land office enforcement efforts and posed the principal obstacle to the Northern Pacific's plan. Stevens held a lengthy meeting with Holmes and secured the latter's half-hearted promise "to prosecute certain trespassers on public timber if he was furnished with information and witnesses." Holmes would take no action to seek out violators of the law. But, to facilitate Stevens's task, he appointed the railroad lawyer a deputy U.S. marshal and provided a stack of blank subpoenas for serving on persons with knowledge of timber theft.[20]

Accompanied by Tuttle, Stevens toured Puget Sound logging camps north of Seattle in August 1871. On orders from Stevens, Tuttle seized logs cut on public land and ordered witnesses to trespassing to appear in federal court. "I think that Capt. Tuttle's seizures of logs," wrote the attorney, "have caused a general feeling among loggers that they must leave, and keep off of, Govt. land." A similar practice was followed on Hood Canal, where timber stealing was most active. Loggers, of course, were angered by the seizure of rafts and there were tense scenes at each of the camps visited. Tuttle, reported Stevens, had "shown them as much leniency and excited as little violence as possible under the circumstances." By the end of 1871, Tuttle had seized, ostensibly on behalf of the government, three million feet of logs. This figure represented, according to Stevens, "one fifth of the amount cut on public lands, during that period." Even this limited success, however, was a considerable advance over previous enforcement efforts of the General Land Office. "I am much pleased with your efforts in Timber matters," Sprague congratulated Stevens, "and must say you are making it lively for the Timber Thieves."[21]

Logs seized by Tuttle were sold at public auction, allowing the railroad to put into play its most effective tactic. Operators who had removed timber from federal land had to reacquire their rafts at the auction or be forced into bankruptcy. To the surprise of no one, loggers agreed among themselves not to submit rival bids. Attending the auctions as private attorney rather than deputy marshal, Stevens countered this ploy by running up the prices on rafts so as "not to let the loggers who cut them bid them in except at ruinous rates." Such actions, as might be expected, aroused considerable hostility toward the Northern Pacific. "Public opinion is against the course of the Company in this matter," Stevens recorded, "and much bitterness is felt and manifested on the subject." But by threatening loggers with ruin, Stevens erected a strong deterrent to further removals of timber.[22]

Although close supervision of the camps continued, Stevens moved toward an accommodation with loggers in 1872, one that would allow them to operate without damage to the value of the Northern Pacific land grant. In doing so, Stevens undertook a remarkable usurpation of authority. The General Land Office credited the railroad with being responsible for the reduction of theft, but antagonism had developed over such actions as the stationing of a deputy timber agent at Seabeck without authorization from the nation's capital. Of more serious import, Stevens and the railroad began to sell government timber. For a deposit of $100 and a rate of 50¢ per thousand feet, loggers could operate on federal land without fear of visits from Tuttle. Sprague believed that such sales were entirely proper, as they would "aid much in preservation of Government Timber and . . . will Cause no loss to Government." The General Land Office, however, took the position that a private corporation had no right to sell public property.[23] When the government refused to recognize the sales, Sprague and Stevens found themselves engaged in what amounted to a form of extortion, selling protection against the seizure of logs.

Hazard Stevens himself was eventually charged with stealing from both the government and the railroad. Sprague informed the attorney in June 1873 that allegations had been made "that you and . . . Tuttle have been defrauding this Co. & the Government out of a large amt of money in Timber matters." A portion of the funds received from the sale of timber and seized logs apparently had not been forwarded to the Northern Pacific or to the land office. Charges were never brought against Stevens—the railroad being unwilling to make public its own involvement in the improper sale of federal timber—but he was dismissed from his position as Northern Pacific attorney.[24] The firing of Stevens ended the rather unique effort of the railroad to protect the public timberland from the designs of lumbermen and loggers.

The Northern Pacific succeeded where the government had previously failed because it had only one goal, the protection of the most valuable resource on the land to be granted to the railroad. The work of Hazard Stevens brought to an end the era in which public timber could be cut with impunity. This was especially so because the government itself stepped up enforcement efforts during the remainder of the 1870s. Under Carl Schurz, who became secretary of the interior in 1877, the responsibility for apprehending trespassers was transferred from easily corrupted local representatives of the government to special agents dispatched from Washington, D.C. Timber stealing, at one time accepted as inevitable, increasingly was treated as a crime. Although trespassing continued, most loggers confined their operations to private lands.[25] With large areas closed to logging, lumbermen were more than ever interested in the acquisition of land.

The Timber and Stone Act

Reformers of the time unhesitatingly called for the outright sale or lease of public timberland as the ideal solution to the problems of all concerned. Legislation expressly designed to facilitate the lawful transfer of timber, went the argument, would enable effective prosecution of those continuing to operate on public land. Substantial revenue would be earned, corruption and waste would be ended, and the legitimate interests of businessmen would be served. Since lumbermen wished to purchase timber, the dictates of reform and business seemed to coincide. Unfortunately, the impact of what Bernard DeVoto later termed "imbecile laws" continued to prevent the working out of a rational approach to the timberland question.[26]

Legislation authorizing the sale of timberland on the Pacific coast was introduced in 1871 by Selucius Garfielde, the delegate from Washington Territory. Although this bill was ignored, other timber laws soon won adoption. The Timber Culture Act, passed in 1873, granted 160 acres of land to individuals planting at least a fourth of that acreage in timber. Based on the theory that forest growth would increase rainfall in semi-arid regions, the act was utilized in eastern Washington by farmers and grazers. The Timber Cutting Act of 1878 allowed timber removal for mining purposes, but was of little use to lumbermen.[27] A second measure approved in 1878, however, dealt specifically with commercially exploitable timberland.

The Timber and Stone Act of that year allowed inhabitants of Washington, Oregon, California and Nevada to acquire up to 160 acres of timber or mineral land at a minimum price of $2.50 per acre. The law's principal purpose, according to supporters, was to end fraud, to "make," as Senator Aaron Sargent of California put it, "an honest business of lumbering."[28] But the restriction of sales to individuals and the acreage limitation meant that the act, at least on the surface, could be of little practical use to large mill companies. Here was another example of the way in which land policy had to be expressed in terms of an appeal to settlers. The main sponsors, Sargent and Senator James Kelly of Oregon, represented important lumbering states and no doubt were aware of the potential uses of the act.

Under the Timber and Stone Act, the only way for lumbermen to acquire large holdings was to engage in subterfuge. Rather than ending fraud, the act increased the sway of corruption, and land reformers soon loudly demanded its repeal. As perceptive critics had foretold, the law could be manipulated with little difficulty. Although not ideally suited for their purposes, lumbermen were able to make substantial purchases under the law.[29]

There was, at first, no great rush to take advantage of the act. By the end of 1881, only 21,000 acres had been claimed in Washington Territory. The hesitancy of lumbermen, though, was short-lived and the Timber and Stone Act soon became the principal vehicle for the assembling of land. Over 216,000 acres were purchased in the territory under the act in 1882 and 1883, the acreage transferred in the latter year representing a third of the national total. By the mid-1880s, the demand for lumber and the population of western Washington were on the increase and timber was a valuable investment. "We think there will be a great increase in sales of timber lands steadily from now on," Charles Holmes, co-owner with William Renton of the Port Blakely Mill Company, observed in March 1882. Timber, he continued, was now "the best property to hold, on this coast & certain to largely increase in value." At Port Blakely, Renton promised to invest all surplus funds in timber and reported that "all the choice Lands within a reasonable distance of the water would soon be taken up."[30]

To the everlasting fortune of the Puget Sound mill firms, the provisions of the Timber and Stone Act were easily circumvented. There was no need to prove residency or improve land, as in the case of the Homestead Act, and claims could be perfected in sixty days. The major lumbering operations maintained agents in Olympia, location of the territorial land office. Often well-connected former government employees, the industry agents were responsible for locating the best timber and for seeing to the filing and perfecting of claims. Persons often applied to be entrymen in much the same fashion one would apply for a job in the sawmill. Although lumbermen attempted to maintain the fiction that those filing on land had no connection with the mills, the actual process was known to all but the most uninformed. In November 1881, for instance, a Seattle newspaper reported the departure on the steamer *Yakima* of "a crowd of men for the Puget Mill Company up to Olympia on land office business."[31]

Great expanses of land were secured at prices so low that lumbermen marveled at their audacity in taking advantage of the law. Between $50 and $125 was normally paid to entrymen for their vital if brief services. Fees for witnesses, commissions and the price of the timber itself increased the outlay. "Our Lands cost us on an average," one of Renton's assistants informed Holmes, "less than $600 for a claim" of 160 acres. Land acquired in a completely upright manner, the same writer noted, would "sell as high as 2,500 to 3,000 [dollars], when handy to water and well timbered."[32] The difference amounted to an unintended government subsidy for wealthy firms and individuals.

Responding to reports that as many as three claims in four filed under the act were fraudulent, the General Land Office suspended entries in Washington in 1883 and again in 1885. As the heaviest purchasers, the Port Blakely and Puget Mill Company operations were the

principal targets of government investigators. "We are informed," John
Campbell of the former firm recorded in mid-1887, "that the land Officials are around hunting up Parties we bought the Lands from, and
getting their affadavits to the effect that they were paid by us to enter
these lands for the Mill Co." Several courses of action were taken to
counter this threat. The dismissal of reform-minded Land Commissioner William A. J. Sparks was sought. To disguise the identity of
the ultimate purchasers, deeds were passed from entrymen to the mill
companies through third parties. Land titles were transferred from the
mills to individuals or to new companies especially created for the
purpose of owning land. Urging speedy action on these matters, William H. Talbot, the president of Pope & Talbot since 1881, noted that
"it is a very serious question & we must act as soon as possible & get
things into shape."[33]

Outright bribery, though, was the principal weapon wielded by lumbermen. In the summer of 1883, for example, Interior Department agent
T. H. Cavanaugh was sent to Puget Sound to investigate fraudulent
entries. William Renton informed Cyrus Walker that the agent was a
close friend of Seattle attorney H. G. Struve and that several of the
mills had paid Struve "to have him use his influence to present the
matter in as favorable a light as possible to the Agent whom he (Struve)
thinks can be fixed." Subsequently, a Renton assistant wrote that Cavanaugh had "been entertained by Gamble and us and we know him to
be all right." Lumbermen took a rather interesting view of such transactions, one that made themselves the innocent victims of corrupt public officials. Will Talbot, for one, wished to "forever put to rest the
thieving of Government agents in blackmailing large timber claim
owners."[34] Mill owners had to maintain a self-image of honesty and
integrity, blaming others for coercing them into the commission of
unethical acts.

The most complex of such schemes involved an effort to persuade
William F. Vilas, secretary of the interior in the Cleveland Administration, to revoke the suspension of Timber and Stone entries and ap-
prove contested land patents. In the spring of 1888, the Northern Pacific requested that lumbermen contribute to a fund being accumulated
to influence the U.S. senate election in Oregon. Cyrus Walker wrote
that railroad officials had "*big* influence with Vilas and through them,
some parties in the Territory, have secured Patents to land which has
been suspended and was in the same position as ours." Cooperation
with the railroad, he suggested, would be "a big card in favor of reaching Vilas through them." According to Walker, Renton—described as
having "very liberal views on such matters"—had already contributed
$1,500 and a similar amount would be forthcoming from the Puget
Mill Company. "It looks like a big sum to pay," observed Will Talbot,

"but if it will help the Co. in getting patents to the land it wont be such a big steal." Vilas ultimately took action favorable to the industry, although it cannot be determined if he had been "reached" by Pacific Northwest lumbermen.[35]

Benjamin Harrison's electoral victory over Cleveland in 1888 also had much to do with the resumption of business-as-usual, or so lumbermen worried about the outgoing president's devotion to law-and-order believed. Shortly after the election, Charles Holmes predicted that the triumph of Harrison would "improve our land matters." Timber and Stone entries during 1889 were in fact three times the rate of the previous year and there was no difficulty securing the perfection of claims. "Our Patents are coming in from Washington very fast," John Campbell reported in midyear. Despite this activity, wholesale manipulation of the land laws by the large Puget Sound mills was coming to an end. Another campaign against fraud and a serious national depression reduced sales after 1890. The creation of the first government forest reserves, moreover, removed public land from entry and meant that private land became the focus of interest. "If any body wants to buy land," Will Talbot noted in late 1888, "they must do so within the next two years if they expect to get it at any decent rate." Faced with the declining availability of cheap public timber, the Puget Mill Company and the other San Francisco–based mills, as Edwin G. Ames later recalled, "decided they had land enough, and stopped buying."[36]

There is considerable difficulty in making an accurate determination of the holdings of the various mill companies. Lumbermen disliked publicity and were especially opposed to publication of their dealings in land. The Puget Mill Company, far and away the largest holder, owned 186,000 acres by 1892, having acquired approximately 80,000 acres in the preceding decade. The Port Blakely Mill Company held 35,000 acres in the early 1880s and made sizable additions under the Timber and Stone Act. "We calculate all our lands are worth [$]*1,000,000,*" John Campbell reported in 1886. Around 30,000 acres of timber were claimed by the Tacoma Mill Company, the only other firm to make extensive use of Timber and Stone filings. George Meigs claimed 55,000 acres—apparently an exaggerated figure—for the Port Madison mill and the Washington Mill Company was reported to own 15,000 acres.[37]

Because of these extensive holdings, the lumberman of the period has traditionally been portrayed as a prime example of the robber baron, the corrupt businessman who supposedly dominated the American economy in the late nineteenth century. The acquisition of timberland, charged a conservation publication in 1903, had been "done in a manner so glaringly fraudulent as to make the rottenness of some of our municipal governments seem mild in comparison." The Bureau of Corporations contended in 1913 that improper use of the Timber and Stone Act and other legislation had allowed the ownership of timber to become concentrated in a few hands. "The amount of fraud in the administration of land laws was almost incredible," a still-outraged Gifford Pinchot recalled in 1937. "It went from one end of the system to the other."[38]

Few would disagree with the essential truth of these allegations. "Our Company," wrote the Puget Mill's Ames, "and I presume others, have oftentimes been accused . . . of coming to Puget Sound, stealing Government timber, manufacturing it into lumber, and getting rich. Technically that is all true." But in the absence of appropriate legislation for the sale of timber, in the absence of effective enforcement, and in the absence of general public outrage over fraud, there was little to prevent the lumbermen from securing land. "Honesty," Mark Twain once remarked, was "the best policy—when there is money in it."[39] When the money was in dishonesty and dishonesty was condoned by government and society, the existence of corruption should provoke little surprise.

Lumbermen, of course, reflected the standards of the time. Cattlemen, mining interests, railroads, and even settlers engaged in widespread violation of the land laws. Mill owners claimed, with some justice, that they were singled out for criticism. This was probably inevitable, as timberland was of great value—James J. Hill contended that an acre of timber was worth forty acres of farmland—and the depredations of loggers were more visible than those of grazers and miners.[40] When such figures as Daniel Bagley, John McGilvra, and Hazard Stevens, usually included among the most distinguished pioneer residents of Washington, involved themselves in questionable activities, it seems clear that corruption was at the center of economic and political life.

The antisocial impact of the Timber and Stone Act, moreover, has probably been exaggerated. Most purchases, especially after the early 1890s, were made by small operators and the act, contrary to its critics, made only a modest contribution to the concentration of timber ownership in the Northwest. Far more important in this respect was the Northern Pacific land grant, the basis for the railroad's own extensive holdings and for those of such firms as the Weyerhaeuser Timber Company. Although Timber and Stone sales were heavy in Washington, the amount of land transferred nationally was relatively small. Finally, by concentrating on the admittedly negative aspects of corruption, reformers overlooked the important connection between timber acquisition and the development of the region's lumber-based economy. Considerations of ethics aside, large-scale lumber manufacturing could not have been sustained over the longrun without extensive private holdings of timber.[41]

Still, just because lumbermen were no worse than most Americans of the period, significant and unfortunate long-range consequences of their actions should not be ignored. Mill owners soon—and conveniently—forgot the critical role played by the government in the gathering of their holdings, preferring to believe that their success was the result of individual initiative. "The lord almighty," a radical writer later complained of this tendency, "didn't make the woods for the special benefit of the Lumber Kings."[42] When government aid was sought by other groups—labor, the unemployed, the impoverished—lumbermen stood out as vigorous advocates of self-reliance. And by simply disregarding inconvenient laws, lumbermen helped fasten upon Washington the notion that economic growth should proceed unfettered by the constraints of public policy.

Depressed Markets

The lumber industry of western Washington made scant economic progress during the years of timber acquisition in the 1870s and early 1880s. True, ten steam-driven sawmills were "in constant operation on Puget Sound," as the territorial surveyor general reported in 1875, and a fleet of fifty lumber-carrying vessels serviced the markets of California. These figures, however, were misleading, as sagging San Francisco demand and an unsteady foreign trade resulted in limited business. Washington lumber production increased only slightly between 1869 and 1879, from 129 million feet in the former year to 160 million feet in the latter. The number of sawmills in the territory decreased during the decade and Washington ranked only thirty-first among the states and territories in value of lumber production. In 1880, output in Oregon actually exceeded that north of the Columbia River. Only one important firm, the Tacoma Mill Company—which turned out its first lumber on Commencement Bay in 1869—was formed on Puget Sound in the two decades following the Civil War.[43]

Much of the stagnation confronting the industry resulted from depressed conditions in San Francisco, where most Puget Sound lumber continued to be sold. The bursting of a speculative boom in mining stocks in 1872 commenced a prolonged period of poor business in the California metropolis. One solution to this problem was to exploit other markets in the Bay Area and in Southern California and mill owners busied themselves locating lumber yards in Alameda, Benicia, Vallejo, and Los Angeles. Within the city, the emphasis was on agreements to control production and prices. Combinations were formed among the major San Francisco–based concerns in 1877 and 1880. Combining the Puget Sound operations with plants in Oregon and California, these associations featured production restrictions, assessments on operating mills, and the payment of subsidies to idle facilities.[44]

Despite the hopes of their sponsors, the combinations faced serious difficulties and quickly collapsed. Weak and debt-hampered operators offered low prices in order to sell lumber and strong concerns were reluctant to limit their freedom of action. "We think," William Renton pointed out in 1878, "that we can make more money to run independent and don't like the idea of dividing profits with some of the Sound concerns."[45] The differing interests of participants stood, in 1877 and 1880 as in later years, in the way of successful combinations. The inability of lumbermen to achieve effective cooperation seemed to be endemic, as did a chaotic market situation in San Francisco.

Conditions in California necessitated renewed attention to offshore markets. "The lumber sent foreign," Charles Holmes observed from San Francisco in February 1878, "can not turn out much worse than it would if brought here last year & retailed for $14. & $15." Trade with Hawaii, fortunately, expanded in the late 1870s as the result of the reciprocity treaty between the United States and the island kingdom. The treaty, allowing shipment of island sugar into the United States free of duty, linked Hawaii to the mainland as never before. As the sugar-based economy prospered, so did the demand for lumber to build plantation buildings and urban homes and mercantile establishments. The value of Hawaiian lumber purchases increased from $48,000 in 1874 to $221,000 in 1880 and reached $344,000 in 1883. This lumber, noted the editor of the *Hawaiian Almanac,* made up the "principal class of imports" at Honolulu and came "principally from Puget Sound."[46]

As enticing as palms and surf to the modern tourist, the demand in Honolulu proved irresistible to the Port Blakely Mill Company. "As so much lumber is used there," wrote Charles Holmes, "it would seem as though the trade should be divided up" and the era of Pope & Talbot domination brought to an end. Holmes and Renton planned their first island cargo in the summer of 1878, endeavoring to keep it a secret so as not to arouse countermeasures. "The Gamble folks are very hostile at our going into the Honolulu trade," the mill reported once the secret was out. Cyrus Walker was so outraged that he refused to speak to Renton at a meeting of Puget Sound lumbermen. Thanks to hurried shipments from Port Gamble, the initial Blakely cargo paid less than expected, but Holmes was determined to persist. Henceforth, the Puget Mill Company, to its displeasure, was forced to share the principal foreign market for Northwest lumber.[47]

That the Puget Mill and Port Blakely firms were the only ones with consistent foreign markets and the strongest operations in the industry was no coincidence. "It is safe to say," reflected Cyrus Walker, "that the lumber market of the Sound may be considered all countries and ports on the Pacific Ocean."[48] But only his mill and that owned by Renton and Holmes were able to effectively capitalize on this truism. The other sawmills continued in reluctant reliance on the depressed markets of California, a reliance that helped account for the lack of progress in the industry. Whether prospering or not, though, mill owners had faith in the future, for they had gathered extensive holdings of timber on Puget Sound and Hood Canal.

By the mid-1880s, lumbermen had augmented their production and transportation facilities with timberland. It was fortunate that they had done so, for the era of cheap land and accessible timber was coming to an end. The population of Washington grew by seventy percent in the first half of the decade and increased at a greatly accelerated rate in the second half. Production of lumber matched the growth in population, increasing from 160 million feet in 1879 to well over a billion in 1889, and a new energy suffused the industry. "Immense new lumber mills have been constructed," a visiting British diplomat reported in 1885, "and the forest has been penetrated with railroads and steam engines which supersede the horse and ox team for procuring timber."[49] In terms of technology, in terms of markets and in terms of demand, the lumbering industry of western Washington was about to be transformed.

6

A Change of Front

There are lots of people getting crazy on the mill question which if they keep on they will burst the business wide open.—Cyrus Walker[1]

It is difficult to give anything like a clear impression of the extent of this business in this northwest corner of the United States. Mere figures convey but an inadequate idea . . . . The lumberman has already made here a beginning on that work which has denuded other parts of the country of the green shelter. But it is only a beginning, and it will be many years before Washington ceases to be the great lumber producing State of the Union.—F. I. Vassault[2]

The last rock was chiseled away on May 3, 1888, opening a long tunnel through the Cascades to the north of Mount Rainier. After years of delay, Puget Sound was linked to the Great Lakes by the Northern Pacific Railroad. A spur had connected Commencement Bay with the Columbia River in 1873 and the Northern Pacific, overcoming bankruptcy and corruption, had completed its main line in 1883. Not until completion of the Cascade tunnel, however, was efficient and direct railroad transportation available to the sawmills on Puget Sound. For years, the inhabitants of old and new Tacoma—the latter a development of speculators connected with the railroad—had lived in a wilderness overlaid with grandiose schemes, waiting for their hopes to be realized. "Avenues, parks, and public squares," a bemused visitor had reported in 1877, "are laid out among burnt stumps, felled trees, and piles of dirt." Residents ignored such condescension, firm in their belief that completion of the Northern Pacific would make Tacoma the center of manufacturing on the coast.[3]

Tacoma's population increased from 720 in 1880 to 36,000 in 1890 and the distinction between old and new towns rapidly disappeared. The Tacoma Mill Company produced great quantities of lumber and a major new firm, the St. Paul & Tacoma Lumber Company, began work on a plant on the mudflats where the Puyallup River entered the bay. Rudyard Kipling, visiting Tacoma in 1889, found it "literally staggering under a boom of the boomiest." The English author "passed down ungraded streets that ended abruptly in a fifteen foot drop and a nest of branches; along pavements that beginning in pine-plank ended in the living tree." Frenzied men gathered on every street corner, "babbling of money, town lots, and again money." No one doubted that Tacoma's domination over the other communities on Puget Sound had at last been established. "Like a new Venice," boasted a local news-paper editor, "Tacoma looks forth over the glassy waters and prepares to handle the commerce of the world."[4]

Only the most demented or inebriated booster, of course, could com-pare Tacoma, with its mud and mills and aroma of sawdust, to Venice. Still, the final arrival of the Northern Pacific was an event of great significance, symbolizing the transformation underway in the region. New markets for lumber were opened to the east, although it would be years before they came to rival the traditional outlets in California and the ports of the Pacific. Riding the trains westward, capitalists from the Great Lakes arrived to build mills and invest in timber, grad-ually forcing San Francisco to loosen its hold on Puget Sound. New technology and the rising militance of labor challenged logging op-erators and mill owners. Lumbermen, wrote George Emerson of Grays Harbor, faced "an entire change of front; an entire change of methods; and an entire change of markets, for all future times; and until one accepts the situation, and re-organizes and re-arranges his business to conform to these new conditions, he cannot expect to be in the modern march of progress, or successfully operate his plants."[5]

Grays Harbor Boom

Even before completion of the railroad to Puget Sound, the geograph-ical scope of the industry had begun to expand. From the founding of the first San Francisco–based mills in the aftermath of the California gold rush, lumbermen had ignored southwestern Washington. Settlers in that region eked out a miserable existence, the lack of transportation facilities making it difficult to ship produce to market or to import supplies. One early pioneer was so depressed by the "primeval forest" that he fled across the Columbia to Oregon. During the 1870s, only two families lived at Aberdeen on Grays Harbor. Michael Luark, mov-ing to the harbor from his Olympia claim, sought shelter from the rain in a tiny cabin with a dozen other disheartened individuals. "Imagine," he recorded in his diary, "some reading some singing; some talking some argueing politics or Religion some Counting and some looking on and wishing Such is Pioneer life on Grays Harbor."[6]

The first sawmill in the southwestern section of the territory was built on Shoalwater Bay in 1853 and the first steam-driven plant was erected at South Bend in the late 1860s. Eventually purchased by Cap-tain Asa Mead Simpson, one of the pioneer San Francisco lumbermen, the latter supplied the California market and was the principal source of employment in the region. Simpson also established a mill at Knappton near the mouth of the Columbia in 1869. On Grays Harbor, a small water-powered plant commenced operations at Cosmopolis in 1880.[7]

These, however, were only preliminaries to the initiation of full-scale lumbering.

Late in 1881, Simpson and George Emerson agreed to build a saw-mill on Grays Harbor. Emerson, who combined in his personality the instincts of the classic robber baron and a sensitivity that bordered on the romantic, arrived at Hoquiam in April 1882 and set to work on the mill. In late summer, the North Western Lumber Company, assuming control of Simpson's South Bend and Knappton properties, shipped its first cargo of lumber to San Francisco and commenced the industrialization of Grays Harbor. Blessed with cheap logs—at least in comparison to those available on Puget Sound—and Simpson's aggressive California sales outlets, the company became one of the leading operations in western Washington.[8]

Other investors were soon inspired by the Simpson-Emerson success to locate on Grays Harbor. In neighboring Aberdeen, A. J. West arrived from Michigan in 1884 to build a sawmill. The following year, John M. Weatherwax, another Great Lakes native, erected a large plant on the Aberdeen waterfront. By 1890, the town—previously dependent on fishing—had become an important manufacturing center. The primitive facility at Cosmopolis was purchased by Los Angeles interests and enlarged with the addition of steam power. The major Puget Sound operators, concerned over the developing Grays Harbor competition, acquired this mill in mid-1888. Renamed the Grays Harbor Commercial Company, its capacity was expanded to match and then exceed that of the North Western operation.[9]

Stimulated by the expansion of lumbering, the population of Chehalis County—the designation was changed to Grays Harbor County early in the twentieth century—grew to over 9,000 by 1893. Although lumber production was only a fifth of that on Puget Sound by that year, it was on the increase and the gap was rapidly being closed. Extension of the Northern Pacific to Aberdeen and Hoquiam and waterway improvements by the federal government would enable local boosters to realize their dream of making the harbor "the Duluth of the Pacific." Grays Harbor, reported a guide to western Washington, "presents a prospective commercial prominence that must soon develop enormously and attain an international as well as national eminence."[10]

Railroad Mills

Completion of the transcontinental railroads to the eastern shore of Puget Sound—following upon the Northern Pacific, James J. Hill's Great Northern entered Seattle in early 1893—coincided with immense changes in the nation's timber industry. American lumber production in the years after the Civil War focused on the Great Lakes, a third of the nation's supply coming from Michigan, Wisconsin, and Minnesota. Between 1873 and 1900, over 180 billion feet of white pine was cut in the three lake states, so depleting the supply of timber that production had to be curtailed in the closing decade of the century. Concerned over the scarcity and high prices of stumpage, lumbermen looked to the South and to the Pacific Coast for future supplies of timber and investment opportunities. Eager to sell land and generate freight, the railroads in turn sponsored excursions of investors to the Pacific Northwest.[11]

Two things were especially impressive about the timber of western Washington: it grew in heavy stands and it was cheap. In the Great Lakes states, merchantable timber averaged 6,600 feet per acre, a fifth of the rate on the coast. The average value of standing timber in 1890, moreover, was 92¢ per thousand feet in Washington, compared to $3.26 on the Great Lakes. High-quality stands could be purchased for as little as 10¢ per thousand feet. These figures were readily translated into greatly reduced costs of operation. A third of the cost of manufacturing lumber in the lake states consisted of stumpage, but only eleven percent on the Pacific coast.[12] Declining supply in the east, vast supply in the west, and the railroad connection made for an influx of lumbermen.

The center of activity was Tacoma, where the Northern Pacific hoped to profit from the sale of timberland and mill sites at its Commencement Bay terminus. In 1887, Frederick Weyerhaeuser, the leading Great Lakes lumberman, considered and then rejected the opportunity to purchase an extensive stand of railroad timber and erect a sawmill. Another group brought together by the railroad, however, accepted this arrangement. Chauncey W. Griggs and Addison Foster of St. Paul joined Wisconsin investors Henry Hewitt, Jr. and P. D. Norton and Michigan lumber manufacturer C. H. Jones under the auspices of the Northern Pacific. These men agreed to purchase 80,000 acres of timber, to construct a railroad from this timber to Commencement Bay, and to build a large mill. Their St. Paul & Tacoma Lumber Company, capitalized at $1.5 million, was formally organized in June 1888.[13]

Managed by Griggs, a heavy-set former Civil War colonel, whose wholesale grocery business had provided the surplus for investment in the West, the St. Paul & Tacoma mill was in operation on the Tacoma tideflats by early 1889. Eleven miles of railroad led into the company's timber and the plant was the only one in the region to have its logs delivered by rail. Outfitted with band saws and assorted labor-saving devices, the mill was the most up-to-date on the Sound. "No lumber is handled by hand in the mill," reported George Emerson after a tour, "until it is ready to load on trucks." The firm's stock, he wrote, "cannot be bought for twenty times its par value." The St. Paul & Tacoma made the city on Commencement Bay the lumbering center of eastern Puget Sound.[14]

This point was never conceded by residents of nearby Seattle, confident in the belief that their community was the "Chicago of the Pacific." Despite its failure to secure the Northern Pacific terminus, Seattle prospered during the 1880s, the city's population increasing from 3,500 to 43,000. Early investors in local real estate—including such firms as Pope & Talbot and such individuals as Cyrus Walker and William Renton—were amply rewarded in these years of expansion. The enthusiasm in Seattle received a severe test in the spring of 1889 when a fire destroyed much of the business district, as well as wharves and sawmills. Walker hurried over from Port Gamble to evict squatters from his property and to discover that 2,000 people were already at work rebuilding the city. "Seattle is famous for *grit* and *pluck*," he informed William H. Talbot.[15] In the rebuilding, there was great encouragement for the expansion of local lumbering.

One result of the fire was the relocation of Seattle's sawmills from expensive Elliott Bay waterfront property to Lake Union and to Ballard, then an independent community north of the city. Leading the way was C. D. Stimson, one-armed son of a Michigan lumbering family, who arrived in Seattle just in time to participate in a bucket brigade during the fire. Acquiring a mill in Ballard, he joined with his brothers to form the Stimson Mill Company in 1891. The Stimson mill was regarded by knowledgeable observers as being among the most efficient in Washington and was far and away the largest producer in the Seattle area. Although the other Seattle mills were small, collectively they could exert considerable influence on the market, especially when they shipped lumber to California. In June 1892, for example, more lumber was shipped from Seattle than from any other point on Puget Sound.[16]

The most important aspect of the Seattle industry, though, was not lumber but shingles. Although in earlier years sawmills had occasionally shipped cedar shingles with their lumber cargoes, no market had ever developed. Completion of the transcontinental lines, however, made it possible to supply the burgeoning demand in the farming regions of the Middle West. Beginning in the mid-1880s, a shingle craze struck Puget Sound, with plants thrown up along the tracks of the Northern Pacific. By 1890, a third of the nation's supply was manufactured in Washington. In May 1893, following arrival of the Great Northern and extension of a Northern Pacific spur to Seattle, 300 cars of shingles were shipped from Ballard, which had become the leading producer in the world.[17]

From the beginning, shingle manufacturing displayed some significant characteristics. Only a small amount of capital was required and the industry tended to attract farmers, lawyers and others whose desire for profit often exceeded their expertise. Shingle operators, complained the *West Coast Lumberman,* "know nothing about their business." Plants were small and numerous and only a few firms were heavy manufacturers of both lumber and shingles. Moreover, the rapid expansion of capacity strained the shipping ability of the transcontinental lines. "They are already manufacturing beyond the carrying capacity of the Railroads," Cyrus Walker observed in May 1892.[18] Exacerbated by car shortages, overproduction caused shingle prices to fluctuate with amazing rapidity.

Lumber and shingles also played an important role in the building of towns on northern Puget Sound. In 1890, Henry Hewitt, Jr., John D. Rockefeller, and other investors, convinced that Port Gardner Bay would be chosen as terminus for the Great Northern, founded the town of Everett. Hewitt began work on a large bayside mill and acquired sizable stands of nearby timber. Visiting Everett, Cyrus Walker discovered that Hewitt and his associates were "laying out a town large enough for a million people." Walker was skeptical of the prospects, noting that James J. Hill had invested no money in Everett and that the railroad would terminate in Seattle. "I fail to see what there is in the place to justify the investment," he concluded. Still, a dozen shingle plants, nine sawmills, and a paper mill were soon operating on Port Gardner Bay.[19]

To the north on Bellingham Bay, the expansion of logging and lumbering created, according to a local settler, "a scene of activity and bustle cheering indeed to the ones who had waited so many years, Micaber-like, for 'something to turn up.'" Between 1890 and 1892, the daily output of shingles in Whatcom County increased from 100,000 to over two million. In the past, the Puget Mill Company's mill at nearby Utsalady had dominated the local lumber market. But as J. J. Donovan, former chief engineer of the Northern Pacific, Michael Earles, and others invested in production facilities, it soon lost its hold to bayside mills. Population grew and the bay towns—Whatcom, Sehome, Fairhaven—coalesced, eventually to merge as Bellingham.[20]

Despite increased capacity and the danger of overproduction, the old California-owned firms expected to benefit from the arrival of the transcontinental railroads and the expansion of lumbering on the eastern shore of Puget Sound. Located on the western Sound and on Hood Canal, they were unable to ship lumber east by rail themselves. Rather, the opening of eastern markets would deprive the new mills of the need to sell lumber in San Francisco or elsewhere in the Pacific. "The eastern business is bound to improve rapidly," Will Talbot wrote in late 1888, "and more and more lumber will be required for shipment east each year." Observing the situation from the perspective of San Francisco, the Pope & Talbot president hoped that the rail mills would be so busy "that they will not be compelled to throw their poor material on this market."[21] The old cargo mills—as those firms shipping by sea were called—would be left undisturbed in their traditional trans-Pacific and California preserves.

Unfortunately for all concerned, high freight costs prevented the rail trade from immediately realizing its potential. The Northern Pacific charged 60¢ per hundred pounds of lumber between the Sound and Omaha and St. Paul. This meant that it cost three times as much to ship lumber to the Twin Cities, a distance of 2,000 miles, than to Australia, a distance of 8,000 miles. The railroads charged by weight rather than by board feet and lumber destined for eastern markets required drying in kilns and storage so as to reduce moisture content. Although a savings in freight resulted, a considerable amount of money had to be expended on construction of the necessary facilities.[22] Frustrated by the heavy costs, lumbermen pressed for reduced freight rates that would open rail markets to full-scale exploitation.

Eager to secure eastbound traffic for the Great Northern, James J. Hill responded to these demands by promising a 40¢ rate to St. Paul. "Should the Great Northern reduce the rates to Eastern points as Mr Hill has promised," Will Talbot wrote from San Francisco, "there will be no question but that a large amount of lumber will be shipped that way, and the Mills now disturbing this market, will have all they can attend to in taking care of their Eastern business." The great moment came in early 1893 when Hill's road entered Seattle and the new rate was officially announced, followed by a comparable reduction in the Northern Pacific's charge. "It is safe to say," asserted the *West Coast Lumberman,* "that a marked increase in eastern shipments will be seen at once and that the railways will be taxed to their fullest capacity in taking care of the lumber business that will be offered in 1893."[23]

Just at the moment when Pacific Northwest lumber became competitive, demand was stifled by the great depression beginning in 1893. Despite the 40¢ rate, the eighty million feet of Washington lumber shipped east in 1893 represented a sharp drop from the previous year. Sales declined still further in 1894. "The condition of the lumber markets in the east," reported the *West Coast Lumberman,* "is not such as to inspire cheerfulness here." Mills built for the rail trade were forced to ship lumber to California and foreign markets, exacerbating competitive conditions at those points.[24] The rail trade would not achieve its potential until the return of prosperity at the end of the decade.

The initial impact of the railroads, then, was less a matter of transforming the direction of the trade than of altering the nature of the industry. Washington lumbering had always been characterized by a small lumber of large mills. By 1900, however, there were over three

hundred sawmills in the state, most of them small affairs strung along
the rail lines like begrimed pearls of industry. The lumber industry,
though, was still dominated by large firms. In 1905, fifty-eight percent
of the state's lumber was manufactured by the twelve percent of the
concerns in the industry earning more than $100,000. Those firms
earning less than $20,000 made up half the industry, but were re-
sponsible for only ten percent of total output.[25] Stimulating the construction of modest-sized mills, the railroads added to the divisions
within the industry a distinction between big and small operations.

A changing focus among the major mills also reflected significant
alterations in western Washington. The rise of production facilities located on the railroads of the eastern Sound and Grays Harbor—the
Northern Pacific reached Hoquiam in the middle of the 1890s—caused
the traditional lumbering centers to lose much of their influence. Kitsap County, focus of the industry in the pre-rail era, gained in population, but the increase was insignificant compared to that on Puget
Sound's eastern shore.[26] In addition, many of the new mill owners
moved to Washington to supervise business personally. This meant
that the industry was no longer exclusively controlled from San Francisco and that decisions increasingly were made in offices in Tacoma,
Seattle and other local centers of lumbering.

Those who relocated on Puget Sound and Grays Harbor, moreover,
differed considerably from the first generation of Washington lumbermen. The latter group originated in New England and came to the
Pacific coast with relatively modest amounts of money. They were
Republican in politics, Protestant in religion, and parsimonious in the
expenditure of money. The new group, in contrast, came from Michigan, Wisconsin, and Minnesota, and some had even been born abroad.
Men of considerable wealth, they invested the earnings of successful
business careers. Some, like Chauncey Griggs, were Democrats and
some, like J. J. Donovan, were members of the Catholic church. Above
all, they were willing to spend money, an important trait at a time
when timber prices and production costs were on the increase.

Confronting new conditions, new competition and new demands for
funds, the older lumbermen reflected on the wisdom of continuing in
the business. "I dont think we shall make much money in saw mill
line for sometime," wrote Charles Holmes of the Port Blakely Mill
Company in mid-1889, "as there are too many new mills & more
building." Cyrus Walker concluded that if the Puget Mill Company's
"mills should burn up and they would not be foolish enough to build
them up, they would be better off than to keep on running with the
present outlook." There was no point in continuing in a business, he
believed, "where there is no hope of satisfying yourself or any body
else."[27] The industry built by Walker and his colleagues had, with the

arrival of the railroads and their sleeping cars full of eager capitalists, become a complex institution. And all of this happened a decade before Frederick Weyerhaeuser, the most famous symbol of the changing front, made his decision to invest in the region.

Efforts at Market Control

Because of the expanded capacity of the industry, the failure of the rail trade to realize its early promise was a profound disappointment. Competition for offshore business became particularly intense with the building of the new sawmills. Fortunately, there was in the late 1880s a considerable demand for Northwest lumber. Hawaii remained the steadiest supplement to domestic business, island lumber imports increasing in value from $197,000 in 1888 to $343,000 in 1890. The Hawaiian trade, especially because of profitable return cargoes of sugar, paid "better than any other business on this coast" according to Charles Holmes. But the mounting demand for lumber attracted many firms to the Honolulu market and led to price cutting. Moreover, Hawaiian purchasers demanded a higher quality lumber than that traditionally shipped, leading a petulant Cyrus Walker to complain that his product was "all right for White people but for Knakas it dont fit the bill."[28]

As Hawaii boomed and changed, the expanding production of coal opened a new and wonderful situation in Australia. After ignoring the Down Under market for a quarter century, the major exporting mills busily shipped lumber to Melbourne and Sydney. Coal provided profitable return cargoes, eliminating the old problem of freight for the voyage back to America. For this reason alone, the trade was preferable to that with Peru or Chile, in the opinion of the Port Blakely Mill's John Campbell, as there was "a chance of getting return freight from Australia, while vessels going to West Coast always return in ballast." Mill company vessels carried coal to Honolulu and California, earning such returns that Charles Holmes concluded on one occasion to send a cargo of lumber to Australia in spite of low prices, "as coal freights back are so good."[29]

Lumberships taking Australian coal to Honolulu normally loaded sugar at that point for San Francisco. The final leg of such a voyage marked the concluding segment of a quadrangular trade pattern, in which mill company vessels sailed to Puget Sound from California with supplies, then carried lumber to Australia, where a return cargo of coal was loaded, the latter to be exchanged for sugar in Hawaii. This pattern retained its importance for only a few years, coal strikes and shipping shortages contributing to a sharp downturn in the Australian economy in the late 1880s. Lumber exports to Australia in 1889 dropped by over a half from the level of the previous year. Reports from Mel-

bourne led Will Talbot to predict in March 1889 that the trade would
be "very limited" in the forseeable future, "the boom having fallen
off." The collapse of the Australian market was so sudden and so severe that Puget Sound mills were forced to curtail production.[30]

Operators attempting to break into the foreign trade preferred to send
lumber to the west coast of South America, where railroad construction
in Peru and Chile generated considerable demand during the decade.
The St. Paul & Tacoma Lumber Company, for instance, contracted in
1889 to supply eight million feet of lumber per year to customers in
Callao. Heavy shipments and the willingness of some firms to secure
business by cutting prices, however, made for an unstable situation.
Occasional Puget Sound cargoes were also dispatched to China and
around Cape Horn to England and Ireland. Sales in South America,
China and Europe, though, fell short of compensating for falling prices
in Hawaii and the collapse in Australia.[31] Declining offshore business
combined with the slow growth of the rail trade to force lumbermen
back upon the time-honored California market, a market that could
absorb the industry's output only in the rarest of instances.

For a brief but glorious period in the mid-1880s, California required
most of the lumber turned out in the Northwest. Stimulated by a rate
war between the Southern Pacific and the Santa Fe railroads, the population of Los Angeles increased rapidly and Southern California became for the first time a center of frenzied real estate speculation.
Responding to this opportunity, Pope & Talbot and Renton, Holmes,
& Company—the California parent of the Port Blakely Mill Company—acquired substantial interests in the San Pedro Lumber Company, in a Los Angeles retail operation, and in a San Diego yard. In
turn, these firms agreed to purchase Douglas fir lumber exclusively
from the mills at Gamble, Ludlow, and Blakely. Pope & Talbot also
moved into the San Joaquin Valley, establishing a distribution point
at Port Costa and yards in Stockton and Fresno to compete with outlets
operated by other Puget Sound mill owners. The new trade offered a
number of problems, including high freights and a shortage of rolling
stock for shipment from coastal ports to interior yards.[32]

The main problem, though, was that booms sooner or later were
bound to collapse, a truism accepted with philosophical resignation by
lumbermen. After three years of phenomenal growth in Southern California, the speculative bubble burst and the demand for lumber collapsed in 1888. "San Diego is dull as the D—, infact dead," reported
Will Talbot, and Los Angeles was "completely fizzled out."[33] With
foreign trade also on the decline in that year, the collapse added to the
need for a sharp curtailment in production and increased the pressure
in the Bay Area.

San Francisco remained the focal point of the industry, the market of last resort for lumber that could not be sold elsewhere. As always, efforts to bring order out of the competitive chaos in the city foundered on the inability of operators to trust each other. Conditions were so depressing, however, that efforts to devise workable production restrictions were always being undertaken. Even before the demise of a short-lived combination in 1886, for instance, mill owners had begun work on an elaborate plan. "This scheme is a very important one," wrote Charles Holmes, "& involves quite a change in our method of business." The biggest change called for the establishment of a common yard in San Francisco. Members of the combine would send all lumber destined for the Bay Area to the yard and receive a commission on sales. Profits of this Pacific Pine Lumber Company would be divided according to the amount of lumber provided by each participant. Organized by the end of 1886, the firm commenced operation at the beginning of the new year.[34]

Mill owners began the Pacific Pine experiment with realization that all previous combinations had failed. There was an expectation that such operators as Charles Hanson of the Tacoma Mill Company—who was likened to "the Black Prince" by a competitor—would go their own way in blithe disregard of contractual obligations. The conflicting interests of members firms, moreover, constantly threatened Pacific Pine with disintegration. Lumber manufacturers, recalled Edwin G. Ames of the Puget Mill Company "would go to these meetings and spend an hour developing a scheme for the general good of the lumber business, and then go home and spend hours and days trying to figure out another scheme to beat it." Still, Pacific Pine held together in its first year and helped to stiffen prices in San Francisco. Better times were encouraged by efforts to reduce competition. When the Washington Mill Company plant at Seabeck was destroyed by fire in August 1886, for instance, William J. Adams accepted a subsidy to relocate his firm through purchase of a mill at Port Hadlock. This was bad for Seabeck, which soon became a ghost town, but it reduced the number of operating units in the industry.[35]

Maintaining control in San Francisco at a time when outside mills sought to penetrate the market was the great problem confronting Pacific Pine. Beginning in 1888, the collapse in Southern California and declining foreign trade forced the dumping of lumber in the Bay Area and prices began to drop. Old-time San Francisco lumbermen, so long the undisputed masters of the industry, confronted a deplorable prospect. "Owing to the great number of competing mills and the small demand in comparison with the capacity for supplying the same," Will Talbot reluctantly admitted, it might be impossible "to continue in the same way that we have been running." In the past, the principal threat to combinations had come from within. Now, a new and sobering obstacle came from without, from the large number of companies refusing to join Pacific Pine.[36]

Members of Pacific Pine met continuously to devise strategy. In late 1888, the mills agreed to run only eight hours a day and to close for a week each month. Although a standard response to glutted markets, these moves were insufficient to meet conditions in San Francisco. Trying another tack, Pacific Pine devised a scheme in 1889 to subsidize mills to halt production. One dollar was to be paid from a fund raised by assessments on mills still running for each thousand feet of capacity eliminated. This, too, failed to bring production into line with demand and the only substantive benefit went to mills unable to sell lumber and thus willing to shut down in return for subsidies. Reflecting this tendency, the Puget Mill Company's Utsalady plant, deprived of its local trade by competition from Bellingham Bay, closed and was awarded a subsidy.[37]

Typifying the strained times, some of the oldest firms in the industry were forced out of the trade. The best-known casualty, George Meigs, ended decades of flirtation with bankruptcy. Amidst reports that Meigs was defrauding his creditors, the Port Madison mill was attached by Seattle's Dexter Horton Bank in January 1892. Initially, he seemed to be extricating himself from the crisis and Cyrus Walker marveled that "no one but Meigs would ever be able to pull through." But the creditors had no patience and the mill was closed, never to resume operation. The Port Discovery mill, founded in 1859, also went out of business, its owners failing in efforts to sell the property.[38]

Pacific Pine failed to restore prosperity in San Francisco because only a small number of mills joined the company. The ideal solution, to those of supple and expansive mind, therefore, seemed to be to form a new super combination of all the cargo mills on the Pacific coast. Such a solution amazed rational and experienced observers of the industry. "I hardly see how a combination can be formed that will include all the mills of the Coast and be held together," contended Cyrus Walker. "The State of Washington has 466 Mills which will give some idea of the extent to which the lumber business on the Coast is overdone." Negotiations commenced in spite of the magnitude of the task, ultimately to founder on the refusal of large firms to allow small manufacturers a real voice in policy and on the traditional obstacle of mistrust. "When selfishness and lying are banished from humanity," observed the *West Coast Lumberman,* "combinations may be successfully worked."[39] For the time being, an increased dosage of Pacific Pine medicine was not desired.

Mills belonging to Pacific Pine, however, were reluctant to abandon the habit and their organization staggered on into the great depression of the 1890s, its bedraggled state symbolizing the fortunes of the industry. A period beginning with booming markets ended with both foreign and domestic trade depressed. Production of lumber soared and the spirits of oldtime lumbermen slumped, their once dominant market position challenged by aggressive newcomers. The demise of the Port Madison and Port Discovery mills reflected the declining influence of the San Francisco–based firms. Increasingly, those remaining in business were on the defensive, even in their headquarters city.

New Technology

The change of front taking place in western Washington at the end of the nineteenth century was not limited to fresh capital, new manufacturing centers, and more complex markets. Change was also rapid in the woods, as operations moved inland from tidewater and logging became more difficult and expensive. In the sawmills, however, the situation was a bit different. Lumber manufacturers thought of themselves as men of initiative, "which explains," contended a leading member of the industry in later years, "why we have Paul Bunyan as our Patron Saint instead of Caspar Milquetoast."[40] Mill owners, though, were often closer to the latter timid standard, bemoaning the necessity and the cost of adapting to a new age.

Since the introduction of steam power and the circular saw in the 1850s, only slight changes had been made in the means of manufacturing lumber. Over the years, technical improvements had allowed the substitution of mechanical for muscle power in a number of areas in the mills. The willingness to undertake incidental alterations, though, did not mean that mill owners were willing to make heavy expenditures, even when such expenditures would result in longterm savings. First introduced in the United States in 1869, the band saw—a continuous steel blade running between two wheels—was a major advance over the standard circulars. Because of its small blade, the band saw chewed less of each log into sawdust and greatly reduced wastage. Of an operator who had installed a bandmill, the *West Coast Lumberman* observed: "He wont cut as much dust . . . but hopes to have more dust in his pocket."[41]

Although conceding their benefits, Washington lumbermen were slow to adopt the new saws. The thin blades, some believed, were not strong enough to handle heavy Douglas-fir timber. More important, manufacturers hesitated to spend money on the conversion of their plants. George Emerson recognized "that no investment of money can pay longer returns than that required for such a change," but still objected "to a move in that direction at the present time." It was better to wait until the usefulness of the saws had been demonstrated by other operators. "You can watch them in other mills," recommended Charles

Holmes, "& let them experiment & when they prove to be the thing we can use them."[42] Timber was still plentiful and cheap and reduction of waste was secondary to the avoidance of expense.

Of more concern to mill owners was the constant danger of fire. Sawdust was piled everywhere, machinery and belting often produced sparks, and sawmills were little more than bonfires awaiting ignition. In February 1888, for instance, the Port Blakely mill burned to the ground in less than two hours. During the remainder of the year, a new mill was erected at Blakely and the industry considered how best to protect itself. Will Talbot besieged Cyrus Walker with hysterical letters, expressing his "mortal terror of fire" and his fear that once a conflagration had started, "all the water in the country is of no avail." Little was done to alleviate these concerns. While he did not "wish to be considered [an] old fogy," Walker opposed the installation of sprinkler systems as a needless expense. Electric lighting would eliminate dangerous oil lamps, but few operators were willing to adopt such an innovation. Various insurance schemes were discussed, but only a small number of mills were actually insured.[43] Mill owners preferred worry to the expenditure of funds for protection of their properties.

Lumbermen were also slow to make use of new organizational methods. Only one lumbering firm, Pope & Talbot, followed the national trend toward systematic business management. Hoping to reduce the burden on Cyrus Walker, Will Talbot had sent a Maine cousin, Edwin G. Ames, to Port Gamble in 1881. A man of conservative temperament, belied only by a fondness for the writings of Mark Twain, Ames quickly won the older man's confidence and became Walker's designated successor as manager of the firm. To increase efficiency, Talbot devised what he termed "the 'System'" in 1888. His scheme parceled out the various functions of the Puget Mill Company—manufacturing, transportation, and so on—to five subsidiary companies. "Should any thing happen to the heads of the Biz.," Talbot informed Walker, "there would be such a thorough System going on that the Biz. would suffer no material loss." The reorganization, he continued, "will give us fuller information, as well as create offices whose heads will be responsible for the workings of their Divisions and Sections." Despite grumblings about "red tape" from Walker, who preferred seat-of-the-pants methods, Talbot's system was implemented and Puget Mill became the first firm in the industry to operate according to up-to-date administrative methods.[44]

Although slow to accept change in their mills, lumbermen readily adapted to new conditions in the woods. The depletion of timber on tidewater and the move of operations inland meant that change was a necessity. "Timber within easy distances of the water-courses," a railroad surveyor reported in 1881, "is becoming comparatively scarce."

Three years later, the census bureau pointed out that lands adjacent to Puget Sound "have been culled of their best trees for a distance inland of 1 or 2 miles."[45] The increasing difficulty and cost of logging required new technology, new investment, and new relations between logging operators and mill companies.

Relations between loggers and lumbermen had become acrimonious by the 1880s. Independent logging firms continued to be little more than sharecroppers of the forest, surviving from year to year on advances from the sawmills and on supplies from the mill stores. For their part, lumbermen believed that most loggers were inefficient and incapable of sound or honest business dealings. Mill owners blamed the increasing price and decreasing quality of logs on the rascality of the men in the woods. In middecade, Puget Sound lumbermen formed a log brokerage company in an effort to force a reduction in log prices. The company failed because individual mills were willing to violate the agreement in order to secure raw material. Tacoma's Charles Hanson, ever the villain in the opinion of his colleagues, "would . . . kiss a loggers backside for Logs," complained an official of the Port Blakely Mill Company. The only solution to these problems, many lumbermen conceded, was for the mills to assume a direct role in the woods.[46]

The means of removing the "round stuff" from the woods were also on the verge of transformation. Logs were still hauled to saltwater by oxen over skidroads at the beginning of the 1880s. As loggers moved away from the water, the need for a better method became manifest. Any operator attempting to haul logs for more than a mile with oxen, Stewart Holbrook once noted, "was heading straight for bankruptcy." Fortunately, California lumberman John Dolbeer, an inveterate tinkerer, invented the steam donkey in 1881. Powered by a steam engine, cables pulled logs in from the woods. Perfected by Dolbeer and other operators, the donkey made it possible to move logs over longer distances, at a more rapid pace and—considering the heavy outlays for oxen—at a lower cost.[47]

After some initial hesitation, the steam donkey was adopted by most Washington loggers. George Emerson was the first to use the new device and became its most enthusiastic advocate. "When one considers they . . . require no stable and no feed," he pointed out, "that all expense stops when the whistle blows, no oxen killed and no teams to winter, no ground too wet, no hill too steep, it is easy to see they are a revolution in logging that will allow a logger to do better at $3.50 per M. [thousand feet] than they have at $5." The high cost of the engines, moreover, would work to the advantage of the mills by driving poorly financed and untrustworthy logging operators from the woods. "The number of loggers will be reduced to but few," observed Emerson, "and those, men of means and safe men to deal with." By 1900, there were three times as many steam engines in Washington as in Oregon and California combined. Two counties—Chehalis and Mason—contained more of the devices than either of the states to the south.[48]

Introduction of the donkey engine solved one problem, but the problem of transporting logs from the deep woods to tidewater remained to be resolved. On northern Puget Sound and Grays Harbor, reliance was placed on river drives until well into the twentieth century. Logs were gathered in booms to await the production of high water by nature or by specially built splash dams. On central and lower Puget Sound, however, streams were unsuited for large-scale drives. Moreover, mill owners believed that by opening up timber away from the watercourses they would increase the supply of logs and relieve the pressure on prices.[49] Nature and markets combined to create a demand for another new technological device, the logging railroad.

A small rail line began hauling logs south of Olympia in 1881, but Mason County soon became the center of Washington railroad logging. Fronting on both Hood Canal and the southwestern reaches of Puget Sound, the county was isolated and sparsely populated. Virtually every acre was covered with timber and the surrounding waters offered easy transportation to the major sawmills. All that was necessary was to build railroads into the foothills and bring out the timber. The Puget Mill Company began work on a road in 1885. That same year, two arrivals from the Great Lakes, A. H. Anderson and I. R. McDonald, commenced building westward from Shelton, the principal town in the county. Their Satsop Railroad snaked for a dozen miles into the nearby hills and earned a handsome profit.[50] The most important of the logging railroads, though, was the work of the Port Blakely Mill Company.

Charles Holmes and William Renton agreed in early 1885 that "railroads are now indispensible to logging . . . & a mill cannot be successfully run without them." Only by opening up new sources of timber could the company counteract the rise of log prices. Work on their Puget Sound & Grays Harbor Railroad, with its Sound terminus on Little Skookum Inlet near Shelton, began in the fall of 1885. Caught up in the western Washington boom, Renton decided to extend the railroad on to Montesano on the Chehalis River in 1888, providing common carrier freight and passenger service between Puget Sound and Grays Harbor. This section of the road was sold to the Northern Pacific, but the impact of the logging portion of the line was of crucial importance. "The new mills being built on this coast will increase competition," wrote Holmes in April 1888, "so that another season our margins will be materially reduced, but we hope even then to be

able to do fairly having our own lands & Railroad & we trust a mill
that can manufacture as cheaply as any mill on the coast."[51]

Small loggers could not afford the expense of railroad construction
and operation, and logging increasingly came under the control of large
firms. The Port Blakely Mill Company, for instance, acquired timber
and absorbed debt-ridden independent logging operations along its line.
The mill soon became a leading force in the logging industry, matching
its position in the manufacture of lumber.[52] However reluctantly, mill
owners had to accept new technology and devise new means of operation. The transformation was most dramatic in the woods, where
the search for timber required the substitution of mechanical energy
for the power of animals and water. Although accepting a greater role
for the machine, lumbermen were less willing to accede to a new status
for the human beings operating the machines.

Rise of the Labor Question

Workers in the mills and forests of Washington were the highest-paid
in the American lumber industry. Sawmill employees earned an average monthly wage of $56.40 in 1890, compared to $35.83 in the
Great Lakes states. Nationally, twenty-six percent of the persons working in the woods received more than $40 a month, but the rate for
Washington was ninety-three percent. The comparable figures for sawmill workers were thirty-four percent for the nation and eighty-five
percent for Washington. Employment, moreover, was steadier in the
Pacific Northwest than elsewhere. Terrible accidents, drab mill towns,
and tumbledown logging camps meant that conditions of work and life
were harsh. Still, a substantial advance had been made over eastern
standards, and laborers migrating from Wisconsin and Minnesota realized an improvement in their way-of-life. Lumbermen recognized the
need to pay high wages and to provide good food and relatively decent
quarters in order to secure workers.[53]

Nevertheless, lumbermen had to come to grips with the rising militance of labor in the mid-1880s. In the early days of the industry, mill
owners had felt some personal affinity for their employees, who were
usually fellow New Englanders. But they gradually came to adopt an
impersonal attitude, especially after the workforce expanded to include
men of diverse origins and allegedly inferior capacity. Lumbermen referred to wages as the "price of labor" in much the same fashion as
they would discuss the cost of logs and animals. When production
cutbacks were instituted, no thought was given to the impact of reduced earnings on employees and their families. Rather than acknowl-
edging the contributions of workers, lumbermen treated them with scorn.
"We appreciate the position you are in," Will Talbot sympathized with

Cyrus Walker on one occasion, "in having to deal with such a poor class of labor."[54]

Employers could neither respond to demands for better wages and working conditions nor deal with unions. To do so would risk undermining the belief that mill owners were superior to mill workers. Manufacturers would go to absurd lengths to avoid even the appearance of granting concessions. Cyrus Walker, according to a well-known story, refused to contribute to a fund for the widow of an employee killed on the job because it might be construed as accepting responsibility for what happened to his workers.[55] For their part, laborers could see little reason for loyalty to employers. Wages were high, but workers faced a tenuous existence, seeing themselves as commodities to be used up and then discarded. Production cutbacks reduced income, and wages were not paid at all if the mill or camp went bankrupt. The accident rate was frightening, and even the best logging operations were isolated deep in the woods. There was scant cause for looking to the future and every reason to seek better conditions in the present.

Founded in 1869 with the goal of uniting all workers in the United States, the Knights of Labor presented the first organized challenge to Washington lumbermen. By the mid-1880s, the Knights had attained a membership of around 2,000 in western Washington. On Puget Sound, the organization directed its main effort toward the expulsion of Chinese workers, whose willingness to work for low wages supposedly prevented the improvement of conditions for whites. The anti-Chinese campaign brought the Knights into direct conflict with mill owners, since the lumber firms employed Asian laborers to perform undesirable and low paid tasks. "I dont know how we could wood up the steamers at all hours of day & night without Chinamen," observed Cyrus Walker.[56]

During October 1885, the anti-Chinese crusade came to a climax with the murder of Chinese coal miners outside Seattle and the holding of demonstrations in that city and in Tacoma. Suddenly, Walker informed Will Talbot, "the Chinese question" had become the "all absorbing topic all over the Sound." Under pressure from the Knights of Labor, the Tacoma and Port Blakely mill companies fired their Chinese employees. The Puget Mill Company, though, avoided the ignominy of surrender because of its isolated position and its ability to control the flow of traffic to and from Gamble and Ludlow. In early November, mobs drove Chinese inhabitants out of Tacoma and Seattle and federal troops were sent to the two cities to restore order. "The boys in blue," wrote Walker, "have a wonderful effect in quieting the disturbing element that this country is now cursed with." Despite this salutary lesson, the success of the Knights posed a frightening prospect for mill owners.[57]

For Walker, the implications of the anti-Chinese movement were obvious. "If the Knights of Labor succeed in driving the Chinese out of the Country," he observed, "their next attack will be on the corporations and capital." The Port Blakely mill had no sooner replaced Chinese woodcutters with whites, Walker pointed out, than the latter informed management "that they wanted four bits an hour for their services." Walker's assistant, Edwin G. Ames, forecast "lots of trouble on this Labor Question" as the result of the Knights' victory. "I dont fear any trouble from the sensible, saving, industrious Laborer," Ames maintained, "but only from the Whiskey Bum, Pokersharp Dead Beat . . . class who would never have a dollar if they got one dollar a minute for their Labor." To Walker, Ames and other worried lumbermen, the future carried visions of crazed and besotted unionists wresting away the property of virtuous Americans.[58]

The forebodings of mill owners soon appeared on the verge of realization. In the spring and summer of 1886, the Knights began a campaign to force a reduction in the standard working day from twelve to ten hours. The organization mounted a strike at the Tacoma Mill Company in early August and made repeated forays to Port Blakely in an effort to spark a walkout at that point. "Whenever we find any of them making any demonstrations," reported John Campbell from the latter location, "they are ordered to leave the place at once. In this way if possible we intend to prevent them getting a foothold." Manufacturers held conferences to concert action and employed detectives to infiltrate the Knights.[59]

In the midst of these doings, industry leaders were shocked by the sudden decision of the Puget Mill Company to grant the ten-hour day in advance of an industry-wide strike. "Mr Walker expects that the men will strike for 10 hours," wrote Charles Holmes in late September, "& that it will be better for the mills to make the change themselves." Walker found it difficult to run fulltime with labor in short supply and determined that the sensible course was to reduce the working day before a walkout would make such a move appear to be a surrender to the union. At first, Walker's erstwhile colleagues in the anti-Knights drive stood firm in defiance of the Puget Mill Company. "We think in two or three months Walker will want to go into the Poor house," Campbell noted in amusement, "he will feel so bad to think that the rest of the Mills did not adopt the 10 hour system."[60]

Strikes at the Tacoma and Port Blakely mills in November, however, forced the remaining Puget Sound concerns to adopt the ten-hour day. At Blakely, several dozen men walked off the job in an effort to close the mill, but the presence of county deputies and the territorial militia enabled production to continue at a reduced rate. "They have been ordered off the premises," Campbell wrote of the strikers, "although a great many of them regret the course they have taken, as it probably turned out different to their expectations." Bravado aside, Campbell conceded that the strike had "retarded" output and the prospect of further disturbances required a reduction in hours. Ten hours therefore became the standard working day on the Sound.[61]

Having made the move themselves, Puget Sound mill owners advocated extension of the ten-hour day to cover all Northwest manufacturers. Otherwise, those firms operating twelve hours a day would be able to produce more lumber and gain a competitive advantage. "It is a difficult task forcing Grays Harbor & Columbia River mills [to] run the 10 hours," noted Campbell. Nevertheless, the new system was adopted in those areas by the spring of 1887. As for the Knights of Labor, the organization began to fall apart soon after the anti-Chinese and ten-hour day campaigns, hampered by internal divisions and financial problems. Surveying the reports of undercover agents at the end of 1886, Campbell was confident that the Knights were "squelched for some time to come."[62] And so they were, bringing to an end the industry's first skirmish with organized labor.

It had been a disturbing experience, hardening attitudes toward workers and unions. What was needed, according to Cyrus Walker, was a return to the old days when mill owners were treated with deference. "The idea of being called a *good fellow*," he complained, "is far too prevalent among some of the heads of the departments here, besides they are too easy with the men to keep up the discipline necessary with a large crew of men." In common with American businessmen of the time, Northwest lumbermen defended their right to form combinations, but denied to labor the right to organize in defense of its interests. They were willing to pay high wages and to improve conditions of labor, but were unwilling to concede the legitimacy of unions. For older operators, the rise of labor was another depressing aspect of the industry's changing front. "There is a strong inclination among all the employees," wrote Walker, "to become high toned as to how they shall live, the time they shall work, the amount of work they do & the pay they receive, which makes things run harder than in times past."[63]

Statehood

Washington's admission to statehood in 1889 reflected the expansion of lumbering during the preceding decade. Washington lumber production increased in value from $1.7 million in 1880 to $17.4 million in 1890—the new state ranking fifth in the nation in the latter year—providing one of the principal arguments used by statehood advocates. Lumbermen, however, feared the creation of a new and uncertain political environment and the likelihood that people attracted to the region would have little sympathy for industry. "We shall, no doubt, have to take more interest in the politics of the new state," observed Cyrus Walker.[64] To the concern of mill owners, their control over waterfront property came under immediate challenge.

One of the most controversial questions before the state constitutional convention, meeting in the summer of 1889, was disposition of tidelands. A majority of the delegates to the Olympia conclave apparently believed that such lands should belong to the new state, as private ownership would retard economic opportunity. "This question involves interests of great magnitude on Puget Sound," reported Walker, "as nearly every mill, every warehouse, every railroad, every manufacturing industry, as well as all other interests concerned with shipping will be directly affected by the settlement of this question." The proper solution, Walker believed, was for the state to sell the lands to those who had in good faith built improvements. But domination of the convention by hostile rural delegates and ambitious "young lawyers" made such a compromise unlikely.[65]

Mill owners met in San Francisco to concert action and Walker worked to form an alliance of interests opposed to the course of the convention. What may have been the most effective response developed from the report of Seattle attorney H. G. Struve that a number of convention delegates "are generally accredited with being there for the purpose of boodle." Recommending that the mills "spend some money with some of the members," Walker set off for Olympia with a $5,000 fund at his disposal. Returning from the capital, he informed Will Talbot of the results: "the feeling among a majority of the members is much more favorable to giving the upland owners, that have occupied and improved the tide lands, a privilege to acquire a title from the State at a nominal valuation."[66]

Ultimately, the convention affirmed public ownership of the tidelands, but encouraged those claiming prior rights to seek legal redress. Creation of a special commission to establish exact harbor lines was mandated and the state was authorized to lease waterfront property. The convention, in effect, left final resolution of the issue to the legislature and to the courts. "It would be better to have the thing settled in the Constitution," noted Will Talbot, "as we would then be free from the blackmail of Legislators."[67] Lumbermen and others interested in protecting tideland improvements, though, had won a significant victory by forestalling complete abrogation of what they believed to be their rights.

Subsequently, the Harbor Line Commission and supporters of public ownership were pitted against waterfront interests in a flurry of emotional hearings and lawsuits. "The people of the town," George Emerson wrote of the proceedings at Hoquiam, "are in much the condition of a cage of unfed rats and have fallen upon themselves to tear themselves to pieces." For his part, Emerson, a longtime advocate of "metallic argument" as the best means of dealing with public officials, tried to bribe members of the commission. The life of the panel expired in early 1893 and its duties passed to the State Land Commission. The latter body, reported a relieved Emerson, "is very favorable to us." Acceptable harbor lines were finally drawn up and longterm leases of public holdings were negotiated.[68]

The tideland affair demonstrated that the industry would have to pay closer attention to political developments. To Cyrus Walker, for instance, it was essential "to prevent the demagogues from ruining the country." Although aware of the new importance of politics, mill owners attained only a limited understanding of the usefulness of political activity. Their goal was to prevent detrimental action by government, rather than to secure assistance. The task of industry lobbyists, wrote Walker, was to defeat "much of the unfriendly legislation, that is likely to occur." Such views reflected a restricted view of government and a contemptuous regard for politicians. When C. F. White, manager of the Grays Harbor Commercial Company, announced his candidacy for the state legislature, to take one example of this attitude, the *West Coast Lumberman* observed that it was "sad to see a first-class lumberman start down hill."[69] A full realization of the potential of politics would be left to later lumbermen.

Western Washington's lumber industry was transformed between the mid-1880s and the early 1890s. New investment and new markets stimulated the establishment of new manufacturing centers. Lumbermen, whether of the old San Francisco variety or the new Great Lakes strain, confronted changing technology, militant labor, and an altered political environment. Although overproduction and stagnant demand had created temporary hard times, most members of the industry joined in the early months of 1893 in confident expectation of a return of prosperity. "Unless a panic or cholera strikes this country," commented the *West Coast Lumberman*, "it is safe to say that the lumber business on the Pacific coast in 1893 will be large."[70] Mill owners were to be spared an outbreak of cholera, but they would soon fall victim to the most severe epidemic of depression yet to strike the United States.

7

Memorial Times

We have not failed yet to *meet every* paym't & every pay day with *Cash* & are the *only* mill on the Sound that can say *that* & we are doing our level best to keep the good start we have, as these times are going to be memorial.—C. D. Stimson[1]

When a lumberman now says 'come share my board,' it carries very little weight to a girl who is posted on the lumber market.—*West Coast Lumberman*[2]

Lumbermen who lived through the depression of the 1890s remembered it as the most devastating experience of their business careers. Coinciding with the return of Grover Cleveland to the presidency, and thus blamed on that hapless politician, the panic of 1893 ushered in the greatest economic crisis to that point in American history. Businesses, banks, and railroads by the thousand failed and an estimated four million workers lost their jobs. The best economic minds, recalled George S. Long of the Weyerhaeuser Timber Company, forecast that "it would only last a year or two, while as a matter of fact it took about six years for it to find the bottom and get an upward turn." Even the strongest lumbering concerns were severely burdened by the prolonged strain. Noting that the value of his holdings had been cut "right in two," Chauncey Griggs of the St. Paul & Tacoma Lumber Company described the depression in 1896 as "a burden which I am hardly able to stand."[3]

Rail and cargo markets collapsed, forcing the closure of many mills and the curtailment of those remaining in business. Washington lumber production increased only slightly during the decade, and the state's share of national output declined by a small margin. Great Lakes capitalists hestitated to invest in the Pacific Northwest, and the industry's changing front was halted at midpoint. Shaken San Francisco–based lumbermen sought to sell their properties, but could find no buyers. The economic collapse exacerbated social tensions and brought on an emotional confrontation between conservative and radical movements centering on the rise of populism and the presidential candidacy of William Jennings Bryan. At decade's end, mill owners sought in new methods of transportation and new markets the means of restoring prosperity to their industry. All in all, said Grays Harbor logger Alex

Polson, reflecting from the perspective of later years, the depression "set the state and the United States back nearly thirty years."[4]

The Panic of 1893

On the Pacific coast, the panic of 1893 struck first in the financial centers of California. "Business is very dull here in all lines," Frederick C. Talbot, Wiliam H. Talbot's brother and assistant, wrote from San Francisco in mid-May, "a great many people are leaving daily for the East, & the people remaining have no money to get away with." Banks, continued a flabbergasted Talbot, were "bursting all over the world." As depositors scurried to remove their funds from the banks, commercial activity ceased in San Francisco and in Los Angeles. The hysteria reached Puget Sound in early June, C. D. Stimson reporting that several Seattle banks were "quivering." Over the next several months, two-thirds of the financial institutions in Tacoma were bankrupted. In Everett, three of the city's five banks collapsed, as did the value of Henry Hewitt's extensive Port Gardner Bay investments.[5]

The disappearance of circulating money characterized the panic in Seattle, Tacoma, and other Sound communities. "Nobody has money," noted Cyrus Walker, "as it is locked up in the Banks, or stowed away in Safe Deposit Boxes, and old stockings." Threatened by depositors, banks hoarded their resources and refused to extend normal financial services to even the strongest customers. "I don't believe I could go to any of the Banks of Seattle today with $50,000. of good securities," observed Walker, "and raise $1000. on it." The *West Coast Lumberman* reported that merchants accepted shingles in lieu of cash, but discounted "the rumor that shingles were being thrown in contribution boxes and that ministers were taking them on salary accounts."[6] Under the financial strain, trade fell off and the Pacific Northwest economy collapsed.

Reflecting the impact of the depression, Washington's rate of population increase dropped sharply from the 375 percent of the previous decade to a relatively modest forty-five percent during the 1890s. Communities dependent upon lumbering were especially hard-hit by the economic crisis. Tacoma, for instance, grew by only five percent between 1890 and 1900. In contrast, Seattle benefitted from a more diversified economy and nearly doubled in size, establishing itself as the predominant urban area on Puget Sound. Once-important towns like Port Discovery, Port Madison, and Seabeck, where sawmills had gone out of business, all but disappeared as viable entities.[7]

When the initial period of panic passed, reflection on the economic future became a widespread and not altogether distressing pastime. The operators of the stronger sawmills actually welcomed the depression, believing that it served their interests. For one thing, the collapse would eliminate weaker competitors and allow the industry to achieve at last effective control over production and prices. "These tight times will have one good effect, in the end," predicted Cyrus Walker, "as it will knock out a whole lot of irresponsible concerns, that have ruined the business." As the weeks went by during the summer of 1893, Walker wrote with pleasure that "the outside Mills . . . are all the time getting into poorer circumstances, and they cant last much longer."[8]

Those firms surviving the crisis, moreover, could anticipate that organized labor would be undermined and that sharp cuts in pay would be possible. Wages were in fact reduced by fifteen percent at all of the major mills in August 1893, followed thereafter by periodic reductions. "The country from the Atlantic to the Pacific," pointed out George Emerson, "is full of unemployed men and hundreds of thousands who formerly received from $3. to $5. a day are anxiously seeking work at present at a $1. a day and boarding themselves." There was no cause for workers to complain, contended Will Talbot: "The men certainly cannot expect us to continue losing money while they are getting the same wages as they have been getting in the past when the Company made big money."[9] Adding to the certitude of this observation was the fact that plenty of willing men were readily available to fill the positions of those who quit in protest.

Labor would be forced to accept its proper and inferior place in the economic scheme of things, or so employers believed. "Thus far," wrote Emerson in September 1893, "it has been a rich mans panic, and before spring, it will be a poor mans empty stomach, which is perhaps the only way to teach a large class of our people that they can not organize against or dictate to capital without suffering consequences." The Hoquiam lumberman, who had become greatly alarmed over the decline of old-fashioned labor-management relations, contemplated the pleasure of again being able to supervise his affairs "with a taut rein and whip in hand."[10] The prospect of such a development was a large compensation for the monetary losses inflicted by the depression.

Unhappily for those welcoming the depression, the collapse proved to be a disaster for all concerned. The rail trade was immediately effected, financial conditions in the east causing demand to disappear by mid-1893. Railroad shipment of Washington lumber dropped by a third between 1892 and 1894, despite the lowering of freight rates by the Great Northern and the Northern Pacific. The shingle industry, with its makeshift financial structure, was also hard-pressed. "The Shingle business on the Sound has collapsed," noted Cyrus Walker in a neat summation of the situation, "as the East has no money, consequently cant buy them." By early 1895, three-fourths of the shingle mills in the state had closed down.[11]

Rail mills located on tidewater compensated for declining trade by shipping lumber to California and foreign ports, driving prices in those markets down to new lows. A sudden influx of lumber from British Columbia, made possible by the removal of import duties in the Wilson-Gorman Tariff of 1894, added to the problems in California. British Columbia operators made quick and abundant use of the state as a dumping ground for lumber that could not be shipped east over the Canadian Pacific, and they were more than willing to cut prices. Competition from north of the line combined with excess shipments from Puget Sound and Grays Harbor to produce what George Emerson described as "fearful" conditions in San Francisco.[12]

The depression also caused the exodus of Great Lakes investors to come to an abrupt end, dealing a serious blow to those old-time operators wishing to sell their property. Despite the fact that timberland could be acquired at bargain rates, eastern capitalists hesitated to invest in the region because of the economic downturn. Frederick Weyerhaeuser, for instance, allowed an option on 850,000 acres in southern Oregon to lapse in mid-1893. Two years later, Weyerhaeuser rejected an offer of 55,000 acres near Olympia, commenting that he "would not want to make any investments at the present time owing to the condition of trade throughout the country."[13] In these depressed years, the only major investment by men from the Great Lakes was the establishment in 1894 of the Pacific Empire Lumber Company at South Bend by O. H. Ingram and William H. Day, two of the leading lumbermen of the Mississippi Valley. To the chagrin of Pacific Northwest boosters, there was a distinct lack of enthusiasm for the region among investors.

Lumbermen expecting to regain their old-time dominance over labor also endured disappointment. Rather than stifling the protests of workers, the depression brought increased unrest and frightful prospects for mill owners. The first sign of this development came from the strike of the 700-member shingle weavers union in the summer of 1893. The strike was broken and the union all but destroyed, but manufacturers anticipated continued trouble from their unpredictable employees. "The shingle element is such an unknown quantity," complained one mill operator, "that one can never tell what they will do next." C. D. Stimson closed his Ballard shingle plant because of constant difficulties with employees, difficulties that made the state of affairs in his adjacent sawmill seem almost placid in comparison.[14]

For men of conservative instincts, the country seemed on the brink of revolution by mid-1894. The Pullman strike halted nearly all railroad traffic west of Chicago, made it impossible to ship lumber from Puget Sound, and, according to Chauncey Griggs, "prostrated" mills dependent on eastern markets. At the same time, large numbers of the unemployed on Puget Sound and Grays Harbor joined the industrial army led by Ohio businessman Jacob Coxey. Planning to march on the nation's capital to protest economic conditions, the army presented a disturbing prospect to owners of property. "The Industrial Army is daily growing larger there," Cyrus Walker wrote after an April visit to Seattle, "and now numbers some 900 men. They are quiet and orderly thus far, but the citizens will be relieved when they leave for Washington."[15] The bands of the unemployed and the railroad strikers brought home to lumbermen the realization that the depression was unlikely to be a beneficial experience.

Such episodes as the organizing of Grays Harbor mill workers to protest wage reductions in 1894 destroyed the gleeful anticipation of George Emerson and other employers that the industry would be able to return to the good old whip-wielding days. American business enterprise was set on a course, feared mill owners, that was likely to end in worker control. Contemplating the Fourth of July, a dejected Emerson observed: "I have a feeling of late that the people of America have no longer any independence to celebrate, and that memorial services are more in order."[16] Workers, in the opinion of lumbermen, failed to learn the proper lesson from the depression and, in their refusal to take upon themselves the burden of collapsing markets, denied the validity of economic law.

New Efforts at Control

Two approaches to the depressed conditions of the 1890s were standard: Individual firms slashed expenditures through wage reductions and installation of labor-saving devices, hoping thereby to sell low-priced lumber at a profit, and, secondly, the industry as a whole created new instruments for the control of production and prices. Along the former course, the Port Blakely Mill Company was the most active in the drive to reduce costs. Brothers John and James Campbell, in charge of the mill since the death of William Renton in 1891, installed new saws, refuse chutes, and an elevated tramway, enabling reduction of the workforce and of the time required to load vessels. Increased efficiency allowed the mill to run fulltime in spite of the depression, to the mystification of old-fashioned observers like Cyrus Walker. "The way they are running the Blakely Mill is a conundrum to me," wrote Walker. "Somebody must be loosing money fast, I think."[17]

Port Blakely's effort to reduce the cost of logs led to formation of the Simpson Logging Company, one of the major firms in the industry.

The mill company's Mason County logging railroad having reached the limit of its timber holdings, the Campbell brothers and Charles Holmes joined with Sol Simpson and A. H. Anderson to found a new woods operation in 1895. A virtual subsidiary of the Port Blakely Mill Company, the logging firm operated eight camps capable of producing a half million feet of logs a day. Port Blakely received first call on these logs, at slightly below the prices charged to other customers, giving the mill a significant competitive advantage. "We always get our logs some cheaper than the other people," boasted John Campbell.[18]

The depression, though, was so severe that collective action, rather than individual efficiency, seemed to be the best remedy for the ailment of glutted markets. Despite the evident shortcomings of the Pacific Pine Lumber Company, which continued in a paper existence, lumbermen worked on combination schemes. Longtime opponents of such arrangements like George Emerson, Chauncey Griggs, and C. D. Stimson now favored an expanded San Francisco pool, demonstrating better than anything else the impact of the depression. The new combination was to include virtually all of the cargo mills on the Pacific coast. Arrangements with retail yards and independent loggers were anticipated, adding to the complexity. Everyone agreed that the plan offered the best hope for the industry, but few believed in its success. "From its general appearance," observed Pope & Talbot executive A. W. Jackson, "I think that it *can be worked* successfully, . . . provided the right kind of men are at the head, . . . men with clear heads and honest. Whether it is possible to obtain such men in the lumber business is an open question to be answered."[19]

An affirmative answer was secured by late 1895 with the signing up of most of the San Francisco mill owners. "There has been less trouble in securing the signatures of the people here than anticipated," reported Will Talbot, "every one being willing to go into the enterprise as they do not think that by doing so they can make matters any worse than they are at the present time." To operate for one year, the Central Lumber Company distributed a thousand shares of stock, the Puget, Port Blakely and Tacoma mill companies, and the Asa Simpson operation becoming the largest holders. "The proposition is," explained Talbot, "that all of the Mill Companies sell their product to Central Lumber Co. at a scheduled price." The lumber would then be sold back to the San Francisco agents of the mills "at an advanced rate, making a profit of from $1.50 to $3.00 per M." At the end of a year, dividends would be distributed on the basis of shares held, and the scheme would be extended for an additional period.[20]

For all the high hopes, the new concern proved to be a crushing failure. California retailers stocked up with low-priced lumber prior to the start of business and thereafter balked at dealing with Central Lumber Company participants. "There are still enough mills outside the pool," noted one lumberman, "to furnish a good deal of lumber and small yards will buy of them when they can get their lumber at a little less price than they can buy of the combination." Within the trust, owners of small mills complained that they were treated unfairly. "While all parties may be living up to the letter of their contracts with the C.L. Co.," wrote the San Francisco representative of the Pacific Empire Lumber Company, "in the spirit they are being violated. The parties here who were active in forming the C.L. Co. did not ignore their own interests when doing so." Cyrus Walker retorted that the major firms should not be expected to make undue sacrifices. "If all the small mills were treated equally liberal by the Central Lumber Co," he argued, "there would not be much business for the large mills to do."[21]

Under these circumstances, the existence of Central was short and painful. To the surprise of few and the regret of many, the combination expired at the end of 1896. The episode once again demonstrated the complexity of the lumber industry. Despite the economic emergency, mill owners were unable to reconcile their differing interests. The result was further demoralization, Chauncey Griggs pointing out that the California market had been "left . . . wide open." Trade in the state, commented the *West Coast Lumberman,* "seems to be in as bad a condition as a bladder with a hole in it or a balloon with a puncture." According to that trade publication, the best hope for the market appeared to be a providential outbreak of "the Bubonic plague" among the lumbering interests of the Pacific coast.[22] Washington mill operators, however, preferred the blandishments of William McKinely to the competition-reducing spread of epidemic diseases.

The Populist Challenge

Throughout the United States, urban reformers, farmers, and organized labor reacted with outrage to the economic collapse and to the inability of the Cleveland Administration to bring an end to the depression. Especially frightening to conservatives was the rise of populism and its fusion with the Democratic presidential campaign of William Jennings Bryan in 1896. In that year, Democrats like Chauncey Griggs joined a conservative coalition of lumbermen in support of Republican nominee William McKinely, believing that the abandonment of traditional party ties was necessary to save the country from revolution. Neither before nor since have industry leaders felt so threatened by political developments.

The focal point of the challenge mounted by the Populist party and by Bryan was the call for an increase in the supply of money through the free coinage of silver at a ratio of sixteen silver to one gold. Such a course supposedly would advance prices and end the depression. In contrast, conservatives supported the gold standard, arguing that only the maintenance of so-called sound money would enable business to recover its long-vanished confidence. Northwest lumbermen were, in fact, divided on the merits of silver. Grays Harbor logger Alex Polson, an influential Republican, was an enthusiastic advocate of the metal and equaled the most rabid Populist in his denunciation of the gold standard as a British plot.[23] Even Polson, though, joined in the prevailing industry sentiment that implementation of free silver would complete the wreckage of the economy and end for all time the possibility of renewed investment in western Washington.

Bryan's nomination on a free-silver platform by both the Democrats and the Populists provided the catalyst for joint action on the part of lumbermen. "The success of the democratic, silverite, populist fusion," wrote George Emerson, "means financial disaster, distress, and perhaps revolt, if not civil war." Cyrus Walker, who actually met and shook hands with Bryan, believed it would be "a hard fight to beat the *free silver cranks.*" Nevertheless, the uniting of conservative Democrats and sensible Republican silver advocates behind McKinley offered the best chance to stave off Bryan and ruin. The campaign, predicted Emerson, "will take none from the solid front the Republican party presents, and many from the mixed and many headed uncongenial mass that is arrayed against it."[24]

Lumbermen threw themselves into the campaign with a determination and vitality rarely demonstrated. The elderly Asa Simpson even journeyed from San Francisco to speak for McKinley among his employees at Hoquiam, South Bend, and Knappton. Labor, went the argument followed by Simpson and other industry spokesmen, was as dependent on Republican victory as management. "If McKinley is elected," summed up Cyrus Walker, "there will be some hope for improvement but if Bryan wins, we shall all become bankrupt." Mill owners throughout the region threatened to close down and retire to California should the so-called "Popocrat" candidate be elected. Although "the best element of the laboring class" was receptive to this chilling threat, noted Chauncey Griggs, most workers were difficult to reach. "They dont read," explained Walker of the inability to deal with such men on a reasoned basis.[25]

Despite the efforts of the conservative alliance, Bryan carried the state, Populist John Rogers was elected governor, and the Democratic-Populist forces secured control of the legislature. McKinley staved off the Bryan challenge nationally, however, more than making up for the local disappointment of lumbermen. Washington's businessmen were "ashamed of our State," according to Emerson, but "proud of our Nation." The discrediting of silver and the rapid dissolution of the Populist party following the election removed the principal obstacles to the restoration of business confidence. National recovery was anticipated and the demand for lumber was expected to revive. Even at the state level, no one expected antibusiness actions from the legislature. Washington had "gone popocratic," observed Emerson, "but not on popocratic principles, only on the free silver question."[26] All in all, the election was cause for the first expression of optimism in years.

Within days of the McKinley victory, Cyrus Walker reported the coming into existence of "a marked feeling of confidence in business generally on the Sound."[27] The return of prosperity was credited to the new Republican president just as the panic had been blamed on Cleveland. Aside from the brief impact on the nerves of business leaders, the election campaign had important, longrange implications for the state and the industry. The decamping of Griggs and other wealthy leaders deprived the Democratic party of financial support and contributed, once the Populists had disappeared from the scene, to Republican domination of the state, a domination continuing until economic crisis again revolutionized politics in the 1930s. The apparent connection between the election of McKinley and prosperity, moreover, taught the importance of close attention to presidential politics. In the future, lumbermen would often see their prospects as dependent on the election of Republican presidents.

Prosperity Returns

Post-election prosperity was especially manifest in the rejuvenation of the rail trade. A sixth of the lumber produced in Washington was shipped east over the Northern Pacific and the Great Northern by 1899, and the rate would double between that year and 1901. Although turning out a quarter million feet of lumber a day, the St. Paul & Tacoma Lumber Company was still "unable to fill the orders we are receiving" from rail markets, according to Chauncey Griggs. Even the old cargo mills at Port Blakely and Port Gamble entered the trade in a limited manner, sending lumber by scow to Seattle for shipment. George Emerson reported that his firm's "Eastern trade is growing to such proportions we now employ two stenographers nine hours per day instead of one for a few hours as in the days long past." Seattle mills, noted a trade journal, spent $5,000 a year on stamps in their dealings with eastern customers. So great was the demand that mill owners complained of a lack of sufficient rail cars to carry away their lumber.[28]

To operators interested in making a profit, the situation in the traditional cargo markets was less enthralling. In early 1899, the Southern

Pacific Railroad reduced freight rates between Portland and California, enabling Willamette River mills to sell low-priced lumber in the Bay Area and in the interior valleys of the state. Coming after the demise of the Central Lumber Company, this was a severe blow to Puget Sound and Grays Harbor firms. "The Portland dealers and the Southern Pacific R.R.," warned Pope & Talbot's A. W. Jackson, "are laying their forces to capture or control the trade of Northern California." To George Emerson, the significance of the Oregon rail shipments was clear. The California market—historically the mainstay of the Washington industry—"is lost for all time," he wrote.[29]

The dependence of Washington firms on outmoded and expensive sailing vessels made it difficult to reduce shipping costs to meet the Oregon competition for California trade. The perfection of ocean-going rafts, however, offered the promise of a new cost-cutting device. Rafting experiments had been carried out in New England as far back as the 1790s, but the technique received its greatest boost when Hugh Robertson developed during the 1880s a method of building rafts in floating cradles. Although Robertson failed in efforts to tow rafts to San Francisco from Mendocino and Coos Bay between 1889 and 1893, he attracted the attention of important lumbermen. In 1897, Renton, Holmes & Company, Pope & Talbot, and several other Washington mill owners joined with Robertson to incorporate the Robertson Raft Company. The new firm immediately set to work binding piling into rafts on the Columbia River and at Seattle. Lumber was secured on top of the piles, allowing cheap transportation to California.[30]

Promising in theory, rafting operations were plagued with practical difficulties. The worst danger, for instance, nearly came true in September 1899 when a Puget Sound raft drifted away off the Northern California coast following the parting of its towline. A disconsolate Frederick Talbot wrote that "the chances are small of ever seeing it again" and even the recovery of the raft after a week of searching failed to down the disturbing implications of the episode. In addition, vessel owners, fearful of the impact on freight rates, sought legislation banning rafts as navigation hazards and unemployed sailors were suspected of engaging in sabotage. Ultimately, disagreements between Robertson and his backers led to dissolution of the raft company. Rafting efforts continued on the Columbia River, but were abandoned by Puget Sound interests.[31] The latter had to locate new water markets to compensate for declining San Francisco trade.

Several major events provided new, if fleeting, opportunities for lumbermen. Reports of the Alaska gold strike reached Seattle in July 1897, setting off a hysterical reaction reminiscent of the glory days in California. "Seattle is all 'agog' with this gold fever," wrote Eugene Semple, former territorial governor, "and the streets are crowded with knots of men so worked up over the news that they can scarcely avoid being run over by the cars and carriages." Visiting the city, George Emerson found that "conditions . . . are the wildest I have ever seen." Hotels were filled, trains deposited hundreds of persons every day, and all manner of vessels were readied for the voyage to the north. An estimated one hundred thousand people were expected to travel to Alaska with the spring weather in 1898, and here was a vast and obvious field to exploit. "In all probability," observed Cyrus Walker, "lumber will be in good demand there for some years to come."[32]

Sawmills in Seattle, the gateway to the north, moved to dominate the trade. The Stimson Mill Company became the leading supplier, shipping nearly four million feet of lumber in December 1897 alone. Mills outside the city, in contrast, were slow to enter the trade and found Alaska overstocked when they finally dispatched cargoes. In January 1898, Will Talbot and Charles Holmes sent a steamer to Skagway, only to discover that the supply of lumber greatly exceeded the demand. A few months later, John Campbell visited Alaska after a Port Blakely vessel ran aground off Skagway, learned that there was "such poor demand that it would not pay to detain [the] vessel longer to discharge" and had the ship and its cargo towed back to Puget Sound.[33] Except for early-on-the-scene Seattle operators, the gold rush proved a distinct disappointment.

A burgeoning foreign trade at the end of the decade overshadowed the frustration of Alaska. Traditional markets in Australia absorbed a great amount of lumber and Puget Sound shipments to the island continent doubled between 1898 and 1900. The discovery of gold in the Transvaal opened up a profitable new market in South Africa. Trade was retarded by the outbreak of the Boer War in 1899, but mill owners anticipated that a rapid British victory would result in increased demand. "Under British rule," predicted the *Pacific Lumber Trade Journal,* "South African progress and prosperity will be greatly enhanced."[34] More than anything else, the splendid little war between the United States and Spain in 1898 expanded the commercial horizons of Northwest lumbermen.

Like most American businessmen, industry leaders initially feared the war's uncertain impact on trade. There was, in ignorance of the decrepit condition of the Spanish fleet, concern that merchant shipping would be swept from the seas and foreign lumber markets closed to American exporters. Practical men, though, realized that a short and victorious war could result in considerable profit. The sending of an army across the Pacific to Manila required construction of support bases in Hawaii and greatly increased the demand for lumber in the islands. Following upon Dewey's victory at Manila Bay, military construction in the Philippines opened a new outlet for Northwest mills.[35] Of most importance, the annexation of Hawaii and the acquisition of the Philippines placed America in position to exploit the fabled China market.

Railroad construction and the development of industry and port facilities had stimulated the demand for lumber in China prior to the war. With United States sovereignty established on the rim of Asia in 1898, further increases were anticipated. "We feel that it is too good a business not to take a little chance on," Will Talbot noted of the prospects. In the summer of 1900, Pope & Talbot sent a representative, James Claiborne, Jr., to Asia in hope of developing a large-scale trade. Reporting from Yokohama, Claiborne found the economic situation bleak and Japanese businessmen unworthy of trust. "They have no principle or honor and cannot be depended upon at all . . . . they are proud to be called what we would term a sharper or a liar." The common perception that "Japs are a very intelligent and honorable race and far ahead of the Chinese," observed Claiborne, "is way off." Moving on to Vladivostok, he confronted language barriers, "nasty mysterious looking Russian grub" and "discouraging" prospects for trade.[36]

On the main portion of his journey, Claiborne arrived in China in the midst of the violent outburst of nationalist elements known as the Boxer Rebellion. Business was totally disrupted, and the military forces of the great powers marched across the land, putting down the revolt. At Tientsin, wrote Claiborne, the only source of water was the river "and the river is full of dead bodies." With commercial activity halted, Claiborne's mission ended in failure. "I am heartily disgusted, disappointed and ashamed of this trip," he informed his employers. "A great deal of money has been spent uselessly. I have accomplished nothing, and the sooner we bring this unsatisfactory trip to a close, the better for all of us."[37] Thus vanished, amidst war and cultural confusion, Pope & Talbot's vision of vast trade with Asia.

The shimmering away of the China market failed to detract from the improving position of the industry. The boom in railroad and offshore markets more than made up for the stagnant conditions in California and the failure in Asia. Business increased on a month-by-month basis and a visitor to Tacoma reported that every mill in the city was "running day & night." Advancing prices reflected the natonal economic climate in the aftermath of McKinley's election. "The lumber trade," observed a trade journal, "is simply a part and parcel of the general prosperity of the country—a prosperity which bids fair to continue for several years to come."[38] The great depression at last over, lumbermen were free to reflect on the future.

For George Emerson, one of the most astute members of the industry, the principal lesson of the depression was that the old-fashioned mills were doomed to extinction. Future profit, he was convinced, depended on fullscale exploitation of the rail market. In contrast,

ocean shipments should be restricted to selected foreign ports and the
traditional San Francisco market should be abandoned as a prime outlet. "California from this on," contended Emerson, "will be but a
dumping ground for surplus stock." Unlike those who looked to the
past, transfixed by time-honored ways of running their industry, Emerson was determined to "get into the procession and march with the
times." Along that route was to be found the "salvation of the lumber
business," but lumbermen traveling the old course faced only continued frustration.[39]

One of those finding it difficult to adjust to new conditions was
Cyrus Walker. Nearing the end of fifty years on Puget Sound, the old
man was convinced that "I have about outlived my usefulness." At
the turn of the century, Walker shipped an automobile to Port Ludlow,
his home since the late 1880s, but in the absence of roads could only
drive it around the mill grounds. When he wished to take an excursion,
Walker was restricted to loading the car on a steamer bound for Gamble, there to putter about that equally isolated port. The aging lumber
baron's efforts to come to grips with this latest product of American
technology symbolized how cut-off the old San Francisco–owned mills
were from the changes underway in their industry and dramatized the
need for what George Emerson termed "new blood, . . . active brains
and energetic bodies, up-to-date in all business transactions, associations and ways."[40]

Surveying the years of depression at the end of 1898, Chauncey
Griggs observed that "we have passed through the hard times and are
on the upward trend, but it has been a herculean task to overcome
obstacles which have been thrust in our path since we came to this
beautiful section of our country." Although the output of Washington's
sawmills registered only a modest increase between 1890 and 1900,
the monetary value of the state's lumber production doubled, indicating that the future was bright with promise. The best indication of this
was the revived interest of Great Lakes investors. "It looks very probable now," wrote Cyrus Walker in March 1899, "that there will be
parties out here during the coming summer to buy up the timber land
west of the Cascade Mountains." Washington, reported the *Pacific
Lumber Trade Journal,* "is the mecca to which all good lumbermen
seem to be traveling."[41] The value of stumpage began to mount and
those who had refused to buy timber in the aftermath of the panic
began to regret the missed opportunity. Stalled by the depression, the
change of front was about to be concluded in a final rush.

8

Imperial Forests

An age of big things calls for big men. . . . The big man when fully
developed is going to be one of the best types of big men, for he will of
necessity be robust, vigorous, daring, bold, cautious, prudent, broad-minded
and generous, for does he not work in Nature's own temple, the heart of
the forest, drawing health, vigor and inspiration from his environment?—
George S. Long[1]

Great central lumber companies like the Weyerhaeusers moved to the
Pacific Northwest, where they secured imperial forests.—Frederick Jackson Turner[2]

On an early February day in 1900, a group of men in broadcloth gathered about a large table in the Tacoma Hotel. A month earlier, they
had completed lengthy negotiations for the purchase of 900,000 acres
of western Washington timberland from the Northern Pacific Railroad.
Meeting in Tacoma to formalize plans for the management of this vast
tract, the investors were led by a slight man with a wispy white beard
and thick German accent, a man whose shyness masked an ability to
bring together powerful men in the interest of mutual profit. With the
signing of the documents, the Weyerhaeuser Timber Company began
its existence as the second largest private holder of timber in the nation
and the dominant force in the forest industry of the Pacific Northwest.
Although assembled reporters were informed that the firm would soon
be manufacturing lumber in a large way, the real purpose was not the
building of sawmills, but the holding of timber for longterm earnings.
"This is not for us, nor for our children," Frederick Weyerhaeuser,
the investment syndicate's organizer, privately observed, "but for our
grandchildren." The arrival of Weyerhaeuser in the Pacific Northwest
was the dramatic culmination of changes underway for a decade-and-
a-half and, in the apt descriptive summary of the federal Bureau of
Corporations, "one of the most spectacular events in the timber history
of the country."[3]

Weyerhaeuser Arrives

Frederick Weyerhaeuser was to lumbering what Carnegie was to steel
and Rockefeller to oil, the symbol of an industry and the principal
target of its critics. Born in the wine country of Germany in 1834,
Weyerhaeuser's boyhood hero was Napoleon, a bizarre choice for a

German, but perhaps prophetic for the ultimate builder of a forest empire. Traveling to America with relatives in 1852, Weyerhaueser settled in Pennsylvania with the intention of becoming a brewmaster. This
ambition "was given up," he later recalled, "when I saw how often
brewers became confirmed drunkards." Sobered by this realization, he
journeyed west to the Mississippi River and secured a job as salesman
for a sawmill in Rock Island, Illinois. When his employers were bankrupted in 1857, Weyerhaeuser took over the mill and made it the base
for expansion in the Great Lakes region. From this episode, *Fortune*
magazine would devise "a formula for getting rich: get a job, save
your money, wait for your boss to go broke. Then take over the business."[4]

With Frederick Denkmann, the husband of his wife's sister, Weyerhaeuser organized the Rock Island firm of Weyerhaeuser & Denkmann. The company "made money steadily, [and] rapidly" according
to a credit report and provided the funds for additional investment. In
the quarter century after the Civil War, Weyerhaeuser built a con-
glomeration of mill and timber companies stretching over six Midwestern states and became one of the wealthiest men in America.
Spreading his personal assets, he controlled these firms in alliance with
the major lumbermen of the Great Lakes: James and Matthew Norton,
W. H. Laird, O. H. Ingram, John Humbird, Peter Musser, and others.
Although a member of the group complained of the "one man power
management of Mr. Weyerhaeuser," most recognized his coalition-
building talent for reconciling divergent interests and accepted his
dominant position. By the turn of the century, the group owned 300
billion feet of timber, produced twelve percent of the lumber in Wisconsin and Minnesota, and constituted, according to its critics, a gi-
gantic timber trust.[5]

More than anything else, Weyerhaeuser hoped to pass on a thriving
business to his four sons. By the 1890s, however, there was considerable doubt that a secure family lumbering dynasty could be maintained in the lake states. A third of the nation's lumber supply had
come from Michigan, Wisconsin, and Minnesota in 1889, but the rate
fell to twenty-five percent in 1899 and to twelve percent in 1909. Rapidly increasing log prices reflected the declining supply of timber. Like
other Great Lakes lumbermen, Frederick Weyerhaeuser looked to the
west and to the south as the focus of future operations. According to
legend, he was repelled by the climate and the social conditions of the
South and concluded that members of his family could not be expected
to live there. This left the Pacific Northwest as the only outlet for
investment. Therefore, Weyerhaeuser and his associates, as the *Pacific
Lumber Trade Journal* informed its readers in January 1897, had "their
eye on the Pacific Northwest."[6]

Weyerhaeuser had been casting his eye beyond the Rockies and the Cascades for over a decade, his vision expanded through lengthy conversations with James J. Hill, a convenient neighbor in St. Paul. His first direct involvement in the region preceded formation of the giant company that would bear his name. Weyerhaeuser & Denkmann was heavily involved in the acquisition of 45,000 acres of timber on northern Puget Sound in 1899. The firm owned half the stock in the Sound Timber Company, formed to manage this tract, although its ownership apparently was not known to the public. Weyerhaeuser also purchased an interest in the Humbird family's lumbering operation on Vancouver Island.[7] These early moves, though, were only sidelights to bigger transactions.

With a number of his longtime associates, Weyerhaeuser had organized the Coast Lumber Company in 1898. Although the company's ostensible purpose was to purchase and market Washington shingles in the Midwest, it seemed likely to become the vehicle for a full-fledged Weyerhaeuser move to Puget Sound. In July 1899, it was offered the Bell-Nelson sawmill and adjacent waterfront property in Everett. The facility could be so improved, argued Coast official W. I. Ewart, "as to become the leading factor in the Western trade." Weyerhaeuser agreed and gave to the project his "full endorsement." Opposition among the stockholders to the expansion of Coast's activities, however, meant that organization of a new company would be required for serious exploitation of the Pacific Northwest.[8]

Weyerhaeuser's intentions became a leading topic of speculation in the region. One trade journal even dedicated a monthly column to detailing the latest movements of the famed lumberman and his colleagues. The rumors centered on a deal with the Northern Pacific, the size of which was so immense as to defy belief. When George Emerson informed him that Weyerhaeuser had secured an option on all the railroad's timberland, an incredulous R. D. Merrill, himself a recent arrival from the Great Lakes, responded that it was "more likely" the arrangement was limited to "the lands in Chehalis [Grays Harbor] County." By November 1899, though, the railroad had removed its western Washington timber from the market and all observers acknowledged that Frederick Weyerhaeuser would soon be the leading lumberman in the Pacific Northwest. "It would seem as though these parties," observed William H. Talbot, "were intent upon capturing a great amount of timber land, and it may be their idea to eventually control the business."[9]

Under conditions of some suspense, a number of Washington firms decided that the time had come to benefit from the resumption of Great Lakes investment by selling out, if possible to Frederick Weyerhaeuser. In March 1899, Cyrus Walker was approached by an intermediary

for easterners interested in purchasing the Puget Mill Company. Although unable to ascertain their identities, Walker did learn that the
investors were "capitalists from Minnesota, Wisconsin and Michigan"
who did not at present own timber or mills on the Pacific coast. Furthermore, he wrote, the offer "comes through Judge [Thomas] Burke,
who is the Attorney for the Great Northern in Seattle, therefore I presume they are friends of J. J. Hill." Since Hill and Weyerhaeuser were
neighbors, Walker was convinced that "the Warehouser crowd" wanted
the company's holdings. The prospective purchasers were interested
in timber, but were willing to buy mills, vessels, and other assets.[10]

The prospect of selling to Weyerhaeuser excited the interest of the
Puget Mill's owners. Walker responded favorably to the possibility of
unloading "this *elephant* of ours, which is gradually killing off all that
have anything to do with the active management of the concern," and
supported a sale "if anything like a fair price can be secured for the
property." Agreeing with Walker "in every particular as far as selling
out is concerned," Will Talbot believed that "it would be best for all
concerned if we can dispose of the Company's property." Firm in the
belief that the rapidly increasing value of stumpage would enable a
more profitable sale at a later date, Talbot and Walker set a prohibitive
price on their holdings, contending that even the logged-off land was
worth $5 an acre.[11] Negotiations, therefore, did not proceed beyond
the preliminary stage.

A more vigorous and direct effort to conclude a deal with Weyerhaeuser was undertaken by George Emerson, who had chafed for years
at the eccentric business methods of Asa Simpson. Lacking sufficient
funds himself, he needed a wealthy partner to carry out his plan to
purchase the North Western Lumber Company. "If we could get the
Weyerhauser interest centered here," Emerson wrote, "we think all our
holdings would come to the front. I have been therefore very anxious
to accomplish that move and sincerely hope I shall be able to get them
to consider the purchase of the property without farther delay." Should
the rumored arrangement with the Northern Pacific be concluded,
Emerson believed, the Weyerhaeuser investors would require several
sawmills in order to produce enough lumber "to meet their interest and
tax account." The Hoquiam mill, with its access to both cargo and rail
markets, was ideal for this purpose.[12]

In November 1899, Frederick Weyerhaeuser and several of his colleagues inspected North Western's Hoquiam and South Bend plants.
The response of the visitors was encouraging and a Weyerhaeuser-
Emerson purchase of the company seemed likely. "I have felt from
the first," Emerson informed a friend, "the Weyerhauser people were
the parties with whom we would eventually deal." But Emerson's optimism lapsed with the passage of time. Meeting with W. I. Ewart of

the Coast Lumber Company in mid-December, he learned the bad news. Although "Mr. Weyerhauser is very strongly inclined toward accepting our proposition," Emerson reported, other members of the syndicate contended that the pending Northern Pacific purchase was too costly an undertaking to allow for additional acquisitions.[13] As the Grays Harbor lumberman reluctantly conceded, the scope of the railroad-Weyerhaeuser transaction was indeed breathtaking.

Convinced that the best investment opportunities were slipping away, Frederick Weyerhaeuser was ready to move by the summer of 1899. Early in the year, he offered $5 an acre for a million acres of Northern Pacific land west of the Cascades. This proposal was rejected, but both parties remained interested in working out an agreement. Weyerhaeuser increased his offer to $6 an acre and the Northern Pacific, for its part, was in need of funds to meet financial obligations. "The great inducement for us to concede such a price," wrote Charles Mellen, the line's president, "is in order to get cash and notes . . . that we can convert, so that the money we receive may be used immediately for the purposes of our General First Mortgage sinking fund."[14]

Two obstacles delayed conclusion of the sale: the price and the railroad's insistence on a clause requiring that it be granted a monopoly on rail transportation of Weyerhaeuser lumber. Negotiations focused on these questions for several weeks, both sides eagerly in search of a compromise. Under the final terms, the transportation clause was limited to fifteen years, the extent of the purchase was reduced to 900,000 acres, and the price was set at $6 an acre. Since there were no plans to manufacture lumber in the near future, Weyerhaeuser's attorney privately pointed out that the shipping section "does not make any more difference . . . than though we had received a clear deed, and made no contract whatever." With agreement on these points, the transfer documents were rapidly put into shape.[15]

A down payment of $3 million was transferred to the railroad on January 3, 1900, the balance to be paid in eight semiannual installments of $300,000. One third of the $5.4 million purchase price came from Frederick Weyerhaeuser and the remainder was provided by a dozen investors. Thirty percent of the stock in the firm created to manage the timber was owned by Weyerhaeuser, a fifth was held by Laird, Norton & Company, and the rest went to O. H. Ingram, Robert L. McCormick, Sumner T. McKnight, and other lumbermen. The long-term nature of the investment was demonstrated by the fact that Ingram's partner, W. H. Day, declined to participate, explaining that he wished to "invest in what seemed more certain of speedy income, as I am more interested in returns to spend and enjoy now, . . . than in laying up an asset for my heirs."[16]

Concentrated in southwestern Washington, the new Weyerhaeuser holding electrified the Pacific Northwest. The transaction "has taken us all off our feet," one lumberman wrote Ingram, "and we have hardly gained our equilibrium as yet, not only the magnitude of the deal, but the cheapness of which it was bought." All observers agreed that the purchase would enhance the worth of timber holdings. "It will immediately stiffen values all over the country," observed Charles Holmes of the Port Blakely Mill Company when the sale was first reported, "and stumpage certainly will advance." All agreed as well that the firm organized by Weyerhaeuser and his associates would dominate the industry.[17] In the early months of 1900, old-time lumbermen looked to Tacoma, where Frederick Weyerhaeuser's plenipotentiary busily assembled the new giant.

The Work of George S. Long

One member of the syndicate, R. L. McCormick, moved to Puget Sound to take an active role in the management of the Weyerhaeuser Timber Company as the firm's secretary. The real operating head of Weyerhaeuser, though, was George S. Long, an Indiana native with extensive experience in the lumbering business of the Midwest. Before accepting the position of manager of the timber company, he had for many years been the sales manager of the Northwestern Lumber Company of Eau Claire, Wisconsin, a frequent participant in the schemes of Frederick Weyerhaeuser. Tall and lanky, with receding chin and prominent adam's apple, Long wore an old-fashioned stiff collar and, except for an ever-present cigar, resembled a country schoolteacher. He combined a reputation for integrity, an ability to view events from the large perspective, and a penchant for humor with the dominating size of Weyerhaeuser to exert for three decades a degree of leadership unprecedented in the history of the Pacific Northwest industry.[18]

Operating out of a small office suite in the Northern Pacific building in Tacoma, Long commenced the business of Weyerhaeuser in March 1900. He worked so hard in these early weeks, reported the *West Coast Lumberman,* that when "it was rumored in the east that Manager Long was working only thirteen hours a day" his employers feared that he had fallen ill. The labor was necessary, for Long readily admitted that he had no firsthand knowledge of the Pacific Northwest or of the problems involved in timber management. "Absolutely without a particle of experience in the land business," he focused on acquainting himself with the company's holdings and with the region's industry. "Every time I get out into the woods," Long noted, "I learn something that I didn't know before."[19]

Long confronted several immediate and important problems. For one thing, he believed that it was imperative to establish an image of cooperation for Weyerhaeuser, to overcome any concern that the company would use its size to the detriment of the public or of other firms. "If you have such a thing as public sentiment . . . to contend with," he observed, "you want to study that very carefully, and find out who really controls public sentiment; trying to get them friendly towards your interests, and avoid doing things that will irritate the public; in other words, a man has to be a good fellow." Thus, when Long unwittingly acquired at an auction a section of state timber desired by the Port Blakely Mill Company, he wrote John Campbell that Weyerhaeuser had "no inclination or disposition to interfere with your plans." He then arranged to sell the timber to Port Blakely at the same price he had paid the state. "The province of a diplomat," Long believed, was "to work out the situation and get the best results possible to obtain."[20]

With this in mind, Weyerhaeuser sold some of its original holdings. Many small mills depended on the Northern Pacific for timber and now feared that their supply would no longer be available. One of Long's first acts was to inform trade journals and newspapers that he would sell timber to operators who "were more or less dependent on the Northern Pacific lands for logs." He did so, Long explained to a member of the syndicate, "for policy, as I thought it much better to have it understood that we are willing to sell, then to have it understood that we would not sell." Otherwise, Long feared, "the little mill men" might influence county assessors to raise valuations on the company's land. Weyerhaeuser sold its first timber in June 1900 and disposed of 19,000 acres by the middle of 1903.[21]

The principal task facing Long, though, was assimilation of the lands acquired from the Northern Pacific. Relying on the railroad's figures, the syndicate made no detailed examination of these lands prior to the purchase. But the Northern Pacific had cruised only a portion of its property, and a fourth of the land sold to Weyerhaeuser had never been surveyed at all. Within days of his arrival on the Sound, Long began sending cruisers into the woods in an effort to secure a more accurate idea of the company's holdings. The results of this preliminary examination were shocking. "I have tried," Long wrote Frederick Weyerhaeuser in late June 1900, "to be not at all prejudiced in this matter, but I cannot get away from the conviction, that on the so-called unexamined lands, we have been imposed upon, and have been equally convinced that there has been given to us a large quantity of isolated, scattering, timberless tracts of land, which it will never pay us to log." The Northern Pacific's estimates, Long concluded, "are padded and excessive." Prolonged discussions between the company and the railroad resulted in agreement for a joint cruise of the unexamined land. Compensation for errors supposedly would be figured into a second

Weyerhaeuser purchase of Northern Pacific timber, then in the negotiation stage.[22]

While dealing with the railroad, Long initiated a vigorous timber acquisition program that focused on the even sections within the boundaries of the Northern Pacific grant. His policy, Long wrote, was "to clean up the individual small ownership as much as possible." This required a great deal of time and expense. "The buying up of these lands adjacent to our holdings," Long noted, "is somewhat of a tedious process, as we have innumerable people to deal with; in most instances buying single claims of 160 acres from homesteaders, or people who have quarter sections they have taken up for timber claims." Spending from $60,000 to $70,000 a month, Long acquired 1.5 billion feet of timber from private owners by the end of 1900. Weyerhaeuser holdings in Washington, including a second transfer from the Northern Pacific, amounted to 1.3 million acres by mid-1903. These purchases filled in the checkerboard squares missing from the original transaction and assured efficient timber management.[23]

Concentrating on its timber holdings, Weyerhaeuser did not immediately enter the manufacturing end of the business. When it became evident that the company would not produce lumber, newspapermen and manufacturers alike clamored for information. "This whole country seems to expect that we will build mills at once," Long complained in January 1901. "The newspapers have been building some more mills for us," he noted on another occasion, observing that "while these edifices are hardly air-castles, they are at the same time newspaper mills." In statements and interviews, Long repeatedly stressed that Weyerhaeuser would not make lumber until market conditions allowed. "You have to say something to these newspaper fellows," he wrote to Weyerhaeuser, adding that he was studying "the art of saying nothing that can be misconstrued, and [trying] to leave the impression that we are ready, willing and anxious to build mills, and apt to build them at any time."[24]

Weyerhaeuser's hesitancy to add sawmills to its holdings was not for want of opportunity. The owners of the Puget Mill Company continued to discuss a sale with persons assumed to be Weyerhaeuser representatives. In early 1902, elaborate arrangements were undertaken for an "accidental" meeting between Cyrus Walker and Frederick Weyerhaeuser. "There may be no material advantage from such a meeting," Pope & Talbot official A. W. Jackson pointed out, "still there might be material benefits aris[ing] from such a meeting." The two lumbermen, one representing the old order in the Pacific Northwest and the other the new, might establish a personal rapport that would lead to a cooperative effort to monopolize the Washington timber industry. Will Talbot, for one, desired the "eventual coming together of the large timber owners, thereby creating a Timber Trust which can control the mkts & prices of the world." Despite the preparations, Walker and Weyerhaeuser did not meet, and the Puget Mill Company continued to operate as an independent concern.[25]

Resisting the blandishments of operators eager to sell out, Weyerhaeuser finally became a lumber manufacturer with acquisition of the Bell-Nelson Lumber Company of Everett in January 1902. The mill was small and inefficient and owned only a small quantity of timber. "I am afraid some of you will not like the saw mill," Long informed the members of the syndicate, "for it is not a first-class one by any means, and one which will either have to be re-built before a great while or else run . . . in a way that may not be entirely satisfactory." But the mill provided an opportunity for experience in the lumber business, as well as the company's first direct involvement in logging Washington timber. Rebuilt in 1903, the mill continued to be a relatively minor factor in the lumber trade and for a decade was the only Weyerhaeuser manufacturing venture.[26] Aside from the relationship between rising stumpage values and prices for logs and lumber, the direct impact of Weyerhaeuser on lumbering remained one of potential.

The original purchase from the Northern Pacific and the subsequent Weyerhaeuser acquisitions dramatized the increasing concentration of Washington timber in the hands of larger operators. "It will not be long before the timber in small tracts, will be used up," Cyrus Walker pointed out in March 1901. Heightened by the Weyerhaeuser dealings, speculative mania produced an abrupt increase in the value of timber. By mid-1901, Weyerhaeuser itself was forced to limit purchases because of high prices demanded by land owners. "There is no question about the sharp advance in the price of stumpage being on account of the large purchases of timber land, made in this country," Walker observed. The boom continued until the panic of 1907, Long reporting at the end of 1906 that the company's "stumpage has increased 50% in value during the past twleve months."[27]

Concentration of ownership raised the possibility that the large holders might achieve control over the manufacturers of lumber. Frederick Talbot welcomed the Weyerhaeuser purchases, for "the fewer hands the timber gets into the better position we will all be in to dictate terms re prices of timber lands etc." Cyrus Walker believed that the time was not far distant when "the large holders will have it in their power to controll the price of stumpage." Long too was confident that control of stumpage was the only course that could result in stable lumber prices. The big firms, he argued, should "continue the policy of adding to their holdings, . . . and thus bring about at an earlier date the time when they can control the market conditions."[28] The arrival of Weyerhaeuser signaled the beginning of the era when ownership of timber, not ownership of sawmills, was the key to success in the Northwest forest industry.

Completing the Great Lakes Connection

Frederick Weyerhaeuser's massive achievement was the most dramatic manifestation of the turn-of-the-century rush to the Pacific Northwest. William E. Boeing, Joseph Fordney, Russell Alger, and other well-known individuals invested in western Washington timber, occasionally bribing telegraphers to keep them informed of the competition's dealings. Previously backward areas like Willapa Bay and the northern Olympic Peninsula were opened up by investors in timber and milling ventures. "Hardly a week passes without the purchase of large tracts of stumpage by Eastern investors," reported one trade journal, and another observed that timber was "without doubt, the best investment a man can make."[29] Speculators were confident in the belief that, as oil and steel had been the key industries of the old century, lumbering would be the focus of the new one.

Thanks to the renewed interest in the region's timber, George Emerson was at last able to succeed in his effort to acquire the North Western Lumber Company from Asa Simpson. Following the failure of his effort to involve Frederick Weyerhaeuser in the scheme, Emerson secured in C. H. Jones of the St. Paul & Tacoma Lumber Company a new partner. The two men spent a week-and-a-half in San Francisco in the spring of 1900, working out the details of an agreement for the purchase of the Hoquiam mill for $550,000, most of this sum to be in deferred payments. No sooner had they returned to Washington than news arrived that Simpson had changed his mind and refused to sign the contract. The sale, the old man now insisted, would have to be on a cash basis, a requirement that was unacceptable. Despairing, Emerson again tried to bring Weyerhaeuser into his scheme. In the midst of discussions with George Long, however, the situation again changed. Eccentric to the end, Simpson suggested a return to the original terms and even offered to loan Emerson and Jones the money for the down payment. A month of haggling remained before the transaction was concluded in April 1901.[30]

North of Emerson's Hoquiam domain, the dominant concern at the beginning of the new century was the Merrill & Ring Lumber Company, a Michigan firm that had been investing in Olympic Peninsula timber for over a decade. Until the late 1890s, only a few modest and short-lived efforts at manufacturing had been attempted along the Strait of Juan de Fuca, including a mill operated by the utopian Puget Sound Co-operative Colony. Timber, though, was cheap, and logs could be formed into rafts for towage to the Sound or to new mills built on the strait. Investing heavily in timber near Pysht, to the west of Port Angeles, Merrill & Ring led the way in the opening up of the rugged region along the strait. The company's holdings soon exceeded 25,000 acres and continued to grow in the years after 1900.[31]

Concentrating on Pysht, Merrill & Ring reorganized its Grays Harbor investments in conjunction with logger Alex Polson, a colorful figure who roasted chestnuts in his office fireplace, shipped Quinault salmon to his friends, and composed poems about his umbrella. Founded in 1903, the Polson Logging Company brought together the holdings of Merrill & Ring and Polson to form one of the largest woods operations in the state. "They got so many goddamn locies [locomotives] over there," observed one employee, "a feller don't dare sit on a rail for fear one will come along and nip off a piece of his ass." From the beginning, though, there were signs of trouble in the boardroom. Bristling over Merrill & Ring's initial one share majority in the firm, Polson and his brother Robert insisted that they be given an equal stake in the business. With profits flowing, this demand was granted and thereafter the brothers were left alone as operating heads of the company.[32]

Renewed change from Great Lakes investors was also underway in the older lumbering centers of Puget Sound. Like their longtime competitors at Port Gamble, the owners of the Port Blakely Mill Company had decided to sell their holdings. Although Weyerhaeuser was the first prospect approached by Charles Holmes and John and James Campbell, the company was finally sold to two Michigan investors, David E. Skinner and John W. Eddy, in February 1903. Relations between the new owners soon began to deteriorate, contributing to Port Blakely's reduced stature in the industry. According to informed sources, the aggressive Skinner resented the fact that the Eddy family's role in financing their venture forced him to endure his plodding partner's interference with the proper conduct of business. For his part, Skinner was regarded by much of the industry as an undependable hustler. "A second thought with him is a mighty good thing," advised Edwin G. Ames of the Puget Mill Company.[33]

New money from the lakes also sparked an industrial renaissance at Everett, where the depression had ruined the speculative ventures of the Rockefellers and Henry Hewitt. David Clough, the former governor of Minnesota, arrived in 1900 to erect the Clark-Nickerson mill, the largest plant in the city. Joining Clough was his son-in-law, Roland H. Hartley, a future governor of Washington. Everett banker William C. Butler, brother of Columbia University president Nicholas Murray Butler, also became an influential figure as financial backer of loggers and mill owners and as investor in timber. Weyerhaeuser's decision to manufacture lumber at Everett enhanced the community's proud if gritty image as the "city of smokestacks."[34]

To the north on Bellingham Bay, one of the most important operations in the industry was assembled in the early years of the century. Peter Larson, J. J. Donovan, and J. H. Bloedel, each involved in a variety of northern Puget Sound investments, founded the Lake Whatcom Logging Company in 1898. Three years later, the associates formed the Larson Lumber Company for the production of lumber. In 1913, Larson having died, the firms were merged as the Bloedel Donovan Lumber Mills. Operating three sawmills, two shingle plants, and two logging camps, the company employed over a thousand persons. As in so many firms, business convenience brought together in Bellingham contrasting personalities. Devoted to Catholic charity work, Donovan was regarded as one of the most honorable men in the Northwest. Resembling a gay-nineties barber in his well-groomed sleekness, Bloedel carried a contrasting reputation. "You can not always put your finger on him," Will Talbot worried of Bloedel.[35]

Compounding the impact of Weyerhaeuser's arrival, the investments of other Great Lakes capitalists boomed the value of Washington timber. The Bureau of Corporations estimated that Washington stumpage tripled in value in the first decade of the new century. The ownership of timber, moreover, was coming to be concentrated in the hands of large holders. Defenders of the industry argued that this would lead to greater control of lumber production and ultimately to improved utilization of timber. Opponents countered that concentration would retard the growth of agriculture, corrupt state politics, and result in artificially high lumber prices.[36] Assessment of this question depended upon individual balancing of good and evil, as both industrialist and reformer made valid points. More certain was recognition that concentration portended the end of the era when sawmills and loggers could expect to operate without owning sufficient quantities of timber. The holders of timberland controlled the destiny of the industry and those who had been slow to recognize this truth would eventually be forced out of business.

Completion of the Great Lakes connection severed the industry's linkage to San Francisco. The presence of wealthy investors and powerful representatives like George Long on Puget Sound and Grays Harbor ended the period when economic decisions were made in the city by the bay. That "the business of the manufacture and sale of upward of 600,000,000 feet of lumber annually should be done 1,000 miles from the base of operations," noted the *Pacific Lumber Trade Journal* in June 1901, ". . . does not appeal very strongly to ninety per cent of the lumbermen doing business in the Pacific Northwest." To their chagrin, old-time operators witnessed the dwindling away of their once

John McLoughlin

William Renton

Cyrus Walker

Edwin G. Ames

Loggers of the St. Paul & Tacoma Lumber Company. Photo by Clark Kinsey.

Oxen hauling logs in the Washington forest

Railroad operation of the Snoqualmie Mill Company

Logging camp near Shelton

Pope & Talbot's Port Ludlow mill

Pope & Talbot's Port Gamble mill

GEO. H. EMERSON,
HOQUIAM, W.T.

George H. Emerson

George S. Long

R. D. Merrill

Frederick Weyerhaeuser

J. P. (Phil) Weyerhaeuser, Jr.

mighty influence over the industry. This was a highly beneficial development in the opinion of George Emerson, himself free of San Francisco. The California-owned firms, he contended, had continuously driven down earnings in short-sighted efforts to ruin competitors. The result was that "the lumber business ran in cycles of about three years, two of these years were of distress, while the third was one of prosperity." Under fresh and imaginative leadership, Emerson and other friends of the new order believed, the industry would secure uninterrupted prosperity.[37]

9

Fits and Starts

There is something fascinating about the lumber business and the unin-
itiated seem always numerous and anxious to enter it. . . . If there could
be some means by which the manufacturer could be held back profits
would be assured to him, but there seems to be none. Even under present
conditions, when overproduction is evident, the building of mills goes
merrily forward, [and] while there is an open movement to curtail the
output of mills new ones are being added. There is no reason to think
these conditions will change or that men will.—*West Coast Lumberman*[1]

The lumber business out in this country goes by 'fits and starts.' For the
last two years everything has been gay and buoyant and prices of all kinds
have soared skyward. Today they are in the dumps, and half of the mills
in the state of Washington are closed down, and those that are continuing
to run are not making a dollar. The trouble is we get too much prosperity
one while and then we get in the dumps too much.—George S. Long[2]

A great earthquake struck San Francisco in April 1906, crumbling
buildings and starting fires that devastated the city. Advancing along
the bayfront, flames destroyed the headquarters of the old Washington
lumber companies. San Franciscans, though, responded to the disaster
in plucky fashion. In the city to recuperate from injuries suffered in a
mill accident, Cyrus Walker found refuge in a suburban mansion, where
his aristocratic wife cooked and washed for refugees. Before the debris
had settled, Pope & Talbot began selling lumber from a makeshift
office for the reconstruction of homes and stores. "Everybody writes
in good courage," noted Edwin G. Ames at Port Gamble, "and seems
to think that the City is going to go ahead and rebuild on a more magnificent scale than ever before." Bringing on one last boom for the
oldtime cargo mills, the San Francisco earthquake dramatized the historical instability of firms dependent for profitable business upon natural disasters and mining bonanzas. The departure of Weyerhaeuser's
George Long from the Bay Area a few hours before the quake, moreover, symbolized the apparent capacity of operations oriented toward
rail markets and timber investments to isolate themselves from the vagaries of uncertain markets. "We came very near to having a little
experience," Long could afford to quip from the safety of his Tacoma
office.[3]

Railroad Frustrations

The return of national prosperity at the end of the 1890s allowed the
resumption of a phenomenal rate of growth in western Washington.

Tacoma doubled and Seattle and Everett each tripled in size in the first decade of the twentieth century. "I guess, the town has come to stay," Cyrus Walker remarked of Seattle, where the value of his early-day real estate purchases now exceeded a half million dollars. Regional boosters, reported George Long, predicted "that there will some day be a continuous city from Everett to Tacoma; another London if you please." On Grays Harbor, Aberdeen surged ahead of Hoquiam thanks to the presence in that community, according to one speculator, of a larger number of "hustlers." South of the harbor, the new town of Raymond, featuring a half dozen mills, was founded on the Willapa River opposite South Bend.[4]

An equally phenomenal expansion of lumbering activity coincided with the population boom. Washington became the leading lumber producer in the nation in 1905. Production neared four billion feet in that year, double the output of 1899, and increasingly centered in communities able to serve both eastern and offshore markets. Tacoma, Aberdeen, and Bellingham, all with rail and cargo connections, established themselves as the leading lumber communities in the state. Among the old ocean-oriented mainstays, Port Blakely dropped to fifth, Port Gamble to eighth and Port Ludlow to eleventh. National demand for lumber peaked in these years and the average price of Douglas-fir increased from $8.47 per thousand feet in 1899 to $14.40 in 1907.[5] At the head of their industry, Washington mill owners anticipated an unparalleled period of prosperity.

Although expecting prosperous conditions, perceptive lumbermen worried about the increase in the industry's capacity. To George Long, one of the most striking things about the Pacific Northwest was the fact that, in comparison to eastern standards and available timber, "the actual business of manufacturing lumber is not conducted on a very extensive scale." Still, Long was concerned from the moment of his arrival in the region by the delicate balance between output and profits. "There are enough mills here at the present time," he reported in late 1900, "and if there was any material increase in the manufacturing capacity . . . the market would soon be in a very bad shape." As if to realize these fears, the number of sawmills in Washington, large and small, increased from 317 in 1899 to 557 in 1905. By the end of 1907, over a thousand plants were engaged in the production of lumber.[6]

Mill owners expected the growth of the rail trade to save them from the consequences of this expansion in capacity. Sparked by a reduction in freight rates to the Missouri River, eastbound shipments of Washington lumber doubled between 1900 and 1902. Twenty-eight percent of the lumber produced in the Pacific Northwest was sent east over the Northern Pacific and the Great Northern in 1905. Six years later, the figure had risen to fifty-nine percent. Those mills oriented toward the rail market secured most of the benefits and thereby enhanced their positions within the industry. In 1903, for instance, the St. Paul & Tacoma Lumber Company shipped seventy percent of its cut by rail and George Emerson's North Western Lumber Company sold twice as much lumber in the east as in California and foreign ports combined. The Tacoma and Port Blakely mill companies, on the other hand, sold only twelve and twenty-one percent of their output, respectively, to rail customers.[7]

Enhanced by completion of a third transcontinental, the Milwaukee Railroad, the eastern trade produced boom times in Washington. Lumber production and timber values had reached such a stupendous level, observed George Long in 1906, that it was difficult "to stay on terra firma instead of floating off into the unknown." Complaints about the railroads, though, grew as fast as the trade. The refusal of the lines to reduce freight rates to points east of the Missouri River gave a considerable competitive advantage to Southern mills. According to one calculation, lumber could be shipped from Louisiana to Chicago at a rate forty-eight percent below that available to Puget Sound operators. Horror stories of railroad inefficiency detailed the abandonment of lumber trains on remote sidings and the snail-like transfer of cars from one Sound town to another, adding to the discontent.[8]

Shortages of rolling stock hampered railroad shipments, a problem that increased in gravity as the demand for Pacific Northwest lumber mounted. Reluctant to haul large numbers of empty cars west in order to service the eastbound traffic, the transcontinentals insisted upon the efficient handling of their costly equipment. "I am in the railroad business, . . . not the lumber business," James J. Hill reminded angry mill owners. Lumbermen naturally refused to concede that there were sound reasons for a shortage that reduced potential profits. "It seems to me criminal that any corporation such as yours should have so neglected their patrons," George Emerson bluntly exclaimed to an official of the Northern Pacific. "The lumber industry of the Pacific Northwest," contended a trade journal, "is being deliberately and wantonly sacrificed." The railroads, went a popular argument, were obviously engaged in a plot to build up the South at the expense of the Northwest.[9]

Lumbermen called upon the federal government to institute strict regulation of the railroads. The selfish operators of the transcontinental lines, mill owners believed, had brought this sorry fate upon themselves. Leaders of the Northwest industry joined shipping interests from throughout the country in support of President Theodore Roosevelt's successful campaign to secure Interstate Commerce Commission authority over the making of railroad rates. Defeat of Roosevelt's leg-

islation, warned the *Pacific Lumber Trade Journal* during the delib-
erations in congress, would leave "government ownership" as the only
alternative. Manufacturers conceded that the railroads had created an
enormous new demand for lumber, but insisted that the transcontinentals had snatched away much of the trade's potential by arbitrarily restricting shipments and access to markets.[10]

While attention focused on the rail trade, the old-fashioned cargo
business languished outside the limelight. The development of large
brokerage firms as middlemen between producers and consumers was
the principal new feature of offshore operations. Controlling shipping
through ownership or charter of steamships, now the favored mode of
transportation in the foreign trade, such firms as Charles R. Mc-
Cormick of San Francisco and Dant & Russell of Portland became
major factors in the lumber industry. Forced to deal with the brokers
because of the heavy cost of modern shipping, mill owners frequently
complained of poor performance and low prices. "Instead of acting as
our agents and telling us they can secure business at a certain price,"
George Emerson observed of his San Francisco agent, "they take the
position of a customer and ask us at what price we will furnish, and
then charge us commission and cash discount, and that cash discount
at the end of 90 days."[11] Here was an additional measure of frustration
for cargo men.

The San Francisco earthquake, welcomed as a nostalgic reminder of
the old days when natural disaster meant good business, provided a
brief respite from frustration. From just over a million feet of lumber
prior to the quake, San Francisco's daily consumption increased to
better than three million feet in the months following the disaster. "Those
mills with transportation," wrote Emerson, "have shipped even the
foundation of their shipping piles, Laths, Shingles and all." Despite
waterfront strikes and high freights, Washington lumbermen prospered
from disaster-enhanced prices. Shipping lumber from Weyerhaeuser's
Everett mill, George Long noted that "we may as well get all we can
out of the market before the re-action sets in."[12]

Angry residents of San Francisco charged lumbermen with taking
unfair advantage of their plight. Investigations of the allegedly outland-
ish prices charged by the mill companies were undertaken with attendant publicity by William Randolph Hearst's *Examiner* and by a grand
jury. Always sensitive to charges of monopoly and pricefixing, Long
insisted that prices were determined by market conditions and would
have been even higher were it not for the compassion of the industry:
"While I do not suppose we will get credit for this, and most certainly
will not ask for credit, the fact remains that our attitude from the day
of the disaster has been one where we felt and have acted on the belief
and the theory that no advantage should be taken of the San Francisco

calamity." If seriously undertaken, these efforts had failed, as Long privately admitted, and the aftermath of the earthquake delivered a severe blow to the industry's image. Shippers so overstocked San Francisco, moreover, that demand collapsed by early 1907.[13] Once again, lumbermen acting on their own had failed to mesh production with demand, a failure that doubtless increased the attraction of a new form of organization.

The Rise of the Trade Association

In the first years of the twentieth century, economic theorists pushed for acceptance of the new concept of the trade association. Within each of the nation's often chaotic industries, associations could work for the resolution of common problems. Without, they could be effective spokesmen before the public and especially before the government. Their industry particularly prone to disunity, Northwest lumbermen became early leaders of the association movement. The industry's first trade organization, the Lumber Manufacturers Association of the Northwest, had been formed in 1891. With George Emerson as its leading figure, the association fought for reduced railroad rates until collapsing during the depression. Similar attempts by Oregon lumbermen, by Seattle lumber dealers, by Puget Sound loggers, and by shingle manufacturers also fell victim to the economic crisis of the 1890s.[14] Facing runaway growth in output, shortages of transportation and erratic markets, lumbermen returned to the association idea at the turn of the century.

Founded in Seattle in 1901, the Pacific Coast Lumber Manufacturers Association brought together the leading mills of Puget Sound and Grays Harbor. Opposition to the railroads and their alleged "stand-and-deliver" policies was the principal motivation behind creation of the P.C.L.M.A. The group pressed the transcontinental lines for reduced freight rates and demanded a federal investigation of the car shortage. Although disputes with the railroads continued to be a central purpose of the association, other activities soon came to the fore. The association information bureau, for instance, assembled and distributed statistics on production and prices. Of most importance, the association quickly became involved in price fixing.[15]

Soon after the founding of the Pacific Coast Lumber Manufacturers Association, members began to discuss the practicality of controlling prices. "It is well to get together and compare notes," the Port Blakely Mill Company's Charles Holmes observed in April 1901, "and [we] may be able to advance prices by so doing." This sentiment resulted in the drawing up of detailed price lists for the various grades of lumber sold in the California, foreign, and rail markets. The initial determination and later revision of these lists was a difficult and time-consuming process. Setting prices at too high a level would undermine the effort by encouraging the building of new mills and the conflicting interests of association members had to be reconciled. "I do not believe it is possible to ever get one up, [that] will suit all hands," reflected Cyrus Walker of the latter problem. Still, adherence to the listed prices was at first fairly widespread in the industry.[16]

Price fixing was carried on in an open fashion until around the middle of 1906, as indicated by the fact that control committee meetings were actually held in the Tacoma federal building. Thereafter, the antitrust activities of the Roosevelt and Taft administrations combined with public antipathy toward the industry to produce a more circumspect approach. "The times are not very propitious for making an iron clad association," observed George Long, "because I think the government is inclined to attack any combination that is gotten up to maintain prices." Explicit public efforts were abandoned in favor of circumspect circulation of closely guarded price lists and information bureau statistical circulars. Mill owners, trade journals, and association spokesmen loudly maintained that prices were dictated by supply and demand and that, legal considerations aside, rigged prices were a practical absurdity.[17]

Although refusing to abandon their efforts to control prices, lumbermen grew increasingly fearful of the Sherman Anti-trust Act, especially following federal prosecution of the Standard Oil Company and other controversial business organizations. Visited by a federal investigator in 1913, Edwin G. Ames reflected this apprehension by simply refusing "to talk about such matters." When asked by the agent for copies of price lists, Ames "told him I had a file of which I was very careful, and which I wished to preserve as a curiosity and as I had only one copy of each list, and desired to keep my file complete, that I had locked them up in the safe deposit vault for safe-keeping." The belief that the trade association's confidential endeavors would sooner or later be revealed to the public and the government, wrote William H. Talbot, "keeps us on the anxious seat a large portion of the time." Federal prosecutors, commented George Long, were "anxious to take a crack at the lumbermen and are only waiting for the right kind of a spot to hit them to show up."[18]

On the public level, the association moved into the inspection of lumber cargoes. Traditionally, purchasers had been allowed to determine how closely lumber received matched up with lumber ordered, a procedure that often resulted in demands for rebates. "The mill man absolutely puts himself at the mercy of the buyer," pointed out the *Timberman*. "No matter how good the grades are," contended the *Pacific Lumber Trade Journal* in a similar vein, mill owners were "sure to be mulcted at the other end." The obvious solution to this problem was for inspection to be made the responsibility of the mills, but customers were as unwilling to trust manufacturers as manufacturers were to trust customers. "How will you be able to carry out the plan," wondered Cyrus Walker of this obvious obstacle, "when you have to deal [with] so many bankrupt institutions, run by parties, many of which, you would not believe under oath."[19]

Ultimately, the answer to Walker's question was establishment of an industry-wide inspection apparatus providing qualified tallymen and certificates attesting that shipments coincided with contract specifications. Headed by Edwin G. Ames, the Pacific Lumber Inspection Bureau was organized in 1906 to perform this service. San Francisco retailers, the prime beneficiaries of the old system, were irate, but mill inspection was quickly accepted because of the bureau's effort to maintain standards. Despite generating a good deal of personal antagonism, for instance, Ames refused to certify the inferior cargoes of such important operators as E. K. Wood and Asa Simpson. Eliminating the troublesome rebates, inspection was a major achievement. "It is a wonder," noted a trade journal, "that mill inspection was not inaugurated twenty years ago."[20]

Because price controls were susceptible to violation and government investigations, the inspection bureau may well have been the most important early result of association work. At the time, however, lumbermen credited adherence to the P.C.L.M.A. price listings with bringing on a period of excellent business in 1905 and 1906. Although booming demand was the actual cause, there was no question but that times were prosperous. "Pacific Coast conditions never looked so good as they do now," reported George Long in March 1906. "They seem abnormally good; but I presume in this respect they partake of the great prosperity that is broadcast in our land." Demand was on the rise and there was no difficulty maintaining prices.[21] The real test of the Pacific Coast Lumber Manufacturers Association, and of the industry itself, would come when new and unexpected crises upset the calculations of mill owners.

Continued Hard Times

Surveying the year 1907, the *Timberman* reflected that its beginning and its end had been marked by "extremes . . . unparalleled in the history of the lumber business of the Pacific Coast." In the opening months, lumbermen had operated under "an unclouded sky" and had sailed "along in the noontide of prosperity." Then, two unexpected and disastrous developments had "settled like a blight and pall on the business." The first was the outbreak of open warfare between the industry and the transcontinental railroads. Before this conflict was concluded, the second, a national economic crisis, dried up the demand for lumber and initiated a lengthy period of depression for the industry.[22]

As if in counter-reward for their years of campaigning for reduced freight rates, mill owners faced a 10¢ increase in charges in the summer of 1907. The railroads thereby continued, charged the *Pacific Lumber Trade Journal*, "their favorite pastime of 'sticking it into the lumbermen.'" Making it impossible to compete with Southern lumber east of the Missouri River, the increase had an immediate and devastating impact. Seventeen thousand carloads of lumber were shipped from western Washington in the twelve months ending in November 1907, compared to 77,000 for the calendar year of 1906. "This advance in the freight rate to the east," wrote George Emerson, "has shut off our eastern trade." Sawmills were idled and empty freight cars, so recently a rare sight, clogged every siding in Washington.[23]

The Pacific Coast Lumber Manufacturers Association secured an injunction against collection of the new rate and brought suit before the Interstate Commerce Commission, which had recently been given the power of review over freight charges by the Hepburn Act. Lengthy hearings were held in the nation's capital, at which the lumbermen argued that the increases were unjustified and were meant as punishment for the industry's support of railroad regulation. In a compromise decision announced in June 1908, the I.C.C. restored the old rate to the Missouri and allowed a five cent increase to points east of the river. Reflecting the depth of hostility generated by the dispute, Emerson feared that the victory would be "an empty one for the railroads have unlimited resources of meanness and draw from those resources very largely when opposed."[24] The belief that the triumph would be hollow proved to be correct, albeit for a different reason.

Northwest lumber was restored to its competitive position in the markets of Chicago and the Twin Cities, but as the result of the panic of 1907, there was scant demand in those or any other communities. The bursting of the speculative bubble on Wall Street in October 1907 brought a jolting halt to the economic life of the nation. In western Washington, bankers restricted payment of drafts and mill owners resorted to fiat money in order to pay their employees. With the freight rate battle, reported George Long, the "financial 'whirligig' . . . has just about taken the snap out of everything." Some lumbermen anticipated only a brief hiatus in the advance of prosperity, but most recognized that fundamental instabilities—especially excess production— were bound to result in a prolonged period of depression. By January 1908, half the sawmills and most of the logging camps west of the Cascades were closed down.[25]

Despite occasional short-lived bursts of improved prices and demand, the trend was ever downward. As Douglas-fir prices dropped
by a third, the number of operating sawmills in the state fell from
1,036 in 1907 to 389 in 1915. Especially hard-hit were those who had
borrowed heavily in good times and now had to cut timber and manufacture lumber in order to generate cash for payment of debt. Reflecting this tendency, industry output continued to expand, adding to
the excess of supply over demand. National per capita consumption of
lumber declined due to economic stagnation and the rapid development
of wood substitutes, accentuating the industry's sorry position. In no
other field of economic endeavor, noted the journal of the American
Forestry Association in August 1915, was "the outlook for the future
more cheerless."[26]

Northwest lumbermen filled their correspondence books with
expressions of despair and puzzlement. The financial crisis passed and
farm prices improved, but there was no recovery in the demand for
lumber. "For some unaccountable reason," observed a mystified George
Emerson in late 1910, "there is no business doing." Prospects on the
Pacific coast, wrote George Long a few months later, "are rather deplorable, and as a matter of fact, getting a little worse, I think, instead
of better." Customers took advantage of a buyers' market to pick and
choose among mills, driving down prices. The Washington Mill Company, after a half century of business, was bankrupted and the Tacoma
and Port Blakely mill companies closed for lengthy periods. At Gamble and Ludlow by early 1914, machinery was idled, the wharves were
deserted and, for the first time in the history of the Puget Mill Company, there was not a single foreign order on the books.[27]

Mill owners looked to a strengthened trade association for defense
against the distressing state of things. Formed in 1911 through the
merger of the P.C.L.M.A. and trade groups in Oregon and south-
western Washington, the West Coast Lumbermen's Association led the
effort to secure control over production. The work of the association,
however, achieved at best a limited success. Partial curtailment was
achieved, but more as the result of bankruptcies than of the willingness
of members to abide by agreements limiting output. Price standards
fell victim to the desperate need of mills to secure business and to the
refusal of operators to provide honest statistical data to the information
bureau. "We are all a lot of liars," admitted Frederick C. Talbot, "and
until that phase of our business can be eliminated, I doubt whether
anything will ever be a success in lumber organizations." The ability
of the W.C.L.A. to cope with the depression was further impaired by
the loss of membership and by the need to cut back on association
expenses.[28]

In time past, the industry had recovered from hard times by locating and exploiting new markets, creating the demand that would absorb excess production. The opening of the Panama Canal in mid-1914 would reduce shipping time between Puget Sound and the Atlantic coast from two months to three weeks and provide just such an opportunity. The long haul around South America had always kept Northwest lumber out of the eastern United States and had limited shipments to Europe as well. For a quarter century, lumbermen had called for the opening of these markets through construction of an isthmian canal. Now, thanks to the big stick diplomacy of Theodore Roosevelt and the big shovel genius of American engineering, that call would be answered and, it was anticipated, prosperity restored. "The lumber business of the United States will be practically revolutionized," enthused the *Pacific Lumber Trade Journal*.[29]

Hopes for the path between the seas were almost immediately dashed when the Wilson Administration, determined to honor treaty commitments with Great Britain, forced through congress the repeal of a law exempting American coasting vessels from the payment of tolls. Combined with the nation's shipping regulations and prevailing import duties, this action, charged lumbermen, would enable British Columbia mills to ship lumber to New York at a cost well below that of operators south of the international border. The advantages of the canal, thus, would be given away to foreigners. "The Panama Canal was built on American soil by American capital and American brains," protested the *Timberman*, "to facilitate commerce between Atlantic and Pacific coasts." And yet President Wilson and Secretary of State William Jennings Bryan, the old nemesis of 1896, had turned their backs on American businessmen. "Doves of Peace perched on halos floating over heads full of sentiment," wrote an outraged Edwin G. Ames, "do not have any good effect on business."[30] The salvation held out by Roosevelt the manly patriot, to adopt the characterizations favored by men of the woods, had been snatched away by Wilson the effete professor and Bryan the radical bumpkin.

Under these conditions, trade with the Atlantic coast was slow to develop. For 1915, the first full year of the canal's operation, 54 million feet of Washington lumber was shipped to eastern ports. This was three times the rate of the previous year, but still only six percent of the ocean shipments from the state's mills. Thereafter, closure of the canal to civilian traffic during the First World War barred the isthmian passage to Northwest lumbermen.[31] The hopes placed in the Panama Canal by mill owners reflected a recognition that the only permanent solution to the problem of overproduction was to develop to the fullest all possible markets. The number of mills and the conflicting interests of their operators rendered ineffective efforts to restrict production and control prices, especially when demand was falling. When markets slumped, as had been the case since 1907, the industry came under an inevitable and severe strain.

Proposed Remedies for the Industry

Northwest lumbermen, wrote Edwin G. Ames in June 1913, "are all looking for a Moses to lead us out of the wilderness." Although the original might have found the task thankless, there was no shortage of would-be guides to the elusive promised land of prosperity. "I scarcely meet anybody now," noted George Long of this phenomenon, "but what they have some plan to suggest." Ames himself believed that he had found the means of bringing production under control, proposing that mill owners secure laws from the legislatures of Washington and Oregon mandating the eight hour day in the camps and mills. In painless fashion, output would automatically be reduced to a level equivalent to demand. The proposal received only bemused consideration from operators convinced that allowing the government a role in determination of hours would be a step burdened with uncertain and dangerous implications.[32] That Ames, one of the most conservative figures in the industry, would consider such a move, however, suggested that labor reforms could be initiated by indirection when in the interest of management.

Other self-styled industrial saviors observed that the American public, schooled by the writings of the muckrakers, believed in the existence of a lumber trust. One such observer, Spokane banker Henry J. Pierce, therefore concluded that he might as well supply the genuine article. Backed by Wall Street financiers, Pierce set out in late 1911 to acquire two-thirds of the cargo mills in the Pacific Northwest, along with enough timber to keep them supplied for a lengthy period of time. By early 1912, he had secured options on most of the Columbia River and Willapa Bay properties, as well as on a number of mills on Grays Harbor. Lining up these concerns, he also began negotiations for purchase of the Polson and Simpson logging companies.[33]

Crucial to the success of the plan was the willingness of Weyerhaeuser to sell 300,000 acres of timber to the new company, providing the raw material for longterm operation and supposedly inducing other operators to sell out. Initially sceptical about the scheme, but reluctant to antagonize its supporters, George Long held out stiff conditions that had to be met before Weyerhaeuser would even set a price on its timber. Pierce, however, secured a judicious letter of sympathy from Attorney General George Wickersham, demonstrated the strength of his financial backers, added to his list of mills and, most incredible of all, offered to pay cash for the timber. Impressed with the promoter's initiative and convinced that the time had come for Weyerhaeuser stockholders to receive a return on their investments, Long became an advocate of the merger and the timber sale.[34]

In February 1912, Long and Pierce visited St. Paul to confer with members of the Weyerhaeuser syndicate and then traveled on to New York for meetings with the bankers. Back on the Pacific coast, an option was secured on the Puget Mill Company. Unfortunately, one of Pierce's financiers drowned in the sinking the *Titanic* and the remainder balked at the enlarged requirements of the scheme. Despite intense lobbying, moreover, the Justice Department declined to provide an ironclad guarantee against Sherman Act prosecution. Taken together, these obstacles eventuated in the collapse of the merger by mid-1912. Long was chagrined over the failure, as he had become convinced that the plan would have introduced a much-needed element of stability to the industry. "The effort which you have made to bring about better conditions . . . ," he wrote Pierce, "has demonstrated that you clearly realize the business folly of the business being conducted as it has been in the past."[35]

The heavy tread of John Sherman's trust-busting ghost also haunted the ambitions of the most aggressive savant of the industry, David Skinner of the Port Blakely Mill Company. A mercurial and controversial figure on Puget Sound, Skinner formulated in 1912 a plan for a joint sales agency among the exporting sawmills of the Pacific Northwest. The company would handle all export sales for its members and send agents abroad to promote the trade. The sway of the Sherman Act, those advocating the Skinner plan contended, should not extend beyond the beaches of Washington and Oregon. Rather than stifling competition, they insisted, their proposal would expand the nation's commerce and increase the earnings of mills and the wages of workers. "We are all agreed that something ought to be done in this particular branch of business," noted Edwin G. Ames, "and that the first thing to do is to assure ourselves of the legality of this move." Skinner arranged a series of conferences with federal officials, hoping to secure advance immunity from anti-trust prosecution. Like Pierce, though, he received only expressions of sympathy, not the desired assurances.[36]

For four years, Skinner worked to achieve acceptance of his plan, despite frequent depression over the failure of the government to see the justice and necessity of his cause. Although "considerably disgusted," Ames reported after one meeting with Skinner, "I could see that he still had his nerve with him, and was on the job." The Douglas Fir Exploitation & Export Company was finally organized in October 1916. With thirty-eight of the region's sixty cargo mills as charter members, the D.F.E.&E.—as the firm was normally called—offered the prospect of stabilized conditions and improved prices. To the mortification of Skinner, however, the members ignored his role in creating the company and chose the experienced and dependable William H. Talbot as president.[37] This was a nice display of ingratitude and a good indication of the uncertain regard in which Skinner was held by his colleagues.

Those of an intellectual bent also busied themselves with plans for the reorganization of the industry. Burt Kirkland, a University of Washington forestry professor, developed a scheme to set aside the Sherman Act and equip the nation's various lumber trade associations with the authority to forthrightly control production and prices. "If no harm can come to the public through complete liberty to cooperate within a given industry," he argued, "that industry should be given the right of cooperation." Advocates of this approach also called for a federal buy-back of timber. Since poorly designed laws had forced lumbermen, virtually against their will, to acquire excessive amounts of timber, went the argument, a rescue of those hard-pressed by the folly of federal policy was in order.[38] Unable at this time to get around the anti-trust obstacle, the concept of industrial self-government was filed away, to be resurrected from time to time in the years after the First World War.

Although their approches differed, the prophets of salvation focused on a central problem: except in flush times, the industry produced more lumber than could be sold for an acceptable return. The failure to implement the Ames, Pierce, and Kirkland proposals and the delayed application of the Skinner plan therefore increased the despair felt by Washington lumbermen. What expectation there was for an end to chaos rested on hope that some giant firm might become a force for unity. In 1914, as if to meet this expectation, the Weyerhaeuser Timber Company began construction of a large all-electric plant at Everett—known as Mill B—to supplement the old Bell-Nelson facility. At the same time, Weyerhaeuser combined its holdings east of Seattle with those of the Grandin-Coast Lumber Company to form the Snoqualmie Falls Lumber Company. The firm's sizable sawmill, located on the Milwaukee Railroad's main line, was designed to sell lumber in the markets east of the Rockies. When completed, the new mills would make Weyerhaeuser the leading manufacturer of lumber as well as the largest owner of timber, and enable it to exert a greatly enhanced influence over the course of the industry.[39]

While Weyerhaeuser, on the eve of the First World War, looked to the future, others in the Northwest looked with brief nostalgia on the final disappearance of the old nineteenth-century order. In October 1912, Edwin G. Ames visited San Francisco for a reunion with Cyrus Walker, living in retirement in that city since the 1906 earthquake. To his distress, Ames found the old man enfeebled and senile: "He had a little 'memo' in his pocket that he had made up and he asked me a great many questions about things that were settled along ago." Ames drove away from the Walker mansion crying, he confessed, "like a little child." A year later, the famous lumber baron was dead, soon to be followed by other reminders of the past. Suffering from diabetes, George Emerson passed away in the summer of 1914. Asa Simpson, Emerson's old nemesis, followed a few months later, after working in his San Francisco office until the final day of his life. "The old-timers and pioneers," reflected Ames, "are passing away gradually one by one."[40]

The crashing boom of the guns of August in 1914 provided a final blow to seven years of lumbering despair. For ill-timed tourist Frederick Talbot, who was trapped behind German lines for a month, the beginning of the war brought personal discomfort. For mill owners at home, the impact was equally distressing, if limited to the financial sphere. Shipping was disrupted, overseas markets were cut off, and commercial life, at least for a time, halted. By fall, Edwin G. Ames calculated that "in all probability less than one half and possibly not more than forty per cent of the capacity of Oregon and Washington mills is now being operated." At least four-fifths of the workers normally employed in the woods and mills, he estimated, had lost their jobs. "I would say so far as business is concerned," summed up Ames, "that it never was so bad before in my thirty-five years' experience."[41] The coming of war destroyed the last vestiges of hope, frustrating lumbermen who had already confronted their full share of upheaval, and not just in the economic arena.

10

Gifford Pinchot Men

It is sometimes believed that the lumberman is the enemy of forest preservation and should be compelled to greater duty to the public. Whether
or not this is true elsewhere, in the Pacific Northwest he is doing more
for the cause than any one else, and the problem, if the cause is to succeed, is to get the public to perform its own duty.—E. T. Allen[1]

I am a Pinchot man in every respect pertaining to forestry matters.—
George S. Long[2]

Small forest fires, ignited by careless humans and the vagaries of nature, smoldered across southwestern Washington throughout the dry
summer of 1902. A month of soaring temperatures and no rain was
followed in the first week of September by high winds, uniting the
small blazes in a giant conflagration. "Forest fires are general all over
[the] country," reported an alarmed Cyrus Walker, "and are worse
than ever before known." Centered between the Columbia River and
Mount St. Helens, the major fire—known as the Yacolt Burn after a
small town that, ironically, escaped without damage—burned over
25,000 acres of green timber alone. Other fires ravaged the forests to
the north and south of Grays Harbor, so threatening Hoquiam that saltwater was pumped into the town's mains in preparation for a last stand.[3]

Great clouds of smoke spread across the Pacific Northwest, eliminating the difference between night and day and disorienting residents
of the region. "At this hour 2 P.M.," wrote George Long, "the City
of Tacoma is practically in a state of semi-darkness, the sky having a
pinkish overcast, and the general effect is exactly that of the reflection
of a big fire at night." Smoke drifted as far south as San Francisco,
where, noted Will Talbot, "mixing with the fog [it] makes a very dreary
day for all of us." Hysteria mounted to such an extent that those attending a camp meeting near Vancouver prepared themselves for the
end of the world. After a week, however, the wind died away, the
first fall rains arrived to douse the flames, and assessment of the damage got underway. An estimated two billion feet of timber, worth $13
million, had been destroyed in Washington and across the Columbia
in Oregon.[4]

The destruction of timber, logging camps and railroad trestles, though,
was the least graphic result of one of the great disasters in the history
of the Northwest. Three dozen persons, shocked residents of the region
learned, had perished, most in a five mile wide firestorm that had swept

the Lewis River valley. The huddled remains of two families were discovered just short of the shelter of a creek. A woman was found next to her charred sewing machine and the corpse of a mailman still clutched the reins with which he had attempted to lash his horse to safety. Watercourses were clogged with the carcasses of drowned animals, wild and domestic.[5] The loss of life, so graphically portrayed in the press, produced considerable public interest in the prevention of future forest disasters. Taking the lead in the drive to secure protection for valuable natural resources, large timber owners found themselves at the head of the Pacific Northwest conservation movement. In this area as in others, lumbermen recognized that, despite considerable antipathy toward the temper of the times, progressivism could benefit private as well as public interest.

The Progressive Era

The defeat of Bryan in 1896 and the subsequent McKinley prosperity seemed for a brief moment to have ended the era of political and social instability. The United States, enthused W. H. Day of the Pacific Empire Lumber Company, had been saved "from the hands of the fifty cent on the dollar repudiationists" with their hateful and un-American aim of redistributing "assets whenever the dissolute and incompetent become necessitous." Unhappily for those of such expectations, the nation was soon overwhelmed by the progressive crusade to restrain the political and economic influence of powerful individuals and organizations. The populists may have been destroyed, observed George Emerson, "but the voters of the country still stand ready to listen to any blather-skite who will talk to them about something he knows nothing of."[6] With other businessmen, lumbermen complained that the making of money was small compensation for being transformed in the public mind from Horatio Alger heroes to enemies of society.

Turn-of-the-century Northwest lumbermen, as Stewart Holbrook once noted, were made of the stuff of villainy. They composed the most powerful economic interest group in the region and were, reformers contended, responsible for the corruption of politics and government. Their timber holdings, went the argument, had been acquired by fraudulent means and were being destroyed with singular unconcern for the needs of future generations. Muckrakers focused their attention on the machinations of the so-called lumber trust, especially on the Weyerhaeuser Timber Company, a tempting target for anti-monopolists. One journalist questioned whether Frederick Weyerhaeuser had built his forest empire "with clean hands" and another discovered that Weyerhaeuser's "hand is in all of the big timber operations of the day." Although outraged, lumbermen recognized that such allegations were well re-

ceived by the American people. "We are beginning to feel the heavy
hand of public sentiment against large corporate interests," noted George
Long in the fall of 1907.[7]

Driven by the charges of the muckrakers, much of the public did in
fact believe in the existence of a lumbering trust. "The most remarkable of all the great business combinations ever known in this country," thundered William Randolph Hearst's *World Today,* "is that now
dominating the lumber industry and vaguely called the Lumber Trust."[8]
Such charges represented a notable failure to understand conditions in
the industry. The huge number of operating units meant that even the
largest firms produced only a tiny fraction of the nation's lumber. The
company most often singled out by trustbusters, Weyerhaeuser, manufactured only a modest amount of lumber prior to the First World
War. In its organization and in its inherent problems, lumbering was
closer to farming than to such industries as steel and oil. To some, the
activities of trade associations, often carried on behind closed doors,
carried the implication that a trust had come into existence. If so, the
failure to stem the tide of falling prices after 1907, made this a sin-
gularly ineffective trust.

Nevertheless, the federal government responded to popular sentiment by launching a series of investigations of the industry. The Bureau of Corporations, the Justice Department, and, after its creation in
1914, the Federal Trade Commission probed the activities of timbermen and trade associations. In a multi-volume study, the Bureau of
Corporations detailed price fixing operations and, while denying the
existence of a manufacturing trust, charged that concentration of timber ownership stifled competition and resulted in artificially high lumber prices. Although the study provided useful information about the
industry and the F.T.C. inquiry resulted in authorization of export sales
combinations, lumbermen strenuously opposed the investigatory impulse. Biased and inexperienced government agents, argued industry
spokesmen, failed to correctly interpret the evidence and threatened
the sanctity of free enterprise. "It certainly looks as if government
investigations and investigators are running mad," observed the *Pacific
Lumber Trade Journal.*[9]

Critics of the industry were especially interested in the means by
which investors had secured timberland. There was nothing in Washington to match the sensational exposures of timber fraud in Oregon
in the first decade of the century. Allegations of illegality by Washington lumbering firms centered on legislation passed in the late 1890s
to allow the Northern Pacific Railroad to exchange its holdings within
forest reserves and the new Mount Rainier National Park for public
land elsewhere. The Northern Pacific had thereby exchanged worthless
tracts for valuable timberland. Pointing to and overstressing the con-

nection between Weyerhaeuser and the railroad, reformers charged that the two had conspired to secure the legislation, the timber company supposedly coming into possession of much of the land acquired by the Northern Pacific as reward for its efforts. Considering the lack of evidence and the fact that so-called lieu land made up less than ten percent of Weyerhaeuser's holdings in Washington and Oregon, however, there was scant substance to the charges.[10]

At the state level, those interested in exposing corruption concentrated on the sale of timber by the State Land Commission. Upon the achievement of statehood, Washington had been granted three million acres of land for the support of public schools and other services. In 1909, a special legislative committee investigated allegations that fraudulent cruises prevented the state from securing a fair return on sales of its timber. The committee concluded that as the result of "the dishonesty, incompetency and inefficiency" of E. W. Ross, the politically controversial land commissioner, the state had "been for years systematically defrauded." Weyerhaeuser and other large operators, it was alleged, had secured timber at ridiculously low rates through the helpful medium of bribery. Ross and industry spokesmen pointed out in response that cruise reports were at best rough estimates and that the price of state timber had kept pace with private sales.[11]

Failure to bring formal charges against individuals and firms targeted by the committee suggested that the investigation had been driven by considerations of politics and a misunderstanding of the mechanisms of timber transactions. Dishonesty did exist and George Long, while denying that Weyerhaeuser was involved in improper activities, privately conceded that "a great deal of fraud and graft has crept into the sale of timber lands belonging to the State of Washington."[12] Corruption, though, was generally limited to dealings between the state and small loggers and millmen. The land commission sold only cutting rights and required that tracts be logged within a specified period. Large firms like Weyerhaeuser, in contrast, preferred to acquire land for purposes of investment and longterm support of manufacturing operations. Such companies possessed the financial means to purchase timber legally and the intelligence to realize that this was the wise course in an era of public scrutiny.

The allegations of muckrakers and reformers were but one reason for disenchantment with the tenor of progressivism. As the population of Washington increased, so did the demand for government services. The property tax, assessed and collected at the county level, was the principal source of state and local revenue, to the exasperation of timber owners. Western Washington timberland bore a heavy burden and was taxed at a higher rate than timber holdings in most other states. The assessed valuation of Weyerhaeuser's holdings, for one example of the upward trend, increased by a fourth in 1906, by a third in 1907 and by half in 1908. The 1906 increase in Chehalis County, where the company owned 195,000 acres, was a full 100 percent. "It is a horrible situation," George Long informed his employers, "and one which we have not been able to prevent, although we have gone at it in every conceivable way."[13]

As far as lumbermen were concerned, this situation was grossly unfair. The new direct primary system, replacing the old closed convention as the mode of nominating candidates for office, supposedly encouraged politicians to curry public favor by placing the burden of government on corporations. "As a rule," Long pointed out, "the average man is inclined to think that the big fellow ought to pay a lot of taxes, but does not like to pay them himself, and there are always office seekers . . . who like to make a noise about what the big fellow is doing." Moreover, the standard practice of holding off assessments until after the harvest season allowed farmers, a group to whom politicians were abnormally sensitive, to escape taxation on the value of their crops. The unfortunate timberman, in contrast, paid year after year on his crop of growing timber.[14]

Efforts to manipulate the tax-gathering system resulted in endless frustration. Some lumbermen called for the splitting of the state along the crest of the Cascades, reasoning that the removal of eastside farming elements would increase the chance of securing tax exemptions for timber. While contemplating this fantasy, though, they had to face present-day realities. County assessors and equalization board members, officials often susceptible to the persuasive powers of wealthy property owners, received considerable attention. One of George Long's first actions upon arriving in Washington in 1900 had been to visit the counties where Weyerhaeuser had purchased timber to meet with assessors and other local officeholders. Thereafter, he was continuously involved in what he described as "missionary" expeditions to the various backwater seats of county government. Although reduced property valuations were occasionally secured, such efforts failed to halt the rise in taxation. Timber owners increasingly resorted to lawsuits, and the situation became so acrimonious that Long himself was briefly jailed in 1910 for refusing to turn over company records to state tax officials.[15]

Looking as well to the receiving end of the revenue apparatus, industry leaders attacked allegedly wasteful expenditures for schools and roads. "The general public," insisted Edwin G. Ames, "does not owe the boys and girls of this country anything better than a grammar school education." The state, though, insisted upon spending taxpayers' money on high schools and colleges and on lavish salaries for "these Socialistic teachers" at the University of Washington. As for roads, George

Long reported that automobile enthusiasts had stirred up "such a furor
. . . for good roads, that they are taxing the country to death to build
them."[16] Most citizens of Washington, to the consternation of those
concerned with the increasing rate of taxation, desired to educate their
children, to send their produce to market, and to relax with a Sunday
drive. Lumbermen, it seemed, were at odds with the extravagance of
modern times.

To the discomfort of those desiring to neatly compartmentalize the
actors of history, however, mill owners and loggers often became pro-
gressives themselves. Many lumbermen, disgusted with the anti-trust
policies of the Taft Administration, supported Theodore Roosevelt's
campaign for the Republican presidential nomination in 1912. Some,
most prominently George Emerson, actually followed the former president into the Bull Moose party. Having campaigned for the regulation
of the railroads, lumbermen favored an enhanced role for government
in other areas when in their interest. The industry, for example, supported financial aid for the nation's farmers in expectation that this
assistance would stimulate the demand for lumber and shingles.[17] These
and other actions demonstrated that lumbermen had moved beyond their
oldtime view of government as an antagonist to a more sophisticated
regard for the value of politics. This was clearly demonstrated by the
Northwest industry's involvement in two key advances of the progressive era: conservation and labor reform.

The Forest Fire Menace

Destruction of the forest was at the center of the European occupation
of North America. Wherever they went, settlers cut timber for their
immediate use and removed what was left, to clear fields or for the
psychic pleasure of subduing an otherwise hostile environment. Devastation, as William B. Greeley once noted, was "part of the price of
democracy" in a land where resources were so abundant that few could
imagine their extinction. By the final decades of the nineteenth century, leaders of the new environmental movement were convinced that
this price had become an excessive drain on the nation's heritage. George
Perkins Marsh and Charles Sargent contended that trees must be regarded as part of a fragile ecosystem instead of as sources of lumber.
As the United States became an urban and industrial nation, Sierra
Club founder John Muir defended the forest as an indispensable refuge
from the pressures of modern life. In the forested wilderness, Muir
argued, "the galling harness of civilization drops off, and the wounds
heal ere we are aware."[18]

Between the Civil War and the end of the century, foresters drew
on the theories of Marsh and Sargent and benefitted from the popu-

larity of Muir's writings. In contrast to these thinkers, however, advocates of forestry sought to convince the public that they were not, as Carl Schurz put it, "a set of amiable sentimentalists, who have fallen in love with the greenness of the woods and [who] break out in hysteric wails when a tree is cut down." Their first great achievement came in 1891 with passage of legislation authorizing creation of government forest reserves, the predecessors of the national forests. Under the law, presidents removed from private entry extensive tracts of timber, the most spectacular such action being Grover Cleveland's withdrawal of twenty-one million acres shortly before leaving office in early 1897. By 1905, three reserves had been created in Washington state, containing 7.4 million acres.[19]

Creation of the reserves was generally unpopular in the Pacific Northwest. The Seattle chamber of commerce, reflecting prevailing opinion, criticized Cleveland's 1897 action as "a galling insult" and "an amazing instance of the indifference of the East to the facts, conditions, necessities, and rights of the people of the West." Timbermen, though, supported the reserves for reasons of direct self-interest. The withdrawal from entry of large tracts, for instance, enhanced the value of timber remaining in private hands and strengthened the position of the major private holders. Establishment of the reserves, George Emerson pointed out, "goes a long way toward the immediate control of timberland in the state by large owners." Among prominent members of the industry, only J. J. Donovan was a vocal opponent of the government's policy. The Bellingham lumberman's opposition, however, stemmed from the fact that creation of a reserve in the northern Cascades interfered with his plans for mining and railroading ventures, rather than from any impact on his timber operations. He was "in full sympathy with all proper conservation ideas," insisted Donovan.[20]

More than anyone else, the lumberman was aware of the wastage typical of logging and milling. Only the best trees were removed, the remainder being left to decay, and the introduction of steam-driven machinery greatly increased damage to the forest. As early as 1885, Cyrus Walker had expressed concern for "the way logs are being slaughtered, when by better judgement in manufacturing lots of money could be saved to the concern." According to one progressive-era calculation, the amount of wood wasted in the state each year, if piled in cords, would suffice for the construction of a wall 3,000 miles long. "If all this is not a record of crazy men, fools and idiots," observed George Emerson, "you will have to go further than a Washington Boom to find them." With timber rapidly increasing in value, lumbermen had to reduce waste through development of more effective manufacturing methods and marketing practices. "The extinction of the Buffalo is a parallel case," Emerson reminded those declining to abandon traditional practices.[21]

In addition to waste, forest fires became an increasingly urgent problem with the expansion of logging activities and the influx of field-clearing settlers at the end of the nineteenth century. Each summer, great clouds of smoke spread over western Washington, often making for hazardous navigation on inland waters. "The smoke from these fires," wrote an English tourist in 1883, "for weeks and for months troubles the air and obscures the landscape." During his entire stay, noted a visiting journalist in 1889, "the sun has not been clearly visible on account of the smoke in the air." State officials estimated in 1905 that more timber had been destroyed by fire in the land west of the Cascades than by logging, a startling indication of the extent of the problem.[22]

Aside from timbermen, few Washington residents exhibited concern over the danger of fire. In an appropriate symbol for the destruction of America's natural resources, some early settlers actually ignited the forest in celebration of the Fourth of July. The near-burning of the territorial capitol by a forest blaze in 1859 failed to arouse interest in the subject. A series of nineteenth-century laws imposed penalties for the setting of fires, but legislators rejected requests for more stringent measures. The opposition of farmers to restrictions on the burning of fields, the belief that the forests were inexhaustible, and concern over the propriety of government regulation prevented effective action.[23]

By the end of the nineteenth century, lumbermen accepted the theoretical necessity for conservation measures. The great obstacle, though, was the expense required for protection against fire and for reforestation of logged-off land, an obstacle accentuated by the growing burden of taxes. The up-and-down nature of the lumber market made it difficult to justify longterm forestry expenditures. Within the bounds of the free enterprise system, moreover, there were, as far as timber owners were concerned, severe limits on how far the government should be allowed to go in forcing private interests to take action. Still, the growing popularity of forestry—dramatized by the transformation of the Agriculture Department's forestry division into the U.S. Forest Service in 1905—and the creation of the reserves offered hope that the public might be willing to cooperate with timber owners in the search for reasonable solutions.

Along with Theodore Roosevelt, his friend and mentor, Forest Service chief Gifford Pinchot made conservation one of the most dramatic issues of the progressive era. To Roosevelt and Pinchot, the "underlying principle of conservation" was nothing more complicated than "the application of common sense to common problems for the com-

mon good." Common sense, wrote Pinchot, "holds that the people
have not only the right, but the duty to control the use of the natural
resources, which are the great sources of prosperity." Directed by a
corps of professionals, forestry would be so applied as to provide for
the needs of all Americans, present and future. "The outgrowth of
conservation, the inevitable result," insisted the nation's chief forester,
"is national efficiency."[24] Here in brief summary was progressivism's
emphasis on the benefits of enlightened management, a technique as
applicable to private as to public endeavors.

Adept in the use of propaganda, Pinchot encouraged the belief that
the nation faced a crisis in its supply of timber. According to available
estimates, a third of America's forest cover had been removed since
1776. Assuming the continuation of present rates of removal, Pinchot
posited "a probable duration of our supplies of timber of little more
than a single generation." Supporting this analysis, Theodore Roosevelt warned the first national conference of governors in 1908 that "we
are over the verge of a timber famine in this country." Soon, conservationists feared, only the wealthy would be able to afford lumber.
Although the talk of famine acted to stiffen the values of their presumably endangered holdings, lumbermen discounted the existence of
a crisis. Conceding that "Mr. Pinchot has the better of the argument
as he has facts and I have fancy," George Long expressed the "intu-
itive belief" that the forest would not be exhausted.[25] Most Americans,
though, were receptive to the argument, increasing the popular support
for conservation.

If a national emergency did in fact exist, Pinchot's forestry prescription was a remarkably mild antidote. Essentially, the chief forester
and his followers called for reduction of the fire menace and revision
of state tax laws. The danger of fire and the burden of the property
tax, they argued, discouraged longterm holding of timber and thus mil-
itated against the application of forestry concepts. "Owners, willing
and even anxious to retain their forests," observed the journal of the
American Forestry Association, "are . . . driven to cut the forests and
dispose of their timber in sheer self-defense." The nation's "forest policy," Roosevelt wrote Pinchot on one occasion, "substitutes the good
of the whole people for the profits of a priveleged few."[26] Actually,
the aim of their approach was to serve both public and private interest
by making forestry a financially viable undertaking.

Pacific Northwest lumbermen were, for obvious reasons, attracted
to the course of forestry laid out by Pinchot. Those capitalists who had
invested huge sums in the region were particularly interested in protecting their timber holdings from destruction. "The more I get acquainted with our investment the better I like it," wrote George Long
in the spring of 1903, "aside from the one great element of fire haz-

ard." The Forest Service, moreover, offered informed, authoritative and apparently disinterested assistance in the campaign to secure tax relief. The efficiency aspect of Pinchot's approach, finally, struck a receptive response among lumbermen who were striving to bring order out of chaos through the medium of the trade association. Under the right circumstances, forestry could be integrated as a legitimate and valued aspect of business operations. Lumbermen, concluded a spokesman for the industry, were "too intelligent . . . not to undertake those methods which will perpetuate their supplies of raw material and thus prolong their business."[27]

Convinced that the support of industry was essential to the successful implementation of forestry, Pinchot portrayed himself and his profession as the unsentimental friend of business. As head of the Forest Service, he worked to increase lumber prices through maintenance of tariff barriers and encouragement of the trade association movement. Every opportunity was taken to speak to lumbermen on the practical merits of forestry. The Forest Service, Pinchot told the Pacific Coast Lumber Manufacturers Association in 1905, wished "to meet you not less than half way." Northwest timber owners, in turn, kept Pinchot informed of their activities and utilized his circulars in efforts to secure fire protection and tax revision measures at the state level. "All of us," one lumberman later recalled, "thought very highly of Mr. Pinchot."[28] The meshing of the public and the private interest became evident in the aftermath of the disaster of 1902.

The Campaign Against Fire

Although dramatic, the losses in the 1902 fires were less than devastating to many timber owners. Much of the timber in the Yacolt Burn area tributary to the Columbia River, for prime instance, was owned by Weyerhaeuser and other large investors able to withstand the financial blow. Most of the green trees, moreover, had been killed, not destroyed, and therefore could be salvaged. "If the timber is accessible," George Long pointed out, "it is possible to get in and log it with approximately only 25% to 30% loss, if the logging is completed at anytime within an epoch of 7 or 8 years after the fire." Long and his employers were soon at work combining the holdings of investors in the Yacolt region and making arrangements for the sale of logs in Portland. "I do not doubt but that all of our burned timber will be made to net us $1. per thousand," Long calculated, "and some of it considerable more."[29] This, of course, was less than would have been earned if the timber had not been burned. The retention of a modest profit, however, remained a closely-guarded secret as timber owners moved to capitalize on public concern over the loss of life by securing protection for their investments.

Under the direction of George Long, who was determined to "strike while the subject is warm," a committee of lumbermen drew up appropriate legislation. Trade journals and newspapers provided outlets for publicity and a lobbyist was employed to press for approval of the program by Washington legislators in early 1903. The result was legislation establishing a state fire warden, deputy fire wardens in the counties, and a system of patrols and burning permits. An additional act in 1905 created a state forest commission and fire legislation was approved in Oregon in the same year.[30] Passage of these laws marked an important step in the direction of conservation of the Northwest's natural resources.

If the legislation was to serve the interest of timber owners, successful implementation was as crucial as passage. Thus, when the state of Washington exhausted the funds appropriated for fire work in mid-1906, George Long arranged for an $8,000 grant from timbermen to finance forest protection through the remainder of the summer. "The only menace there is to our property in this country is fire," he reminded those questioning the private financing of government activities. The amount of money being spent on the fire problem, he pointed out in a similar vein in 1909, "is a mere bagatelle compared with what should be expended in that direction." Still, effective cooperation between the state and the industry greatly reduced the fire danger in Washington. Across the Columbia, in contrast, Forest Service researchers later concluded that a more lackadaisical approach by public officials and private interests was responsible for continued devastation of Oregon forests.[31]

To step up the campaign against fire, Weyerhaeuser and other large timber owners formed the Washington Forest Fire Association in 1908. "It is quite evident that the owners of timbered land out here will have to watch their property," Long informed one of his employers, "and the territory is so large which is to be covered, that it seemed more economical to all go into one association and have the work done under one hand." The amount of money invested in the state's timber was too great for fire protection to be entrusted solely to public officials. "It is folly to depend upon the State to handle this matter," Long pointed out. Initially, 138 firms owning 2.6 million acres joined the association. Assessments of a penny an acre supported the employment of seventy-five men to patrol the forests in the 1908 fire season. Thereafter, membership and expenditures increased and the W.F.F.A. was accounted the first great success of forestry in the region.[32]

Fire patrolling, though, was only one of the important functions assigned to the association by its organizers. The designation Washington Forest Fire Association, explained Long, was merely "the high-sounding name that will go before the public." A confidential tax division was created within the W.F.F.A., its identification with the industry kept secret, so Long wrote, "for prudential reasons." The task of the division, Long reported, was "to nose around in all the nooks and corners of each county to investigate in regard to road tax, and school taxes, and watch the spending of money." Issued under the auspices of the State Tax League, information gathered in these forays publicized the allegedly wasteful expenditure of public money.[33]

For a broader-based effort, George Long led the way in formation of the Western Forestry & Conservation Association in 1909. The organization served as a clearing house for the various regional forestry associations and, most importantly, as a public relations agency. Its assignment, observed Long, was "to circulate the right kind of literature upon the necessity for re-forestry and conservation, and for reduced taxation; and to attend to the details of trying to get improved laws on our statute books for fire prevention and for taxation." To further these aims, Long secured the employment of E. T. Allen, Pinchot's district forester and a former newspaperman, as executive director. In its early years, Long provided much of the W.F.&C.A.'s financial support and arranged for publication of Allen's book on what became known as practical forestry.[34] Controlling the two most important forestry associations in the region, Weyerhaeuser and other timber owners were able to direct public opinion along courses beneficial to their interests.

Above all else, lumbermen were determined to maintain control of the conservation movement, a determination that required constant attention to the scope of state and other private efforts. The University of Washington forestry school, organized in 1907, received considerable financial support from the state's lumbermen. In turn, it became one of the first institutions in the nation to train foresters for industry rather than for government service. Until 1909, the Washington Conservation Association "had for its membership," wrote George Long, "a whole lot of long haired people who had a lot of crazy ideas and who were issuing all kinds of manifestoes and whose every word and sentiment was freely printed and commented upon by the press." During that year, however, timbermen made liberal financial contributions, assumed membership and took control of the society. It was successfully redirected, reported Long, "into channels which were sane and sensible and . . . what the lumbermen wanted."[35]

Although successful in attaching public and private institutions to the cause of practical forestry, lumbermen found it increasingly difficult to maneuver between those desiring to preserve and those desiring to develop the resources of the region. State officials and development groups opposed the creation of additional forest reserves as inimical to the economic growth of Washington, an especially potent argument in the stagnant aftermath of the panic of 1907. "The people up here," J. J. Donovan wrote from Bellingham, "feel that we have been getting most too much of a good thing in the way of forest reserves." Unable to obtain cheap stumpage, small loggers and mill operators charged that large private owners were responsible for the refusal of the Forest Service to sell timber in the reserves.[36] Conservation, it was clear, was becoming a matter of intense controversy.

In the midst of a dramatic dispute with Secretary of the Interior Richard Ballinger, Gifford Pinchot was dismissed by President Taft in January 1910. Divided in their emotions, lumbermen defended Ballinger, with whom they were well acquainted as a longtime political figure in Washington state, but also praised Pinchot. "I confess that I am all up in the air and do not know where I am at in my views in this controversy," wrote George Long. "The lumber interests never had a better friend," the *Timberman* observed of the former chief forester. The appointment of Henry Graves as Pinchot's successor was applauded by Long, "as it reveals a disposition at least, not to abandon the Pinchot ideas in forestry."[37] This was reassuring, but the departure of Pinchot proved to be a watershed in the relations between lumbermen and foresters.

Already, many in the forestry profession had begun to question whether fire protection and tax revision were sufficient remedies for the nation's timber malady. Increasingly, a stiff dosage of federal regulation became the favored prescription. Following his dismissal, Gifford Pinchot himself emerged as the most effective advocate of the new course. Ostensibly, this development resulted from the slowness of lumbermen to adopt the moderate measures recommended in the years prior to 1910. Of practical forestry, Pinchot wrote in later years, "something like ninety percent is propaganda, and a good deal of the rest the assumption of good intentions."[38] Considering Pinchot's oftenvoiced appreciation for the practical difficulties facing timber owners, his change of approach probably had more to do with politics than with frustration over the shortcomings of industry. Like Theodore Roosevelt, Pinchot was more radical out of government than in and appreciated the truism that the methods for attaining power differ from those required for the maintenance of power.

Lumbermen now confronted in their former patron a formidable enemy, a fact made evident at the national conservation congress held in St. Paul in September 1910. Northwest timbermen feared in advance that the meeting would become a forum for a Pinchot-Roosevelt assault on their industry as a prime beneficiary of the allegedly deficient conservation policies of the Taft Administration. "I think our duty there,"

George Emerson quipped of the conference, "would be more by way of self protection than it would be protection of national resources." At St. Paul, to an audience packed with enemies of Taft, Roosevelt called for federal regulation of "predatory corporations." Pinchot, in his address, contended that "the private ownership of forest land is a public trust, and the people have both the right and the duty to regulate the use of such lands in the general interest." Protestors, including several leaders of the Northwest industry, were drowned out in the general enthusiasm for regulation.[39]

Lumbermen were outraged by what Governor Marion Hay of Washington termed the "tommy-rot and twaddle" put forward at St. Paul. "'Teddy' has gone to far," Edwin G. Ames complained of the former president. Everett Griggs of the St. Paul & Tacoma Lumber Company, who had attended the conference, deplored the "tendency of would-be statesmen to arouse class hatred, [and] to encourage communism and anarchy." George Long was convinced that Pinchot had "lost his head" and wrote that the forester was "certainly entitled to high praise from every body for the many good things he has done, but candidly, I do not like his attitude as a politician." Because of his "meddling interference in matters which are outside of his duties as a forester," Long feared that Pinchot had damaged the practical conservation movement.[40]

The events of 1910—Pinchot's firing and the St. Paul conference—initiated a period of intense enmity between timber owners and regulation-minded conservationists. Gifford Pinchot, marching arm-in-arm with Northwest lumbermen for nearly a decade, had branched off down a path overgrown with the tangle of federal supervision. Continuing along the same practical forestry course, timber owners increasingly complained that their problems and their contributions failed to receive proper appreciation. "Eastern people," lamented George Long, "do not regard us as anything other than ordinary wood butchers who are so careless that we not only waste our timber in cutting but made no effort to keep it from burning."[41] Critics of the industry charged that lumbermen, out of cynicism, favored limited conservation measures because of self-interest rather than from any concern for the public welfare. This was an accurate assessment, but it was also all that had been required by the initial brand of Pinchot forestry.

Reforming the Conditions of Labor

Laborers, even more than regulation advocates, were a prime source of aggravation for Washington lumbermen in the first years of the twentieth century. This was so even though working conditions in the Pacific Northwest were, in the opinion of Edwin G. Ames, "better than anywhere else on earth." The typical laborer, Ames continued, was "well fed, dresses neatly and warmly, and can go to the moving picture show once a week." Placed in perspective, this observation loses a good deal of its apparent absurdity. Daily wages in the western Washington lumber industry averaged $2.50, a figure that George Long considered "fully 20% higher" than in the Great Lakes. The small number of women employed in lumber and shingle mills were paid at a rate well below men performing the same tasks and were the only ones with a legitimate grievance on this score of pay. Wages, of course, were reduced during hard times, but not—at least as far as employers were concerned—to a level commensurate with economic reality.[42]

Those calling for labor reform focused not on wages but on the conditions of work. A study conducted during the First World War found that half the logging camps in the Northwest lacked adequate bunks or showers and that a third were without toilet facilities. "As a community," Rexford Tugwell reported after a tour of the region, "a lumber camp is a sad travesty at best." Such reports and observations in actuality applied more to small logging outfits than to large operations, which employed most of the workers. These latter firms possessed the financial means to give reality to the view that improved living conditions would encourage the development of a stable labor force. Better camps, pointed out logging operator Thomas D. Merrill, "are as much to our interest as to the men's." Of the money expended by the Puget Mill Company in this area, Edwin G. Ames calculated in 1918, "we have more than gotten back indirectly twice over."[43] Like conservation, the right sort of labor reform could be implemented as a sound business practice.

Even radical writers conceded that food, the most important aspect of camp life, was "fairly substantial and plentiful." Loggers were legendary eaters, one study finding that dinner, the most leisurely meal, was consumed in the average time of thirteen minutes. "Table manners were non-existent," recalled a camp owner, "and the loggers speared food off the platters with their forks. If one reached for a piece of bread and the lights went out you would pull back your hand with four forks stuck in it." It was not unknown for entire crews to walk off the job because of poor food, complaining that they had been "starved out." The most parsimonious logging operators devoted careful attention to the quantity and quality of food provided for their men, especially because other aspects of life in the woods were not so amenable to improvement.[44]

Whether in camp or mill, in large operation or small, the companion of work was the terrifying accident. Each year, five times as many workers were killed in the mills and woods than in any other sector of Washington industry. Because of saws, whirring cables, and other machinery, moreover, injuries were particularly gruesome. Cases investigated by the state in 1905 and 1906 included those of a Centralia logger crushed by a falling tree, a Tacoma sawyer who became entangled in machinery, a King County mill worker who had his arm severed, and a Ballard shingleweaver who slipped and fell on a saw. There were 142 accidents at the Port Blakely mill alone in 1904, four of them fatal. The industry seemed, as a labor leader in Robert Cantwell's *The Land of Plenty* charged, to have "made a science" out of killing and maiming its workers. From the perspective of the employees, accidents meant loss of wages and life. From the perspective of the employer, they disrupted production and resulted in expensive damage suits. The hazardous nature of the industry, observed a trade journal, was "nothing short of a calamity" for management and labor alike.[45]

While accidents contributed to ill-feeling, the principal source of worker discontent was the contempt openly expressed by mill and camp owners. Employees were "things," "cattle" and "rif-raf," to cite expressions uttered, respectively, by Cyrus Walker, Frederick Talbot, and George Long. The foreign-born, making up half the workforce by the turn of the century, were the most likely to be the targets of abuse. Swedes were characterized as dense and those from southern and eastern Europe were lumped together as "black men." Products of a Social Darwinist age, lumbermen believed that to complain about conditions was to exhibit unmanly tendencies. Mill owners, it was argued, had pulled themselves up from the ranks and similar advancement was open to all willing to engage in hard work. Modern workers, though, "all want rooms with bath, and Waldorf-Astoria fare," complained Edwin G. Ames.[46] Such hostility was readily transferred to labor organizations.

Workers responded with an equivalent amount of enmity, exacerbating the relations between office and bunkhouse. To employers, stated a radical union organizer, the logger was "nothing but a living machine." Feelings about the unfairness of the system were reflected in logging slang. Workers were "slaves" and the employment office a "slavemarket." The superintendent was a "bull" and the timekeeper a "pay cheater." George Long once remarked that the key to labor relations was whether or not the men liked the boss.[47] When boss and men despised each other, a volatile situation had come into being.

A sure sign of this mutual hostility was the revival of the labor movement from the depression-induced stupor of the 1890s. The International Shingle Weavers' Union of America, organized in 1903, led numerous strikes over wages and recognition, notably at Everett in 1904 and at Ballard in 1906. Efforts to organize the sawmills and logging camps made less headway. An American Federation of Labor–

chartered sawmill local in Everett collapsed and the Royal Loggers, formed on Puget Sound in 1906, foundered when its leader absconded with the organization's treasury. Although alarmed over the shingle situation, lumbermen were initially confident that the lack of initiative among their workers would for once work in the interest of the industry. "I dont believe we shall have any trouble about the mills," Cyrus Walker reported in mid-1901, "as the men we have employed, are so low in character."[48]

The threat posed by the Industrial Workers of the World, however, had to be taken in a serious vein. Founded in Chicago in 1905 by western labor leaders, socialist intellectuals, and others disgusted with the American way-of-life, the I.W.W. called for the creation of "a universal working class movement" based upon the "irrepressible conflict between the capitalist class and the working class." The organization sought nothing less than the overthrow of capitalism and its replacement by a workers' utopia. "We are forming the structure of the new society," declared the I.W.W. constitution, "within the shell of the old." The Wobblies soon found in the forests of the Pacific Northwest one of their greatest opportunities.[49]

Wobbly violence was for the most part limited to rhetoric, and most workers, especially in the sawmills, were hostile or at least indifferent to the organization. Nevertheless, lumbermen were driven to hysterics by the I.W.W. "They go wild, simply wild over me," the radical poet T-Bone Slim quipped of employers. Wobblies were regarded as shiftless scoundrels, the initials I.W.W. supposedly standing for "I Won't Work" or "I Want Whiskey." In their rejection of capitalism, the Wobblies scorned the system that had enriched lumbermen, committing in the process an effrontery that was considered to be both gross and dangerous. I.W.W. members, finally, were without roots or obligations and had nothing to lose, accentuating their unpredictability and willingness to confront management. The radical challenge was so serious and so obvious, argued lumbermen, as to justify all manner of retaliatory measures. "When people threaten what you have," pointed out Archie Binns's fictional timber beast, Charlie Dow, "you crush them any way you can."[50]

The Wobblies—characterized by the *Timberman* as "unscrupulous fakers, who have no interest in the lumber business"—made their initial appearance in the region in March 1907, closing all but one Portland mill for three weeks in a strike over wages and the nine hour day. Edwin G. Ames, alerted to the new organization, reported that the I.W.W. was made up of "a lot of socialists and anarchists and they only take in the lowest class of laborers, and . . . anyone no matter what nationality may become a member." Washington lumbermen feared that, if successful, the strike would spread across the Columbia to Puget

Sound and Grays Harbor. The walkout collapsed, however, when strikebearers were imported from the Sound and Portland A.F. of L. locals refused to cooperate with the strikers.[51] Although a victory for employers, the disruptive capacity of the Wobblies had been made clear to mill owners.

Washington lumbermen, forewarned by the experience of their Portland colleagues, moved with alacrity to "destroy," as George Emerson put it, "the caterpillars before they breed." Detectives were employed to infiltrate Wobbly locals north of the Columbia. The Grays Harbor Commercial Company patrolled the fence around its Cosmopolis mill with armed guards and dogs, leading to the firm's designation by workers as the "Western Penitentiary." To drive down wages and divide labor along racial lines, mills north of Seattle hired Japanese workers and the Bloedel-Donovan lumbering operation at Bellingham employed large numbers of Hindus. The latter experiment, though, ended when outraged white residents drove the unwanted newcomers from town.[52] Of all the tactics utilized, the most effective and the most sophisticated proved to be the reform of working conditions. Lumbermen had joined with conservationists to influence the direction of forestry and they would join with conservative labor leaders to implement reforms in the mills and logging camps.

The value of such cooperation to employer and employee alike was most evident in the area of accidents. For years, mill owners and logging operators had complained that the courts awarded damages to workers injured as the result of carelessness, rather than company negligence. These improper decisions resulted from the apparent fact that the courts were prejudiced against large business enterprise. "Our juries are always 'Pops,'" noted George Emerson during the heyday of the populists, "and 'pop' it to corporations on all occasions." A number of solutions had been tried by lumbermen, without achieving success. Workers were for a time forced to sign agreements absolving their employers of any responsibility for accidents. At the turn-of-the-century, mill companies bought liability insurance and set up hospital funds, deducting money for these purposes from pay envelopes.[53]

In spite of these efforts, the problem of damage suits grew worse, especially after state supreme court decisions increased the liability of employers. Mill companies charged that they had become easy prey for crooked attorneys eager to manipulate the sympathies of jury members. "Hardly a day passes," asserted the *Pacific Lumber Trade Journal*, "that a mill man is not mulcted of a sum in many cases ridiculously high, and on evidence that would not be accepted in the ordinary civil or criminal action." The industry, wrote Emerson, was "very close to a volcano in connection with the dangers incident to the milling business." With organized labor pressing for legislation that would make employers financially liable for all injuries, no matter what the cause, something had to be done.[54]

Facing this final anti-business prospect, most industry leaders gave their support to compulsory workmen's compensation as protection from liability for accidents. American Federation of Labor officials also backed the concept, since damage suits were uncertain of outcome and much of the money awarded to plaintiffs actually went to cover attorneys' fees. In 1911, a special labor-management committee appointed by Governor Marion Hay developed a compensation bill for presentation to the legislature. After securing pro-industry amendments, a committee of mill operators headed by Edwin G. Ames was instrumental in securing passage of the bill. Creation of one of the nation's first workmen's compensation systems was a triumph for the progressive movement and a benefit to business. The law was "without doubt advanced legislation," conceded Ames, but it would result in "a great saving to the manufacturers" through elimination of lawsuits and reduction of the cost of operating the courts. All in all, summed up Ames, workmen's compensation demonstrated "what can be done when we are inclined to be fair."[55]

Although workmen's compensation proved a prime benefit to workers, it failed to down discontent. Following a lengthy free speech fight, the I.W.W. led a walkout on Grays Harbor in March 1912, closing, the Grays Harbor Commercial Company excepted, every plant on the harbor and at Raymond on the Willapa River. Mill owners responded to the strike with uncommon fury. Pickets were scattered by firehoses and deputy sheriffs rousted strikers. Finnish and Greek workers, considered the instigators of the strike, were rounded up and deported from Hoquiam and Raymond in boxcars, with American stock family men hired as replacements. "The mill men are inclined to import men with families . . . and do away with the Finns and the Greeks," reported logger Alex Polson. "I feel that we owe it to our country to do the same." The strike was over by mid-April, but the Wobblies had made clear that reform would not destroy their movement. "Where they will strike next," wrote a worried Edwin G. Ames, "I do not know."[56]

Proving its resilience, the I.W.W. laid plans for an industry-wide strike in the spring of 1913, encouraged by a visit from Big Bill Haywood, the most famous Wobbly leader. "Most everybody has a detective in or around their plant, as we have," Ames informed Frederick Talbot, "trying to find out who the I.W.Ws. are, and quietly let them go and get them out of the way." When it came, the strike closed only a few camps and mills. With the industry in the midst of a depression, there was a surplus of men eager for jobs. The strike-induced reduction in production, moreover, actually helped mill owners cope with glutted markets. The walkout was "a grand fizzle," observed a relieved Ames as the I.W.W. entered a period of quiescence. The beginning of a new campaign to organize the lumber industry, sponsored this time by the American Federation of Labor's shingleweavers union, meant that there would be no relief for the frazzled nerves of mill owners.[57] The battles in Europe following the outbreak of war in August 1914 made interesting reading for lumbermen, but were as nothing to those being fought in their own camps and mills.

Washington, George Emerson had forecast in 1904, was on the brink of "a state of affairs that is nearer civil war than prosperity." This had proved to be an accurate prediction with respect both to the state of the lumber market and the state of relations between management and labor. Easterners must think, reflected Edwin G. Ames in April 1914, that "we are a crowd of wild ones out here on the Pacific Coast, and that we have lost control of the situation . . . to the disadvantage of business and law and order."[58] Lumbermen had attempted to maintain control through advocacy of conservation and labor reform. But foresters had moved beyond their original probusiness concepts and workmen's compensation had failed to stifle unrest. The lumbering depression added to the overall gloomy sense of failure. When the United States entered the world conflict, however, the debate over conservation would be set aside, prosperity would return, and victory in the labor war would be achieved.

11

The Spawn of Hate

The mob spirit is displaying itself here and there in this country.—Woodrow Wilson[1]

No game has ever been played with more consumate skill to poison the well springs of patriotism and loyalty to a country than the insidious, slimy and soul-destroying doctrine of the Industrial Workers. Without respect for either God, government or men, these traitors have spewed the spawn of hate and sung their hymns of venom all over this fair country. . . . The flag of our fathers and of our common country has been defiled and vilified—and all in the name of labor and the sham pretense of the uplift of the workers.—*Timberman*[2]

A hundred logging operators and lumber and shingle manufacturers gathered in the quarters of the Seattle chamber of commerce in the second week of July 1917. Angry and fearful, they confronted what all believed to be the greatest crisis in the history of their industry. The Industrial Workers of the World, striking in apparent alliance with American Federation of Labor timberworkers, had closed most of the camps and mills in the Pacific Northwest. Summoning their courage, those present at the Seattle meeting formed the Lumbermen's Protective Association and ratified an agreement pledging an all-out fight against the major demand of the strikers, the eight-hour day. "Practically all they did," John Dos Passos once wrote of the Wobblies, "was go to jail."[3] Actually, the I.W.W.'s achievement in 1917 was unprecedented, if unfortunate from its point-of-view. The apparent revolutionary situation spawned by the organization was so threatening that large and small mills and loggers were able to overcome their traditional divisive interests. The initial willingness of reform-minded lumbermen to go to eight hours, moreover, was set aside in order to maintain a common front against labor. United as never before, the industry moved to destroy the Wobblies as an active force in the Northwest, capitalizing on a shrewd feel for the temper of wartime America.

The War Boom

Beginning in August 1914, the First World War created economic uncertainty and shipping shortages and upset foreign lumber markets. Lumber that could not be sold abroad was dumped onto the already stagnant domestic market, glutted with overproduction since the panic of 1907. By the end of 1914, more Northwest sawmills were closed than at any time in the industry's history, and Edwin G. Ames reported that conditions were "fast getting down, . . . to a matter of the survival of the fittest." Meeting in emergency sessions, Washington and Oregon lumbermen concluded that 1915 would be even worse than 1914 and that there was no hope for recovery until the end of the war.[4]

To the surprise of mill owners, however, the war soon proved to be the medium of welcome benefits. With the fighting in Europe bogged down in stalemate, orders arrived for lumber to be used in construction of trenches, cantonments, and refugee housing. The British blockade of the North Sea increased the demand for American lumber in markets normally serviced by Scandinavia. Domestically, farmers stepped up their purchases, Midwestern lumber yards replenished their stocks and the rail trade was restored to prosperity. "Not in months, possibly years," noted the Seattle correspondent of a trade journal in the spring of 1916, "has the trade been in such excellent spirits." Washington lumber production for that year was only slightly below the alltime high.[5]

Some aspects of the war boom were of particular importance, altering the focus of the industry. Spruce from the rain-drenched Northwest coast was demanded for use in construction of airplanes because of its fortuitous combination of strength and lightness. Although the *Pacific Lumber Trade Journal* had earlier quipped that the "usefulness of aeroplanes in war depends on whether or not the enemy can be induced to use them," the fighting provided quick and ample evidence for the utility of flying machines. Mills on Grays Harbor sent an estimated 750,000 feet of spruce a month by rail to Atlantic coast ports for export to England. As demand accelerated, the British government dispatched confiscated enemy merchant vessels to haul lumber from the harbor and assigned air officers to keep watch on production. Airplane lumber was also sold to Germany, despite the cumbersome transfers through neutral middlemen forced by the Allied blockade. Mills producing spruce, the *Timberman* reported in March 1915, were "loaded up with orders" from both the Allies and the Central Powers.[6]

The revival of the lumber industry and the opening of numerous shipyards on Puget Sound and Grays Harbor placed a severe strain on the supply of labor. Lumbermen feared that workers, pointing to the high rates of pay offered by shipbuilders, would take advantage of the situation to demand wage increases and recognition of their unions. Driven by these fears, employers continued to stress the hope that improved working and living conditions would somehow defuse discontent. The industry, in its prime effort along this line, was instrumental in securing first-aid legislation from the Washington legislature in 1917, providing for state payment of the medical expenses of injured workers. While continuing to hope for the best, many industry leaders concluded that labor would never be satisfied with reform. American Federation of Labor organizers, complained Edwin G. Ames, made use of every concession "to show the objects and benefits of organized labor to the individual worker."[7] As for the I.W.W., nothing would mollify that band of hopeless revolutionaries.

Violence exploded with a flash at Everett on the first Sunday in November 1916. Wobbly demonstrators, arriving on the steamer *Verona* to protest the brutal treatment of striking shingle workers, were met at the dock by a posse organized by local mill owners. In the resultant gunfire, seven were killed and forty-seven wounded—among the latter, lumbermen Herbert Clough and Joseph Irving—before the vessel pulled away from the wharf to return to Seattle. Employers and their spokesmen were elevated over the punishment dealt out to the Wobblies. "Law and order must be upheld at any cost," the *Timberman* pontificated of those who had taken the law into their own hands, "or civil government is a farce." When Mayor Hiram Gill of Seattle criticized the townspeople of Everett, Edwin G. Ames angrily retorted that when a public official referred to "the better element of citizenship . . . [as] 'thugs and murderers', one almost loses faith in human nature." So as not to lose his faith, Ames became involved in the organization of a vigilante group in Seattle.[8] To their satisfaction, lumbermen had won the initial stage in their confrontation with revolution, a stage that confirmed the argument that a firm resolve was the best course to take with the I.W.W.

The Wobblies Strike

In April 1917, America, after two-and-a-half years of tortuous neutrality, entered the war, a decision that brought at first only slight diversion from the desire of Northwest lumbermen to continue business-as-usual. Although confident that military service would be good for the nation's pampered youth, Edwin G. Ames sought draft exemptions for industry executives. Managing a business, he contended in succint fashion, was as important "as many military duties which can be attended to by others." Southern mill owners, more attuned through geographical affinity and political connections to affairs in the nation's capital, eagerly saw to the drawing of specifications for lumber to be used in war construction, specifications that supposedly worked to the advantage of yellow pine and to the disadvantage of Douglas-fir. Leading the country in production of a vital defense material, however, Washington operators could not long avoid the implications of the war.[9]

Appointed to a special National Defense Council lumber committee in May 1917, Weyerhaeuser's George Long became an important figure in the mobilization of the nation's economy. Among other difficult assignments, he had to deal with the reluctance of lumbermen to go along with government plans for construction of a thousand wooden vessels, half on the Pacific coast. At first refusing to supply lumber, mill owners complained of red tape and of prices pegged well below prevailing market rates. An emotional meeting in early June, at which Long engaged in "quite a little maneuvering," finally resulted in the agreement of Washington and Oregon manufacturers to sell to the shipyards. Clearly preferring the civilian trade, where prices reflected supply and demand, lumbermen regarded this decision as a patriotic sacrifice.[10]

Concentrating on production of fir lumber for ships and military cantonments, most Northwest lumbermen ignored spruce. There were ample reasons for this lack of interest. "Fully 75% of the mills . . . cut no spruce at all," George Long pointed out, "because there is no spruce in the timber adjacent to their locality." Spruce was usually interspersed with other species that would also have to be logged and marketed, increasing both expense and difficulty. Of the spruce actually removed from the woods, moreover, only about fifteen percent was suitable for airplanes. The construction of expensive railroads, finally, was required before most stands of spruce could even be opened to logging. "All of this," summed up Long, "means that an immediate quick response to the increased demands of the Government is almost an impossibility."[11] Ultimately, spruce would become important to the industry as a whole more as weapon to be wielded against labor than as a weapon of war.

Focusing on the region east of the Cascades, where working conditions were the most squalid in the Northwest, the Industrial Workers of the World revived on the twin currents of war-related prosperity and labor shortages to plan strikes in the spring of 1917. Nervous Inland Empire employers awaited the inevitable walkouts, which occurred at various points during the month of June. The Wobblies closed all but one logging camp and cooperated with striking harvest workers to drive the selfstyled better element into a state of panic. "They are on the verge of revolution over there," reported Edwin G. Ames from the safety of Seattle.[12]

West of the mountains, the I.W.W. and the smaller American Federation of Labor timberworkers and shingleweavers, bitter rivals mutually fearful of losing the initiative, independently concluded to walk out. In the second and third weeks of July, strikers closed most of the operations in western Washington. Among the major firms, only the Grays Harbor Commercial Company, secure behind an electrified fence, and the isolated Puget Mill Company mills at Gamble and Ludlow continued to produce lumber. "Things have been jumping around so fast," wrote George Long, "it is hard to tell where to commence or where to quit in telling the story." Lumber workers paraded through the streets of Seattle and Tacoma, joining with striking teamsters and streetcar employees to create an atmosphere of public support for the walkout. By the end of the month, at least seventy-five percent of the capacity of the state's mills and camps had been shut down.[13]

Considering the emotion generated among both employers and employees, the strike was notable for its lack of violence. Although soldiers were sent to Grays Harbor in late July at the urging of sawmill owners, the dispute was almost entirely peaceful. Rather than starting forest fires, Wobblies helped fight those that were ignited by nature and careless individuals. "They are a bad lot every way," George Long privately observed of the I.W.W., "and there is no telling what they will resort to, but I have been watching their capers pretty carefully . . . and while there have been a whole lot of suggestions and threats about their burning up sawmills and burning up timber land, I do not know of one authentic case where this has occurred."[14] The absence of conflict proved an embarrassing inconvenience for lumbermen requesting the stationing of troops in their mills and camps.

On the major demand of the strikers, the eight-hour day at ten hours' pay, lumbermen were sharply divided. One group, centering on Everett Griggs, Alex Polson, and the Simpson Logging Company's Mark Reed, supported a reduction in hours as a concession to labor. Loggers like Polson and Reed were well aware that lack of daylight limited hours in the woods during much of the year anyway, so there was little to lose in such a move. A second group, led by Edwin G. Ames, William C. Butler of Everett and other operators in that town, was violently opposed, believing that a reduction would increase costs, reduce profits, and encourage labor to make further demands. Adoption of the eight-hour day, insisted Ames, would be "practicing philanthropy with . . . stockholders' money." The hardliners heaped scorn on advocates of conciliation, charging that Reed, a Republican leader in the state legislature, and Griggs, a prominent Democrat, were motivated by a desire to placate organized labor for political reasons.[15]

Despite their differences, lumbermen were able to maintain a united public front. Whether favoring concessions or a fight to the bitter end, mill owners and loggers joined in arguing that the eight-hour day would place the Pacific Northwest at a further competitive disadvantage with Southern mills already favored by the railroad freight structure and a low-paid black workforce. "Until the yellow pine industry agrees to an eight hour day," argued the *West Coast Lumberman,* "the Pacific Coast industry cannot grant it, and long survive." Northwest manufacturers insisted that they were opposed only to a regional eight-hour day and would support a nationwide reduction. Only mills dependent on sales in the Midwest rail markets were threatened by Southern competition, however, and the regional differential served for the most part as convenient pretext for portraying the industry as the victim of discrimination.[16]

Of more importance, all agreed that the industry could not be placed in the position of appearing to capitulate to labor. Surrender on this point, asserted George Long, would be "the hardest blow our lumber industry has ever had, and one which will tie us to a machine that will drag us through the mud for years to come." Conceding the eight-hour day in the face of union pressure would be the first in a series of inevitable defeats by labor. "You can no more meet them half way," contended Long, ". . . than you can meet a burglar half way who enters your house and insists on looting all the valuables." For lumbermen, then, the eight-hour day was not, appearances aside, the central issue. Reform-minded operators backed off from their initial desire to compromise when they realized the consequences. Edwin G. Ames, who had earlier called for a regional eight-hour day as a means of limiting production, claimed for his part that he was willing to implement the reduction when the unions were defeated and it would appear to be a voluntary gesture.[17]

The determination to maintain a solid public front led to formation of the Lumbermen's Protective Association. Members pledged not to grant the eight-hour day without the association's approval, under a stiff penalty for violations. An executive committee headed by E. S. Grammer of the Admiralty Logging Company, a subsidiary of the Puget Mill Company, was selected to speak for the industry and to buck up operators thought to be wavering on the matter of hours. Although the militantly antilabor stance of Grammer's committee was blamed by some members of the industry for needlessly prolonging the strike, the L.P.A. was able to carry on an effective campaign. Only one mill in ten, mostly small shingle plants, abandoned the ten-hour day during July and August.[18]

At first, the industry seemed, through the efforts of University of Washington professor Carleton Parker, on the verge of a great victory. An authority on the social basis of labor unrest, Parker was engaged by the federal government to make sure that the strike did not interfere with delivery of cantonment lumber. Some mill owners were convinced that the professor was a dangerous radical, a conviction encouraged by his ready admission that Samuel Gompers was responsible for his appointment. Fantasies aside, Parker was in reality sympathetic to the problems of lumbermen. "He seems like a fair minded man," noted George Long, who went on to report that the professor had privately asserted "that he did not see how we could go to eight hours."[19]

Working with Long and other industry leaders, Parker developed in late July a plan under which the camps and mills would go to eight hours in return for aggressive government action against the Wobblies. "The mill men all insist on one thing," wrote J. P. Weyerhaeuser, who had succeeded his late father as president of the Weyerhaeuser Timber Company, "that the Government will grant the manufacturers protection from the lawless element of the I.W.W.'s." Parker wired the terms acceptable to lumbermen to the War Department: "they have patriotically agreed to the very serious sacrifice of granting the eight hour day . . . if the Federal Government agree in turn to suppress the agitation and strike activity of the I.W.W." Awaiting the reply from the nation's capital, industry leaders were confident that the strike would soon be over. The Wilson Administration, however, rejected the scheme and it remained for lumbermen to hold fast and await a change of position by federal authorities.[20]

Holding fast was not difficult, since the strike in many ways worked to the advantage of the mill companies. The sharp reduction in lumber output stiffened the market at a time when civilian orders were falling off. "If it was not for the strike," J. P. Weyerhaeuser pointed out, "lumber on this coast would decline in value." The *West Coast Lumberman* observed that "the present enforced curtailment" was "a splendid thing for the market and comes just at the time when most needed." Mill owners took advantage of the idle period to make much-needed repairs and had no difficulty supplying defense orders from stocks on hand.[21]

Attempts by government representatives to secure a compromise between management and labor provided the only obstacle to the industry's course. In early August, Henry Suzzallo, president of the University of Washington and chairman of the state defense council, called a conference in Seattle in an effort to end the strike. Aided by Carleton Parker, Suzzallo shuttled between lumbermen in one room and labor leaders in another. George Long, Mark Reed, and J. H. Bloedel demonstrated "some weakness" in these sessions, according to Edwin G. Ames, but were reprimanded by their militant colleagues. Neither employers nor workers were willing to give in and the conference ended in failure. Lumbermen also fended off those who urged acceptance of the A.F. of L. as a means of luring workers away from the I.W.W., preferring instead to hold out for victory over both organizations.[22]

Resisting efforts to settle the strike through compromise, lumbermen pressed for government intervention on the side of employers. The I.W.W. was portrayed as a violent and unpatriotic organization in the service, at one and the same time, of the kaiser and the Bolsheviks. Ignoring the facts, Alex Polson sent eastern friends lurid accounts of "the burning of timber and grain fields" and urged that they "use their influence with the Government to arrest and interne every I.W.W. leader." In issue after issue, the *Timberman* contended that "the sticky, slimy doctrine of disloyalty has been sown in the minds of the worker." Labor unrest, insisted the *West Coast Lumberman,* had been "fomented, fostered and encouraged by alien enemies." Arguing that the eight-hour day would reduce production of lumber needed for the war effort, employers skillfully took advantage of wartime hysteria to malign the loyalty of their opponents.[23]

In the effort to discredit the Wobblies, lumbermen eagerly grasped the weapon of spruce. The Wilson Administration had announced a greatly expanded airplane program, requiring, according to the estimate of George Long, "about 200 per cent over that which has ever heretofore been produced." Soon, government officials arrived in the Northwest to discover why production lagged behind the requirements of the program. Although the shortfall resulted from the isolated location and corresponding difficulty of opening up spruce, employers shrewdly blamed the problem on the I.W.W. "It seems to me this spruce proposition is the one to work," Clark Ring advised his associates in the Merrill & Ring Lumber Company. "The demand of the government [for spruce] is reason enough the authorities would be excusible in protecting the operation, and the best of our men, that really wanted to work, would also have a reason and excuse to stand by the government during this particular time."[24]

For a moment, there appeared to be no need for government intervention. In early September, the Wobblies returned to work and lumbermen briefly exulted over the apparent crushing of the strike. It quickly became evident, however, that rather than giving up, the I.W.W. had adopted a new and sobering course, the so-called "strike on the job." Workers abided by safety regulations, which were normally ignored in the interest of attaining maximum production, and practiced other means of restricting output, throwing the industry into complete chaos. As the ultimate blow, payment of wages to strikers on the job meant that employers were now, in effect, financing their enemies.[25]

Faced with this crisis, the industry stepped up its campaign for federal intervention. Mill owners and logging operators had enjoyed the upper hand during July and August, but now the initiative had passed to the Wobblies, who could disrupt production at will and continue their protest for an indefinite period of time. "Unless the government closes up every [I.W.W.] hall, and takes some means of arresting them for depredations done," wrote Alex Polson, "this country is in a bad way." Henry Suzzallo, agreeing with employers, concluded that the stationing of troops in the camps and mills was "the only way out." Pointing to the slumping production of spruce as an obvious indication of how the dispute was interfering with successful prosecution of the war effort, lumbermen demanded the drafting of strikers and the augmentation of the workforce with soldiers.[26]

Rise of the Loyal Legion

Reorganization of the government's war mobilization machinery in the fall of 1917 created new opportunities for lumbermen. To enable a more efficient management of the economy, the National Defense Council was replaced by the War Industries Board. On behalf of the W.I.B., George Long assumed direction of a Douglas-fir emergency committee in the Pacific Northwest. Application of price controls to lumber greatly enhanced the impact of government on the industry. The army signal corps, given responsibility for procurement of spruce, established an office in Portland to expedite delivery of airplane lumber. Finally, and most importantly, Colonel Brice P. Disque, a professional officer of uncommon energy and eccentric temperament, was assigned the task of resolving the labor dispute that supposedly hampered production of spruce.

Arriving in Seattle in the second week of October, Disque met with Henry Suzzallo, Carleton Parker, officials of the American Federation of Labor, and leaders of the lumber industry. Determined to achieve the maximum production of spruce and completely flexible in the selection of the means to achieve this end, he was impressed by the argument that the way to end the strike and destroy the Wobblies was to reform the conditions that produced discontent. Disque's mandate to restore order to the spruce program, moreover, brought the industry within reach of its goal. "He has full authority to go to any extreme that may be necessary to bring about its [production]," noted George Long, "even to the extent, if in his judgment it is essential, to put troops into logging camps and do the logging." Reporting to his superiors, Disque acknowledged the plight of lumbermen, endorsing the contention that high wages and the reforms of recent years meant that the strike was something more than a conventional labor dispute. "The activities of alien enemies and domestic agitators," he wrote, had resulted in "a very material decrease in production of lumber required by the Government in the conduct of the war."[27]

In the course of his October conferences, Disque worked out the details for two organizations that would dominate the industry for the remainder of the war: the Spruce Production Division and the Loyal Legion of Loggers and Lumbermen. The former, established within the signal corps, would assume control over all aspects of the spruce question, among other things assigning soldiers to relieve the labor shortage and overawe the Wobblies. The latter, an organization of employers and workers, was supposed to increase production through patriotic appeals and the reform of working conditions. "The idea," observed Edwin G. Ames, "is that the men are going into the woods, and are going to work and do their best." Following meetings in the nation's capital, Disque was ordered to take command of the Spruce Production Division, with headquarters in Portland, and to assume the leadership of the Loyal Legion.[28]

The news that troops would at last be made available was received by lumbermen as a great triumph, "completely overshadowing all of the recent developments in the lumber industry" according to the *West Coast Lumberman*. Should any of the soldiers "be approached by an I.W.W. or a pro-German and asked to quit work or slow up on their work," wrote Timothy Jerome of the Merrill & Ring Lumber Company after a meeting with Disque, "the fellow doing the talking will be immediately arrested by the officers in charge upon a charge of inciting mutiny and will be severely dealt with." Companies that had previously avoided war business now scrambled to secure government orders. "We must be in a position to call upon the authorities from time to time for such assistance in the way of protection as we may require," observed Ames, "and I am under the impression that this protection can be had if we go about it in the right way." Impressed with Disque as a man of action sympathetic to their position, lumbermen pressed upon him the unofficial status of wartime czar of their industry.[29]

Disque's organizations quickly demonstrated their value to lumbermen. The 4L—as the Loyal Legion was commonly designated—served as a rallying point for employers and workers opposed to the I.W.W. from the moment of its organization in late November 1917. With mill and camp employees required to join the 4L and known-Wobblies barred from membership, the Legion went a long way toward eradicating radical influence. Such patriotic devices as daily flag-raising ceremonies encouraged loyalty to the government—not to mention loyalty to employers—and only a small number of workers refused to join. Although a shortage of funds slowed establishment of locals, an estimated 75,000 laborers, employers, and Spruce Production Division soldiers belonged to the 4L by war's end. "The actual result of this organization," Disque claimed, "has been a new spirit in industry in the entire Pacific Northwest."[30]

Focusing his energy on establishment of the Legion and the assignment of soldiers, Disque realized that resolution of the eight-hour day controversy was required as a necessary concession to labor. The colonel supported a reduction of hours, but also recognized that practical considerations called for a cautious approach. Although his close advisers among the employers—George Long, J. J. Donovan, Alex Polson, and Mark Reed—were ready to go to eight hours, the hardline lumbermen still feared the implications of appearing to give in on the issue. Under pressure from the latter group, the Lumbermen's Protective Association resolved, in an emotional mid-December meeting, to continue the fight. "The 8 hr question," Disque concluded, "would be best left as it is for the time being." Nevertheless, few disagreed with the prevailing sentiment that the standard working day would soon be abandoned. "Our industry will sooner or later have to assume an eight hour attitude," wrote George Long, "and the only question that we have to confront is when."[31]

The major sticking point was that many operators distrusted Disque's reliability. Their suspicions aroused by the colonel's apparently close relations with Parker and Suzzallo, conservative employers feared that he would continue what William C. Butler scorned as the government's "feather duster" labor policies. Disque's well-publicized intent to follow any course that would stimulate production, they worried, could lead him in the direction of an accommodation with labor, a course directly contrary to the interest of management. As if to give behind-the-scenes confirmation to these concerns, Disque began several weeks of confidential discussions with Wobbly leaders at the end of December 1917, hoping to "start a conciliatory movement in the I.W.W." leading to compromise on the issue of hours and even participation in the Loyal Legion.[32]

In mid-January, an excited Disque informed J. J. Donovan, one of his closest friends among the lumbermen, of "some very interesting developments" on the labor front resulting from the negotiations. The colonel explained that "by somewhat changing our attitude towards them," the Wobblies could be transformed into good citizens and workers. Disque concluded with the confident expectation that "all employers will be glad to show them every consideration shown any ordinary employee." Horrified by this news, Donovan responded that the Wobblies "are no good and are out for the complete overthrow of our present social system and government and at the present time look to the Bolsheviki of Russia as their model." Any compromise with them, the Bellingham mill owner continued in a tone suggesting that lumbermen would thereby lose their enthusiasm for defense production, would result in total chaos and serious damage to the war effort. Recognizing that he was dependent on employers for the success of his spruce program, Disque broke off the talks with the I.W.W.[33] Henceforth, the colonel's conduct reflected his ultimate reliance on the owners of the camps and mills.

Once Disque had learned his proper place in the scheme of things, the debate over hours moved to a rapid conclusion. During the early part of February 1918, operators on Grays Harbor and southern Puget Sound and the Bloedel-Donovan Lumber Mills at Bellingham adopted the eight-hour day. Meeting in mid-month, the Lumbermen's Protective Association concluded that the time for an industry-wide decision was at hand. "The consensus of opinion," reported Edwin G. Ames, "was that we have made a long fight and are probably up against the eight hour day." Although many lumbermen grumbled that there was little sense or patriotism in a reduction of hours in the midst of war, they agreed to publicly defer to Disque. To encourage acceptance, the colonel announced that firms going to eight hours would be given preference in the assignment of Spruce Production Division soldiers and apparently hinted that war requirements might necessitate by summer a return to ten hours or even to the nineteenth-century twelve-hour standard.[34]

Meeting in Portland at the end of February, mill men and loggers acceded to Disque's request that they adopt the eight-hour day without reduction in daily wages and that they pay time-and-a-half for overtime. The industry had decided, noted George Long, that it would be best for the decision "apparently [to] come as a suggestion from the Federal Government through the activities of Col. Disque." The prestige of the colonel—"who we all believe," observed Long, "can do more to handle labor than any one else out here"—would be enhanced, and lumbermen would appear to be acting out of a spirit of patriotic sacrifice rather than self-interest. Employers privately complained that reduced hours would increase costs and damage their competitive position, and some even insisted that the eight-hour proclamation was illegal. There was a general feeling, though, that tension would be reduced and efficiency increased, as well as recognition that high wartime prices would enable absorption of extra costs.[35] Besides, the grumbling masked the real achievement of the industry.

To a great many observers, including subsequent historians, the strike resulted in a triumph for labor. Through the instrumentalities of military power and persuasive personality, Colonel Disque, went the argument, had forced a reluctant industry to adopt a great reform. Even for hardline lumbermen, however, the eight-hour day had never been of crucial importance. Essentially, the industry concluded the same arrangement first proposed in the summer of 1917: the eight-hour day for a government crackdown on the Wobblies. Lumbermen traded something of modest significance for something of great consequence and emerged as the real victors in the strike. "Their [the lumbermen's] victory was complete," observed the *West Coast Lumberman* after manufacturers agreed to reduce hours. Colonel Disque himself, not an immodest man when it came to his own accomplishments, conceded as much in later years. "Looking back on that measure now," he wrote of the eight-hour day, "it does not seem a very great step."[36]

The Spruce Production Program

Northwest lumbermen enjoyed in the climactic months of the war what a trade journal aptly termed "Crimson Prosperity." Restrictions on private construction brought a sharp reduction in civilian demand for lumber, and most sawmills without government contracts were forced to close down by the fall of 1918. Firms providing lumber for the war effort, however, enjoyed the best business in a decade. Although controlled by the government, prices increased, more than compensating for the falloff in non-defense trade. Washington lumber exports, as one instance of the trend, declined from 41 million feet in the first quarter of 1916 to 29 million in the initial four months of 1918, but increased in value from $540,000 to $684,000. The industry became so dependent on war orders that many lumbermen feared the end of the war would bring on an economic collapse.[37]

Despite the hope that the eight-hour day would mollify discontent, the labor problem continued to be of great concern to employers. Welcome relief came in the form of Spruce Production Division soldiers. Although the dyspeptic William C. Butler complained that the soldiers sent to his logging camps "had Sherman's Bummers backed off the map," most operators praised the quality and enthusiasm of the troops. "It is not always the most efficient [labor]," noted T. D. Merrill, "but you know that it will stay with you." The soldiers provided security against radical labor organizers, relieved the worker shortage, and, by their presence, insured that defense orders would be promptly handled. "If we do not treat the Government right (which means, give their requirements first call)," Edwin G. Ames pointed out, "they will not give us any men, and then we will not be able to run."[38]

By the spring of 1918, the Wobblies were little more than disturbing irritants and the main threat to management came from a resurgent American Federation of Labor. In March, the shingleweavers and timberworkers united as the International Union of Timber Workers and launched an organization campaign. Through sympathetic Wilson Administration officials, the union sought to use the Loyal Legion to further its drive. "It looks to me," reported a worried Ames, "as though organized labor, working thru the authorities, have finally entered the wedge, and are continually tapping it, until finally it will be driven in out of sight, and the damage will have been done." Angered by "the selfishness and lack of patriotism of the union leaders," J. J. Donovan warned Colonel Disque that they were "determined to rule or ruin."[39]

Convinced that the A.F. of L. was "plainly taking advantage" of the Loyal Legion, Disque needed no encouragement to move against the Timber Workers. He considered the union's demand for wage increases to be "treasonable" and its criticism to be "actually doing us more harm than several German Divisions." To counter any takeover attempt, Disque reorganized the 4L in the summer of 1918 to increase the role of employers. Under the slogan "Fifty-Fifty," a central council was established with one worker and one management representative from each of eight districts in Washington and Oregon. Ordered at this time to become members of the 4L, Spruce Production Division soldiers diluted the strength of the Timber Workers.[40]

Outraged A.F. of L. leaders helped bring on a congressional investigation of Disque's activities, an inquiry that probed the basic integrity of the spruce production program. Displaying his best quality—energy—the colonel had erected a spruce sawmill at Vancouver on the Columbia River and had accelerated the opening of timber by granting cost-plus contracts for construction of logging railroads and production facilities. Prominent along the latter line was the letting of a contract in the spring of 1918 for the building of a mill at Port Angeles and a network of feeder lines to link up with the Milwaukee Railroad west of that community. On the surface, the results were impressive. Washington spruce production increased from 198 million feet in 1917 to 276 million in 1918, an alltime high.[41] Behind the figures, however, lurked mismanagement and apparent corruption.

Although Alex Polson was so carried away by the war that he fantasized a personal duel to the death with the kaiser, patriotic fervor did not prevent him from taking advantage of Disque's production mania. In early 1918, Polson secured a cost-plus contract for the extension of a railroad into spruce timber supposedly owned by his logging firm. The Polson land opened up at government expense, however, contained valuable fir timber rather than spruce. "Some flannel mouth investigator," feared Timothy Jerome, "might twist things around so it would look suspicious." The affair was well-known within the industry, but Disque declined to take action against Polson that would expose his own administrative failure to the public. The colonel failed to avoid embarrassment, though, as the congressional investigation, dragging on into 1919, aired serious allegations regarding his activities on the northern Olympic Peninsula. Pointing to the fact that John Ryan, head of the government's airplane program, was a director of the Milwaukee Railroad, a firm standing to benefit from its linkage with Disque's spruce logging network, critics suggested the existence of a plot to defraud the government. No proof was offered and no charges were instituted, but Disque emerged from the episode with a damaged reputation.[42]

A major reason for the bad light settling upon the spruce railroad was the sudden and unexpected end of the war in November 1918 before a single log could be hauled, creating the impression that the government's money had been wasted on a useless enterprise. "The war closed about six months before anyone imagined," wrote George Long, "and one of the immediate results of it is to put all kinds of business more or less in the air." Disque canceled contracts, planned the sale of surplus lumber and the dismantling of the Vancouver sawmill, and ordered completion of the spruce road's mainline in hope of selling it to private interests. The cancellation of contracts was hard news for a number of firms that had, under prodding from the colonel, opened their spruce holdings just as the warplane market collapsed. Patriotic lumbermen had to temper their complaints over this unjust development with satisfaction that they had made a significant contribution to the war effort.[43]

Referring to the Portland office building where the Spruce Production Division had its headquarters, the *West Coast Lumberman* called for the casting of a special medal for participants in the "Battle of the Yeon."[44] In recognition of his apparent accomplishments, Disque was promoted to general. The reality, though, was that Disque was not a homefront hero but a captive commander whose course had been directed by industry. Lumbermen manipulated him into destroying the Wobblies, holding out the eight-hour day as the reward. At best, the Loyal Legion and the Spruce Production Division only accelerated the reform of living and working conditions already undertaken by employers. Spruce production was increased, but this was more the result of natural response to war-generated demand than of Disque's inefficient if energetic efforts. Convinced that many lumbermen had misled and betrayed him, Disque for his part left the Northwest an embittered man, rejecting postwar offers of employment from his friends in the industry.

For those of sufficient awareness, the war taught some valuable lessons. When confronted by a major threat to their mutual interest, lumbermen could overcome their traditional divisions and take effective industry-wide action. Under the proper conditions, moreover, government could be the servant rather than the antagonist of business. As the nation began its perilous return to what Warren Harding would soon term "normalcy," however, lumbermen would set aside the lessons of war and revert to disunity and to scepticism about the value of government. Instead of pondering the reasons for their wartime success, they were overwhelmed by uncertainty over the future. During the period of reconstruction, they feared, civilian orders could not possibly make up for the loss of war business. They worried that the removal of troops from the camps and mills would open the way for a revival of the labor menace. "We don't know what is ahead of us," George Long summed up the prevailing mood, "for while the war is over, the chaotic condition of the world is just about as big a problem as the war."[45]

12

An Epoch of Unrest

It is an epoch of universal unrest and I suppose nothing else could be expected from the causes which led up to the world war and the universal upheaval that followed it. It will take some time to get back on solid ground.—George S. Long[1]

But while they walk the thorny way, They're oft times heard to sigh and say: 'Dear Saviour, come; oh, quickly come, And take thy mourning logger home.'—Alex Polson, "The Poor Loggers' Prayer"[2]

The world war came to an end with the signing of an armistice to take effect on the eleventh hour of the eleventh day of the eleventh month of the year 1918. Although Americans had celebrated four days earlier when premature reports of peace spread across the land, they roused themselves upon receipt of the official news for a repeat performance. Throughout western Washington, mill whistles blew, church bells rang, and workers, soldiers, and sailors paraded through the streets. Outside the Puget Mill Company's Seattle office, men dressed in pirate costume danced around a cabbage-covered effigy of the kaiser. Inside, Edwin G. Ames sat at his desk pondering the immediate future. "Right now," he observed, "is the time for men of brains . . . to use them and keep cool." Intelligence and patience would in fact be needed by lumbermen in the first years of peace, as they confronted matters left over from the war or set aside for the duration of the conflict. "'War is hell,'" George Long wrote of the reconstruction period in 1921, "and we were all in the swirl of the cess-pool not only during the fighting epoch, but in the two years following."[3]

Labor in Peacetime

Everyone in the lumber industry expected a prolonged period of depression following the armistice. Economic uncertainty gripped the country, the wartime bureaucracy was rapidly dismantled, and the Wilson Administration, focusing its attention on the making of peace, devoted scant attention to the problems of reconversion. The increased wages and production costs of wartime prevented a reduction of prices to stimulate civilian demand. Trade journals urged a lengthy curtailment of production while the nation adjusted to peace. Other problems, however, meant that lumbermen could not afford the luxury of brooding over their plight. "These are chaotic days," observed George Long, "and what was a live issue yesterday is frequently supplanted by another one tomorrow."[4]

Foremost among those live issues was the future of the Loyal Legion of Loggers and Lumbermen. With the recall at war's end of Spruce Production Division soldiers, employers concluded that only continuation of the 4L would prevent a revival of the union movement in the camps and mills. "On a foundation such as the Loyal Legion," argued the *Timberman,* "there can be reared an organization which, if properly guided, can be made beneficial to the entire industry." Meeting in Portland in January 1919, management and labor formally organized the Legion as a peacetime operation.[5] Wage reductions combined with the hysteria of the Red Scare to put the 4L to an immediate test.

Hard-pressed by post-war recession, workers mounted strikes across the nation. In Seattle, 25,000 shipyard employees walked off the job at the end of January 1919, demanding wage increases. Noting that the prevailing rate of pay in the yards was 86 1/2¢ per hour for skilled labor, Edwin G. Ames was flabbergasted by the effrontery: "Only think of it—men earning such wages as that striking." The conviction that the unions were out of control was confirmed when the Seattle central labor council proclaimed a general strike in support of the shipyard workers. Beginning on February 6 and lasting for four days, the general strike terrified conservatives, Ames, for one, insisting that those leading the protest "believe in the Soviets ruling the State." Lumbermen joined with other business leaders to urge Mayor Ole Hanson to stand firm and station national guard soldiers in the street to intimidate workers, despite the lack of violence and the continuation of vital public services under the direction of a workers committee.[6]

Although rejoicing when the strike was called off, businessmen could not down the fear that the Northwest was on the verge of revolution. "This general strike," wrote Ames, "was organized more along the lines of a revolutionary activity than to do anything for labor." Seattle labor leaders "really believed that they were going to get control of this City, and probably the State, and that eventually the same would spread all over the United States, and they would get control of the Government itself." Devoting much of his time to reading about the Bolshevik Revolution, Ames was convinced that the failure of the general strike had merely driven the anti-American conspiracy underground.[7] The episode suggested to lumbermen that their wartime victory over labor had been only temporary and that they could generate public support by equating defense of their traditional prerogatives with defense of Americanism.

In the aftermath of the general strike, employers faced a briefly resurgent labor movement in the woods and mills. "We find indications all over the camps of the presence of I.W.Ws.," Ames reported in mid-February 1919, "and their literature is posted up here and there." Dispatching walking delegates from their Tacoma headquarters and devising elaborate security precautions in an unsuccessful effort to prevent infiltration by detectives, the Wobblies closed a number of Puget Sound logging camps in the summer of 1919. Irate operators demanded federal and state protection and the institution of such devices as identity cards for workers.[8] Crippled by the war, however, the I.W.W. was unable to do more than disturb employers with nightmare fantasies of revolution.

The renewed threat of the American Federation of Labor to organize the industry also aroused concern. In July 1919, timberworkers forced closure of Bellingham and Grays Harbor mills. By September, though, the strike had collapsed and the timberworkers began a period of rapid decline that would culminate in their disappearance in 1923. However shortlived, the revival of labor increased the interest of lumbermen in shoring up the Loyal Legion. Employers provided funding, distributed literature, promised to abide by the 4L wage scale and the eight-hour day, and announced that members would receive preferential treatment in the event of layoffs. Providing evidence for claims that unions were unnecessary, the postwar Legion was built on a foundation of self-interest.[9]

Despite the 4L, lumbermen lived in constant apprehension over the activities of labor. The Centralia Massacre of November 1919, for instance, occurred in the midst of an anti-I.W.W. campaign mounted by local mill owners. An armistice day gun battle between American Legionnaires and Wobblies in Centralia ended with four of the former dead. That night, vigilantes invaded the town jail and lynched Wobbly leader Wesley Everest. This violent action and the subsequent convictions of seven I.W.W. members for murder were hailed by employers as indicating that the "good citizens of the state" had at last awakened to the danger. "If we are to establish law and order and make life and property worth living for," wrote Alex Polson, "we have got to about face and get busy."[10]

The events of 1919 demonstrated that camp and mill owners were capable of both great obtuseness and unexpected sophistication. In their contention that they were beleaguered patriots and that organized labor was un-American, lumbermen were guilty of a massive self-deception. The sponsorship of the 4L, on the other hand, revealed an intelligent appreciation that it was better to give a little in order to avoid the risk of losing all. Part hysteria and part calculation, the approach to labor was greatly influenced by the economic health of the industry. The collapse of demand in the immediate post-armistice period and the swelling of the labor force through the return of soldiers and laid-off shipyard workers strengthened the hand of management. A return of prosperous conditions, in contrast, could well eliminate the labor surplus, destroy the leverage of employers and bring a renewal of the union threat.

Postwar Readjustment

A sudden boom in the demand for lumber in the spring of 1919 provided a dramatic and unanticipated manifestation of the return to peacetime. Civilian consumption had "piled up behind war-time conditions like water behind a dam," the *Timberman* pointed out. "And like water it rushed out in a wonderful stream when these obstacles were removed." The rush produced startling production and price levels. Washington lumber output set records in 1919 and again in 1920, reaching 5.5 billion feet in the latter year. The average price of Douglas-fir increased to $24.89 per thousand feet in 1919 and to $34.94 the following year. The national lumber price index, with the 1910–14 period equalling 100, stood at 323 by early 1920. "I think that we are running wild and running amuck," observed an amazed George Long.[11]

Significant alterations in the industry's traditional market patterns became evident during this wild period. Careful analysis of prices and demand dictated, among other things, final abandonment of San Francisco as western Washington's principal market. Leading the way in this development, in a fine irony, was the San Francisco–owned Puget Mill Company. The firm cut loose from Pope & Talbot's Bay Area yard and established a Seattle sales office to go after local and railroad trade. The company also abandoned, at the urging of Edwin G. Ames, its longtime position in the Hawaii market to such price-cutting competitors as Bloedel-Donovan. "We want a price and we do not want to sell lumber in any market that is not going to give us a fair price," summed up Ames.[12]

Although increased rail sales in the months after the armistice provided a partial replacement for traditional outlets, the industry's future lay in the development of new markets. The reopening of the Panama Canal to civilian traffic meant that Washington and Oregon lumbermen were at last able to take advantage of the path between the seas. Northwest lumber shipments to the Atlantic coast increased from 1.9 million feet in 1918 to 48 million feet in 1919 and continued to grow at a rapid rate. In the foreign trade, Japan offered a bright prospect, and shipments to the island nation became, for the first time, of major importance to manufacturers on Puget Sound and Grays Harbor.[13]

As far as the industry was concerned, the economic meddling of the Wilson Administration provided the only negative factor in the situation. A shortage of rolling stock, making it difficult for mills to fill rail orders, was blamed on the government's wartime takeover of the railroads. Charges by the Federal Trade Commission that the lumber prices of 1919 and early 1920 were the result of a trade association conspiracy caused additional aggravation. Reflecting their distaste for his policies, lumbermen developed an intense personal hatred for Woodrow Wilson. Alex Polson, for one, took crude pleasure in comparing the president's physical collapse while campaigning for the League of Nations in the fall of 1919 to the spoilage of salmon shipped over the government-run railroads. "It seems almost impossible to have anything go or come without being spoiled," Polson chortled, "even the President when he went out on a trip was spoiled before he got home."[14] While discomfited by Wilson's personality and actions, lumbermen agonized over the impact of another governmental intrusion: the federal income tax.

The Challenge of Taxation

Instituted through approval of the sixteenth amendment in 1913, the income tax had caused little concern prior to America's involvement in the world war. During the conflict, however, assessments on personal and corporate income mounted dramatically, to the predictable dismay of those in the higher brackets. Rates were so high, it was claimed, that initiative was stifled and businesses driven into bankruptcy. The Internal Revenue Service, moreover, represented in the opinion of unhappy businessmen a particularly odious example of confusing government regulations and arrogant government bureaucrats. The I.R.S., complained T. D. Merrill, was "a law unto itself and as unjust in its decisions as any autocrat of the old Russian regime not to mention the present unspeakable Lenin and Trotsky." In theory and practice, Edwin G. Ames summed up, the income tax was "arbitrary, iniquitous, unfair, unjust."[15] With the I.W.W., it amounted to one of the horrors of modern times.

Of particular concern to lumbermen was the method to be used for determining investment in timberland. This concern mounted when the Supreme Court ruled in 1918 that the book value of timber holdings as of March 1913 must be utilized for tax purposes. The industry thereafter had an enormous stake in the means chosen to ascertain 1913 values, for the higher the rate set for that year the lower would be subsequent earnings and taxes. "Whatever earning is made on timber over and above what it was worth on March 1st, 1913," explained George Long, "is income" and thus "subject to the income tax."[16]

In early 1919, the Internal Revenue Service established a special timber section to handle the job of setting values. Forester David T. Mason, recently serving in France as a major in the army's forestry regiment, was appointed to head this office. Because of the lack of data in government files, Mason turned to the industry for all possible information about private timber holdings. "No one might be expected to know as much of the industry," explained a colleague, "as those immediately and constantly in contact with its problems." Meeting with trade association representatives and with Weyerhaeuser attorney Stiles Burr, Mason drew up a detailed questionnaire to be submitted to timber owners. "Much to the surprise of the lumbermen," noted a participant in the negotiations, "Major Mason conceded every point brought up."[17]

Requiring complete disclosure of 1913 timber holdings and of subsequent dealings, the questionnaire generated an initially unfavorable response. The cooperative hand held out by Mason failed to down a conviction that the government had no right to such information. "Before they are through," observed Merrill & Ring's Timothy Jerome, "they will make everyone who has anything to do with the filling out of these Questionnaires a full fledged Bolshevist." Those adopting a calmer approach were staggered by the complexity of the forms. "I find to answer the questionnaire intelligently," George Long pointed out, "is going to take a colossal amount of work in the preparation of maps and schedules and information as to prices."[18]

To mollify discontent and explain the questionnaire, Mason toured the country in the fall of 1919 for meetings with lumbermen. Despite his inability to understand the technical discussion of accounting procedures, Edwin G. Ames was impressed with Mason and as a result of the conference held in Seattle was no longer anxious about the questionnaire. "I am sure that when you have met him and listened to his story and explanations," Ames informed William H. Talbot, "that you will say that if your questionnaire was in Major Mason's hands for his decision you would be satisfied that you were getting justice."[19] Convincing the lumbermen that at least one federal bureaucrat was their friend, Mason assured prompt submission of the necessary information.

Work on the actual determination of values got underway in the winter of 1919–20, eliciting a variety of responses from lumbermen eager to secure favorable results in their negotiations with the I.R.S. "We should spare no Expense," advised T. D. Merrill, "even to the extent of re-writing the books." Recognizing the value of personal relations, some firms hoped to secure the goodwill of W. T. Andrews, head of the timber section's Pacific Northwest division. "It would seem to me," suggested a Pope & Talbot representative, "that it would be a good idea to cultivate this expert and keep in close touch with him." Adopting this approach, the Merrill & Ring Lumber Company supplied Andrews with an automobile and arranged for him to attend a prestigious golf tournament in British Columbia.[20] A great deal of money was at stake and nothing could be left to chance.

Even before arrival of the questionnaires, George Long had held a meeting in his Tacoma office to establish a common front among Washington timber owners. Long's proposal, reported Edwin G. Ames, was "that it would be particularly advisable to work in concert if we could agree on common grounds." Timbermen realized that the values submitted by one owner would influence I.R.S. receptivity to the figures received from holders of adjacent timber. Of especial concern was recognition that the largely undeveloped Weyerhaeuser holdings would be valued at a lower rate than timber already opened to logging operations. To assure maximum cooperation, many firms engaged Weyerhaeuser's attorney to act as their representative in negotiations with tax officials.[21]

The result of countless visits to the nation's capital by lumbermen and tours of the Northwest by timber section officers—the entire process was completed in 1923—was that in virtually all cases the I.R.S. accepted the figures provided by timber owners. "We were all inclined to think that we got very reasonable treatment," wrote George Long after approval of Weyerhaeuser's values, "and our attorneys especially were very much elated." Contacts established during the valuation work also proved beneficial for the men of the timber section. David Mason resigned at the end of 1920 and became a forestry consultant in Portland. His successor, Carl Stevens, joined him in 1923 and the two went to work for the same firms they had dealt with while in the government. When Stevens pressed R. D. Merrill to retain Mason & Stevens, for example, the logger privately observed: "In view of his past services I thought it was up to us to engage him to do this work for us."[22]

Aside from reducing their potential tax burden, the valuation affair provided lumbermen with valuable statistical information and forced them to adopt modern accounting methods. In the opinion of most, though, the benefits were far outweighed by the continued imposition of excessive taxes. The industry had to confront the unwelcome fact, George Long pointed out, "that a great big chunk of money will have to be raised annually for a number of years ahead of us." The tax problem, contended Clark Ring, was "more serious than the world war."[23] Much of this concern over the allegedly stifling consequences of taxation stemmed from the sudden blow of a severe lumbering depression.

New Water Markets

Taking place in the spring of 1920, the collapse in demand was as unexpected as the boom of the post-armistice period, if a good deal more unwelcome. A response to the national economic slump brought on by reduced government spending, curtailed foreign trade and depressed farm prices, the lumber downturn was exacerbated by an increase in railroad freights. Washington production fell to the lowest point in six years in 1921 and prices dropped by half from the record levels of early 1920. The situation confronting the industry was readily summarized by the sales manager of the St. Paul & Tacoma Lumber Company as "very little buying and what little there was, going at ever decreasing prices."[24]

Sensitive since 1896 to the economic implications of presidential politics, the industry looked forward to Republican victory in 1920 as the best hope for a return of prosperity. "We cannot expect any immediate satisfactory conditions," argued Edwin G. Ames, "until our income tax law has been revised, our tariff laws revised, and our immigration laws revised." The landslide triumph of Warren G. Harding apparently assured rapid achievement of these goals. "I have seen more smiles on the faces of business men than I have seen for some time," Ames reported on the day after the election, "in fact, I myself really enjoyed my breakfast this morning." Timothy Jerome observed that the election of Harding would "create a feeling of confidence . . . which ought to be of some assistance in resuscitating business."[25] Looking forward to the departure of Wilson, lumbermen anticipated a sharp upswing in demand.

Final completion of the reconversion to a peacetime economy by mid-1921 released the full force of pent-up demand. "All at once, in every direction all over the United States," observed George Long, "people began to build homes, and there was an extraordinary and consistent demand for lumber, and under conditions where it was easy to advance prices from time to time." Issuance of building permits mounted on a monthly basis and sparked a new boom for the Northwest lumber industry. Beginning in 1922, Washington's sawmills set production records for five consecutive years, finally reaching 7.6 billion feet in 1926. Prices never approached the inflated levels of early 1920, but remained steady at double the rate of the prewar period.[26]

Burgeoning demand on the Atlantic coast caused the trickle of lumber through the Panama Canal to burst forth in a wooden stream. Housing construction and low ocean freights resulting from the postwar shipping glut enabled lumbermen to take full advantage of the canal. Pacific Northwest shipments to eastern ports increased by more than 400 percent between 1921 and 1923. "The New York market," observed Edwin G. Ames of the impact on the industry, "is certainly a godsend." This was especially so because of the refusal of the railroads to reduce their rates. Shipments through the canal more than compensated for slumping rail trade and allowed mill owners to engage in the rare pleasure of thumbing their noses at the imperious transcontinentals.[27]

Weyerhaeuser dominated the new trade through establishment of the most extensive marketing facilities. The firm's Baltimore terminal, the most advanced lumber yard in the world with wharves, sorting sheds, and railroad connections, was in operation by mid-1921. To enable efficient connections between its Everett shipping center and Baltimore, Weyerhaeuser acquired a pair of surplus steamships from the federal government in 1923. "It has become necessary," explained George Long, "for us to have a dependable movement of lumber at regular intervals . . . and we have not felt at all comfortable in facing the possibility of having to charter ships for every voyage that should be made." As trade increased, so did the company's eastern investment. Additional steamers were purchased in 1925, and work commenced on yards at Portsmouth, Rhode Island, and Newark, New Jersey.[28]

Mills shipping cargoes to the Atlantic coast had to devote considerable attention to sales policy. Eastern customers were unfamiliar with Pacific Northwest timber species and lumber grades. Declining per capita consumption, resulting from increased use of wood substitutes, caused additional difficulty. Applying to sales what a trade journal termed "more brain power," Northwest manufacturers entertained delegations of Atlantic coast retailers and undertook their first experiments with advertising. Encouraged by Secretary of Commerce Herbert Hoover, advances were made in the gathering of statistics and the adoption of standard grades.[29] Thus did the industry enter the era when salesmanship would become as important as production.

The flow of vessels carrying Washington lumber south to the Panama Canal barely exceeded the westbound traffic across the Pacific. Japanese demand smashed all previous records, Northwest shipments mounting to 591 million feet in 1922 and to 862 million in 1923. The latter figure almost matched the year's trade to the Atlantic coast. "Once a nation takes the important stand that the people of Japan are now taking," noted the *West Coast Lumberman* with a hint of condescension, "namely, providing themselves with real houses built of wood, it will be impossible to stop such a movement." Shipped as so-called "Jap Squares"—rough lumber destined for final processing in Japanese mills—most American lumber bound for the island empire was produced on Puget Sound and Grays Harbor. Washington lumbermen credited the economic revival of 1921 to trade with Japan.[30]

Conditions in the new markets caused a euphoric feeling among lumbermen that, as Edwin G. Ames put it, "the Northwest has come into its own." Even the normally cautious George Long rated 1922 as "rather a remarkable year" and 1923 as "one of the very best lumber years that the Pacific Coast has enjoyed." The source of this pleasure, the Atlantic coast and the Japanese trade, absorbed the industry's increased output, to the relief of manufacturers wary of overproduction. In the first half of 1923, as one instance of the importance of the new business, thirty-eight percent of Washington's domestic cargo shipments went to eastern ports and fifty-three percent of the state's foreign shipments went to Japan.[31] While filling their ledgers with the happy figures, lumbermen also had to confront in the union peril the negative companion of prosperity.

The Collapse of Labor

Employers responded to the collapse of the postwar market in early 1920 by slashing wages. Beginning in the fall, manufacturers secured a series of reductions in the Loyal Legion pay scale. By late 1921, the minimum daily wage paid by the Puget Mill Company, for instance, had fallen from $4.60 to $2.40. Having the 4L make the cuts made them seem to be a joint labor-management decision and tended to limit protest. Moreover, the closure of mills and camps and the corresponding rise in unemployment meant that workers were more concerned with retaining their jobs than with the rate of pay. "It is surprising," J. P. Weyerhaeuser pointed out, "to see the difference between operating a concern where there are less men than wanted, or where there are more men than are wanted."[32]

The Loyal Legion was initially welcomed by employers as an invaluable ally in the effort to both reduce the cost of production and prevent the revival of discontent. "They are afraid," Edwin G. Ames wrote of mill owners, "if we withdraw and do not stand by the 4L that with falling wages and no work the situation will play into the hands of the Timberworkers of the World and the I.W.W." The failure of the organization's president, Reed College professor Norman Coleman, to move rapidly enough on the wage-cutting front, however, soon led many operators to abandon this view. The 4L, Ames concluded in mid-1921, was unable to "properly function on a falling labor market." The disappearance of the A.F. of L. union and the moribund state of the I.W.W. added to the impression that the Legion had outlived its usefulness. A number of firms dropped out during 1921 and membership declined by half from the peacetime peak of 1919.[33]

That the reform impulse was on the wane was demonstrated by the Puget Mill Company's adandonment of both the 4L and the eight-hour day in April 1922. The decision to return to the ten-hour shift was justified by Ames as necessary to reduce the cost of production. Although he expected employees to welcome the opportunity to earn an extra two hours' wages, over a hundred workers quit at Gamble and

Ludlow and several months passed before full-scale production was possible at both locations. The firm also hoped that other mill owners would follow its example, but all Ames secured was private admiration for his "nerve." By the spring of 1923, after a year's experiment, Ames was forced to admit defeat and the need to "gracefully" revert to eight hours.[34]

When the industry revived, so to did the union threat and the problem of dealing with contentious workers. Eliminating the labor surplus, the boom of 1922 and 1923 presented the remnants of the Industrial Workers of the World with a final opportunity to frighten lumbermen. With rumors of sabotage spreading throughout the industry, the Wobblies called a walkout for May 1923. Ignoring wages and hours, the strike focused on a demand for the release of imprisoned radicals and amounted, reported Ames, to "more of a demonstration against our form of government than anything else." Operations at some logging camps were reduced, but generally the strike had a limited impact on production. The I.W.W., summed up the *West Coast Lumberman*, could generate little headway against the fact that the average Northwest worker enjoyed "the shortest working day, the best living conditions and the highest wages paid anywhere in the lumber industry."[35]

In the aftermath of the strike, employers congratulated themselves on the end of the I.W.W. as a force in the Northwest and on their complete and apparently final triumph over the labor movement. Deprived by this victory of its principal reason for existence, the Loyal Legion was allowed to lapse into a state of decrepitude. Dependent on the financial backing of its remaining employer supporters, the 4L moved close to the position of management after the selection of efficiency expert William C. Ruegnitz as president in 1926. Legion field organizers, who considered themselves to be "social doctors" engaged in "the evangelism of cooperation & brotherliness," were frustrated by the apathy of employers and by the indifference of workers.[36] One major antagonist, labor, had been eliminated, but another returned in full force in the years following the world war.

The Revival of Conservation

During the war years, interest in conservation faded in favor of all-out production of lumber. Nevertheless, the wartime experiment with economic controls and the belief of reformers that the conflict would redound to the benefit of their varied causes increased enthusiasm for the regulation of private enterprise. "A resource which promises to prove a decisive one in winning the war," asserted the journal of the Society of American Foresters in January 1918, "will never again be allowed to be destroyed or devastated."[37] Spurred by the experience of war and by renewed fears of a timber famine, conservationists again captured the attention of the public and made their cause a major national issue.

A committee of the Society of American Foresters chaired by Gifford Pinchot reported in late 1919 that only federal regulation of the lumber industry could prevent final destruction of the nation's timber supply. "The strong hand of the community and nothing else," argued Pinchot, "can prevent the devastation of the uncut lands which remain." Advocates of regulation contended that it would stabilize lumber production and prices and save manufacturers from the folly of their refusal to voluntarily adopt forestry. The industry, wrote a supporter of the Pinchot view, "is badly in need of such control for its own good." Lumbermen, though, regarded Pinchot as a dangerous radical who had abandoned reality in his ceaseless quest for personal publicity. "He has proven himself," observed George Long, "both impractical and not of a type that plays the game with anything like the fairness of the man who did the most to make Pinchot possible—Theo. Roosevelt."[38]

Regulation advocates, for their part, had developed an intense dislike for lumbermen. Pinchot recanted the "innocent belief we all had some twenty years ago, the belief that the lumbermen would gladly cooperate with us foresters whenever a good chance really offered." Now, he wrote, it was no longer possible to "trust the lumbermen, who have got us into this trouble, to pull us out." When Wilson Compton, secretary of the National Lumber Manufacturers Association, published "Fourteen Points" in defense of reliance on practical forestry, an anonymous author responded with a denunciation of Compton's "pseudo-economic sophistries." Defenders of the industry, argued the *Journal of Forestry,* exhibited "sublime confidence in the credulity of the people of the United States."[39]

In its defense, the Northwest industry received valuable support from William B. Greeley, who became chief of the Forest Service in early 1920. Finding the prospect of government regulation distasteful, Greeley preferred to rely on the goodwill of businessmen and on the stimulus of financial assistance. Forestry, he believed, could be advanced only "as the result of business foresight and initiative." Upon assuming office, Greeley released the Capper Report—prepared in response to a senate resolution introduced by Arthur Capper of Kansas—proposing increased federal support for state forestry programs. The report and Greeley's views corresponded with the position of lumbermen. The new chief forester, wrote George Long, "really understands some of our problems about as well as we do."[40]

Essentially, the debate resolved itself into an argument over the merits of federal as opposed to state regulation. To supporters of the former course, lumbermen would always dominate state governments and be able to weaken regulatory legislation. Thus, as Gifford Pinchot wrote, "the fact of the matter is that the fighting lines are being drawn between National control and no control." This view lay behind the introduction of legislation by Arthur Capper in May 1920 providing for federal regulation of private cutting practices. The Kansas senator, noted Long, was "very much under the influence of Mr. Pinchot in some of his rather ultra and extreme views pertaining to the handling of forestery matters." The prospect offered by the bill was a fearful one for believers in the sanctity of private property. If passed, cried Will Talbot, the Capper Bill "would simply mean confiscation."[41]

Orchestrated by George Long on behalf of the National Lumber Manufacturers Association, the industry's response centered on the drawing of legislation to counter the Capper Bill. Devised by Greeley and introduced by Representative Bertrand Snell of New York, the industry program basically involved increased federal cooperation with state-controlled forestry programs. As such, it was for lumbermen a satisfactory alternative to federal regulation. "The Snell Bill is considered about the best situation we can be in under existing conditions," Edwin G. Ames warned, "as it is recognized we are going to have control, and if we have regulation it must be State regulation."[42] Like the Pinchot group, timber owners recognized that state control was unlikely to be harmful to their interests.

Ultimately, both the Capper and the Snell bills failed of passage. Industry and Forest Service joined to kill the former and a strange alliance of Pinchot supporters and reactionary timber owners ended the chances of the latter. During 1923, a special senate committee chaired by Charles McNary of Oregon held a series of hearings around the country and worked on a compromise forestry program. The recommendations derived from this process were embodied in the Clarke-McNary Act of 1924. Among its key provisions, the legislation provided federal matching funds for state forest fire programs and authorized an investigation of the impact of property taxes on reforestation. Focusing on the twin scourges of flame and taxation, the act was a great victory for supporters of practical forestry. "The foundation," enthused a lumber trade journal, "has been laid for the formulation of a national forest policy."[43]

Adding to the sense of triumph was the accelerating movement of trained foresters into private employment, a movement that divided the profession and ended the era when it stood at the leading edge of the conservation movement. At the beginning of the 1920s, the privately employed forester had been compared by disdainful colleagues to the "lawyer retained by a corporation to help them evade the law." By 1930, however, those working in industry had become so numerous as to earn respectable status within their profession. Foresters moving into the private sector were generally conservative in their points-of-view and had never shared the progressive faith in the social benefits of conservation. Leaving the Forest Service, they commenced a process in which the agency would serve as a training ground for industry.[44]

The most dramatic manifestation of this development was the decision of William Greeley himself to become secretary of the West Coast Lumbermen's Association in April 1928. Although Gifford Pinchot rejoiced over the departure of Greeley's "malign influence" from the Forest Service, the appointment was a major public relations coup for the industry. At a time when most Americans still equated conservation with the Forest Service, the presence of a former chief forester at the head of a major trade association suggested that lumbermen were genuinely interested in forestry. In Greeley, George Long pointed out, "we have brought the right man into the lumber industry and one whose voice and recommendations will carry great weight, not only to the lumbermen but to all open minded people in the United States who are interested in forest problems."[45]

At the state level, Washington lumbermen found the public unsympathetic toward tax revision, for years the mainstay of the industry's forestry program. Thrown back on their own resources, several firms in the state commenced studies of their holdings to determine how best to implement reforestation. Great attention focused on the new concept of sustained yield. Promoted in tireless fashion by David Mason, the idea was to limit each year's logging to the amount of timber replaced by natural regeneration, a restriction that would allow perpetual operation of the industry. Sustained yield, Mason argued, "will enable the operator to get the best returns in the long run out of his timber." Mason's approach, in theory, complemented the desire of investors to maximize earnings. Most lumbermen, though, believed that it would be impossible to restrict annual output until relief was provided on the tax front.[46]

Protection of the nation's forests became something of a public obsession in the years following the world war. "If words would make trees grow," noted the *Journal of Forestry,* "the United States would be the most thickly wooded country in the world." Northwest timber owners argued that forestry had made greater strides in their region than anywhere else in the nation and that such figures as George Long and William Greeley deserved ranking among America's leading conservationists.[47] Although the idea that forestry, if it was to be applied in the real world, must be made into a business proposition appeared to have triumphed, lumbermen continued to worry over the issue of regulation and over the plans of those who would impose regulation.

Uninformed supporters of the Gifford Pinchot view, their gaze focused on the forest wasteland of the Great Lakes states, complained timbermen, refused to acknowledge the advances made in the Pacific Northwest. Rather than at an end, the struggle over regulation had entered a period of armed truce.

As the national economy entered the long and dramatic boom of the 1920s, Northwest lumbermen seemed to be in an enviable and impregnable position. The opening of the Atlantic coast market and greatly expanded shipments to Japan provided the basis for the greatest production in their industry's history. The threat of organized labor had been destroyed, the threat of the federal income tax had been reduced, and the threat of federal forest regulation had been sidetracked. Still, there were distressing signs to trouble perceptive lumbermen. The building of new sawmills and the expansion of old ones threatened disastrous overproduction in the event of a collapse in demand. The increasing value of stumpage destined mill owners and logging operators who had not purchased timber in the days of cheap raw material for eventual ruin. These and other problems would have to be met by skillful and informed management. Reminding its readers of the traditional jest that "the Lord took care of the lumbermen, even though they were not able to take care of themselves," the *West Coast Lumberman* pointed out that the Lord's help would no longer be sufficient.[48]

13

The Law of Supply and Demand

The law of supply and demand never sleeps. . . . The law of evolution governs industrial development just as it does plants and animals. Most industries must, from time to time, re-shape themselves to meet the changed conditions which control their success.—William B. Greeley[1]

The lumber manufacturer has been first and last a woodsman, an axman, a logger, a sawmill man, a producer of lumber, oftentimes a graduate lumber jack, but rarely an economist or business man in the modern connotations of those terms—rarely an advertiser or salesman.—National City Company[2]

Meeting in midsummer of 1924 by the rivers where the Cowlitz flows into the Columbia, the leading lumbermen of the Pacific Northwest witnessed the opening of the largest manufacturing facility in the region. Capable of producing over 300 million feet of lumber a year, the Long-Bell Lumber Company's plant astounded observers with its vast size and up-to-date machinery. "Everything is on such a mammoth scale," reported Edwin G. Ames, "that it knocks the ordinary lumberman completely off his feet." Adding to the astonishment, the mill was only the centerpiece of a huge project designed by Kansas City lumberman R. A. Long. Behind protective dikes, a model city destined to provide homes and employment for at least 20,000 persons was being erected according to what a trade journal termed "the best and latest ideals in civic development."[3] Built on faith in the unlimited expansion of America's economy, Longview reflected the speculative mania of the 1920s. The self-styled lumber capital of the nation, it represented a new source of capital for the Northwest. And in the ultimate disappointment of its owners, the venture dramatized the problems facing the region's lumbermen, for the Washington forest industry was in a fundamentally unhealthy state.

Money from the South

Despite the prosperous years of the early 1920s, change marked the course of lumbering enterprise in the Pacific Northwest. Sale of the Puget Mill Company to the Charles R. McCormick Lumber Company of San Francisco in 1925 attracted considerable attention within the industry. William H. Talbot and Edwin G. Ames were failing in health, the old mills at Gamble and Ludlow required modernization, and tired management was no longer up to dealing with the burdens of modern times. Lengthy negotiations resulted in sale of the production facilities and 92,000 acres of timber, the initial payment made in the form of McCormick stock. The new owner merged the property with its steamship fleet and its Oregon sawmill to form a vast lumbering concern.[4] With the disappearance of the Puget Mill Company, seven decades of Pope & Talbot involvement in the Northwest came to an apparent end.

The major change in the industry, though, came in the form of Southern investment. Although yellow pine mills in the South cut forty-two percent of the nation's lumber in 1920, regional production was falling because of a declining supply of timber. Just as Great Lakes manufacturers had looked to the Pacific coast at the turn of the century, so did Southern lumbermen in the decade following the war. "Apparently," noted William Greeley in early 1920, "a wholesale migration of lumber operators is taking place in the Southern States into the Northwest." Attracted by the extensive stands of timber in Oregon and Washington, Charles Keith and other important Southern manufacturers made large investments.[5] Oriented toward San Francisco and then toward St. Paul in earlier eras, the industry underwent the first development of a distinctive Dixie orientation.

North of the Columbia, the most dramatic manifestation of this changing front was the arrival of the Long-Bell Lumber Company. Founded in Kansas City in 1875, the firm operated 130 lumber yards and a dozen sawmills and was, according to Edwin G. Ames, "easily the third in importance and size of the lumber operations of the United States." With his Southern mills running out of timber, R. A. Long (not related to George Long), the company's founder and dominant stockholder, decided by the end of the war to transfer operations to the west. In early 1919, Long began negotiations with Weyerhaeuser for the purchase of timber along the Washington side of the Columbia. Only that firm, he reasoned, possessed the extensive holdings and the financial strength to extend liberal terms required by his undertaking. Ultimately, Long-Bell acquired over six billion feet from Weyerhaeuser.[6]

A longtime advocate of city planning and scientific farming, Long complemented erection of a huge manufacturing complex with the building of a model city on the Columbia. Hiring the best urban planners and buying up land claims, he had work on Longview underway before the end of 1922. Homes for workers, a business block, and a first-class hotel were built and vacant lots were landscaped. "They are certainly going at this thing in no half-way manner," wrote an amazed Ames. Sales of real estate and store space would presumably compensate for the cost of building the town. When lumber production commenced in mid-1924, over 8,000 persons lived in Longview and the company seemed to have achieved a splendid success.[7]

R. A. Long's city building experiment and the arrival of other Southern investors were taken by the *Timberman* as an indication that "this is the day of doing big things." Here was dramatic confirmation of the Pacific Northwest's national lumber leadership. Yet, perceptive observers found forbidding implications in the Southern influx. New investment, for one thing, increased stumpage values to the detriment of operators lacking timber holdings. The building of new sawmills, moreover, portended overproduction, price cutting, and demoralization in the event of declining markets. "It looks to me," observed William Greeley, "as though the lumbermen were setting the stage for a reproduction of their favorite little comedy, or tragedy, entitled 'Killing the Goose.'"[8] While a source of pride, the Southern interest in the region was also a source of apprehension.

Scrambling for Timber

Southern money set off a new wave of speculation in timber, with investors scrambling to conclude deals before values reached a prohibitive level. The scramble was especially intense in the western Olympic Peninsula, where distance from markets and rugged terrain had previously retarded development. Between 1921 and 1924, Bloedel-Donovan acquired 32,000 acres of timber south of Clallam Bay as a new source of supply for its Bellingham manufacturing center. In 1927, Merrill & Ring, Alex Polson, and William E. Boeing purchased a half-interest in the Irving-Hartley Logging Company, operating near Lake Crescent. Renamed the Crescent Logging Company, the firm took over the old Spruce Production Division railroad in 1929. Through these and other transactions, a few investors achieved practical control of the peninsula's resources. Their railroad networks controlled access to markets and meant that owners of isolated tracts, unable to exploit their own holdings, would have to sell out at low prices. "The Peninsula," Polson noted in 1928, "is as well tied up now, . . . as it is necessary that it should be."[9]

Threatened by this development, small mills and loggers counted on the availability of federal timber for maintenance of their operations. Sensitive to the political influence of large private holders and unwilling to contribute to potential overproduction, the Forest Service declined to make sales on the peninsula. This forced small operators back upon the Quinault Indian Reservation, a 190,000 acre wedge of heavily forested land extending westward from the Olympic foothills to the Pacific shore. Beginning in 1905, the government allotted eighty-acre tracts of reservation land to individual Indians under terms of the Dawes Act of 1887. These tracts, ostensibly meant to foster the development of agriculture among members of the Quinault and associated tribes, contained extensive and valuable stands of timber. Even before the allotment process began, George Emerson and other Grays Harbor investors ordered surveyors into the Quinault country to locate the best timber for eventual purchase from Indians.[10]

White plans for the Quinault were temporarily thwarted following an inspection of the reservation by Bureau of Indian Affairs forester J. P. Kinney in 1910. Kinney concluded that breaking up the Quinault into a myriad of eighty-acre segments would militate against the practice of forestry and the efficient sale of timber. The allotment system, William Greeley later recalled, prevented the "orderly utilization of Indian Reservation timberland as a whole in line with the most desirable economic policy." At Kinney's urging, allotment work was halted in 1914 and a detailed survey of reservation timber was commenced to provide data for management. Backed by white loggers, Indians who had yet to receive land sued the B.I.A. and forced resumption of the allotments. The decision opened the way to white exploitation and, in Kinney's view, created "a condition which made a conservative management of the forest practically impossible."[11]

Determined to circumvent the problems presented by the allotment system, the B.I.A. decided to sell timber in large units rather than on a tract-by-tract basis. Such an approach would result in higher sales prices and, by eliminating the need to deal with individual allottees, make for more efficient logging. "The Quinaielt . . . is very heavily timbered," reservation forester Henry Steer pointed out, "and presents . . . peculiar problems from the standpoint of a logging concern principally because of the great amount of railroad that must be built to advantageously log the timber." Despite opposition from Indians who wished to manage their own timber and from small loggers who lacked the capital for large purchases and railroad construction, the policy was put into effect in the early 1920s.[12]

Five large units containing a fourth of the reservation's timber were sold at auction between 1920 and 1923. Because of the heavy financial requirements, there was little competition for most of the units, resulting in sales at low prices. The Ozette Railway Company, a subsidiary of the Polson Logging Company, did have competitors for the Quinault Lake Unit, however, and had to bid $3 per thousand feet for hemlock and $5 per thousand for Douglas-fir, both record prices for Grays Harbor. "If my partners don't all fall dead when they hear what I did," quipped Alex Polson, "they will live a long, long time, and they won't need to have any examination for heart failure."[13] The Quinault sales demonstrated that large firms were willing to buy timber to maintain control over the resources of the peninsula and that small loggers and mills could not rely on the government to rescue them from their timberless plight.

Pulp and Paper

As timber increased in value, reduction of waste in logging and manufacturing became a vital concern. Traditionally, large amounts of timber had been left in the woods to rot after logging and sawmill burners worked overtime to dispose of refuse lumber. More wood was wasted per acre in western Washington, concluded a Forest Service study, than anywhere else in the nation. The need to make better use of timber resources was especially evident in the case of hemlock, an allegedly inferior species making up as much as a fourth of the holdings of such firms as the Simpson Logging Company. To provide a market for their hemlock, timber owners like Simpson and Polson acquired milling facilities in the early 1920s. Although production of hemlock lumber in Washington increased from 286 million feet in 1919 to over a billion feet in 1925, this was not sufficient to absorb the supply at a profitable rate.[14]

For serious thinkers, the correct solution to the problem of efficient timber utilization involved the development of new forest industries. As one such instance, the manufacture of plywood entered its first important phase in the years following the war. Centered in Tacoma and on Seattle's Elliott Bay, the industry benefitted from the introduction of improved glues and pressing machinery. Plywood, however, was not ideally suited for resolving the difficulties of timber owners. Consumption was limited until the 1930s, when plywood began to be used in home construction. Manufactured from small Douglas-fir logs, moreover, plywood offered no assistance to holders of hemlock.[15]

Happily, the rapid postwar expansion of the pulp and paper industry provided a market for hemlock. Manufactured by hand from rags, paper had for centuries been a luxury item. In the mid–nineteenth century, however, German researchers discovered a process for making paper from wood pulp. Because this paper deteriorated rapidly, further technical refinements were required. In the 1860s, an American scientist developed the sulphite process, in which wood chips were cooked in sulphur-based liquor. When bleached, the pulp was suitable for use in high quality paper. The sulphate process, invented in Germany in the following decade, did not preserve the wood fiber as well as the sulphite method. These processes, with new papermaking machinery, made it possible to produce great quantities of paper at low cost. Reflecting these developments, American paper output increased by 500 percent between 1879 and 1899.[16]

Abundant supplies of timber and of fresh water for use in chemical processes and waste disposal made western Washington an ideal site for pulp and paper mills. Nevertheless, mills at Camas on the Columbia River and at Everett were the only manufacturing facilities built in the state prior to the world war. Distance from markets and the financial requirements necessitated by expensive technology were responsible for this slow development. High costs meant that expansion of pulp and paper manufacturing lagged far behind lumbering and that the industry tended to be dominated by financiers.[17]

Several factors combined at the end of the war to produce a rapid expansion of the industry in western Washington. American consumption of paper was rapidly increasing and the Panama Canal made it possible to serve the Atlantic coast market. The huge quantities of relatively cheap hemlock on the Olympic Peninsula, moreover, attracted mill promoters to that rugged region. According to Olympic National Forest superintendent R. L. Fromme, the peninsula's western slope contained "the most extensive and densest stands of timber of pulp wood value in this country." Foresters believed that the peninsula could provide a perpetual source of supply for pulp mills and that development of the industry would enable efficient utilization of timber resources. Here, in the connection between hemlock and pulp and paper, was to be found validation of William Greeley's dictum that "the useless trees of one generation become useful material to the next."[18]

More than any other individuals, Edward M. Mills and Isadore Zellerbach were responsible for the opening of the pulp and paper era in the region. The former, a junior partner in a Chicago investment house, came to the Northwest during the war to supervise one of his employer's ventures, the financially troubled Elwha River power project near Port Angeles. The latter, head of a major San Francisco paper distributing firm, desired to secure a source of supply for his rapidly growing business. Zellerbach was therefore receptive to a proposal that he join with Mills in purchase of the Elwha power plant as the nucleus of a new manufacturing venture at Port Angeles. Beginning operation in 1919, the Washington Pulp and Paper Company was the first in a series of Mills-Zellerbach projects in the state.[19]

Because of the impact on hemlock, lumbermen became heavily involved in the building of pulp and paper mills. In 1927, Mark Reed of the Simpson Logging Company joined with Mills and the Zellerbach family to open the Rainier Pulp and Paper Company next to Simpson's hemlock sawmill at Shelton. The following year, the Grays Harbor Pulp and Paper Company began operation at Hoquiam, with Alex Polson joining Mills and the Zellerbachs as major stockholders. Operating out of the old Spruce Production Division mill and bringing the Mills-Zellerbach group together with R. D. Merrill and other loggers, the Olympic Forest Products Company began producing pulp and paper at Port Angeles in 1930. For timber and sawmill owners, the benefits were the same in all cases. "The beauty part of it is," wrote Reed of the Simpson-Rainier connection, "that we are utilizing every inch of sound wood that comes into the [saw]mill and we are taking at least 20% more material off the ground than we were before we had the pulp mill."[20] Earnings from stock in the pulp companies were an added bonus for those primarily interested in efficient utilization of timber.

The founding of these and other mills, observed a trade journal, meant that the Pacific Northwest was "now in the pulp age." Between 1923 and 1928, the number of pulp and paper plants in Washington alone increased from seven to twenty-one. In the latter year, the state produced sixty-eight percent of the sulphite pulp and eighty-nine percent of the sulphate pulp on the Pacific coast. The merger of the Zellerbach and Crown Willamette companies in 1928, a joining of the dominant concerns on the coast and a transaction so complex that it was not finalized until 1937, symbolized the new era. Because of the impact on timber, a significant advance had been made in conservation of natural resources, although pollution of water with chemical waste soon dampened enthusiasm on that score.[21] All doubt about the future stability of the lumber industry and of communities dependent on timber was apparently resolved.

Problems of Overproduction

Although new capital and the expansion of pulp and paper manufacturing seemed to guarantee a solid future, an unexpected market slump brought demoralization to the camps and mills in early 1924. Washington production continued at a high level, exceeding seven billion feet each year through 1929, but prices suffered a sharp decline. The average price of Douglas-fir, for instance, fell from $27.26 per thousand feet in 1923 to $19.39 in 1928. "Everything has been going down, down, down," lamented Edwin G. Ames at the end of 1924. Nationally and regionally, lumbering became in the midst of the booming 1920s a depressed sector of the American economy. Admitting that he had "not so much faith in the future . . . as I would like to have," George Long compared the industry to "a large army in disorderly retreat."[22]

Firms that had made large investments during the prosperous years of 1922 and 1923 were especially hard-hit by the slump. Expending over $40 million on its Columbia River project, Long-Bell had engaged in such heavy borrowing that George Long made discrete inquiries about the company's financial standing. The completion of Longview, unfortunately, coincided with the beginning of the industry's depression and both lumber prices and real estate sales failed to meet expectations. In early 1926, Edwin G. Ames informed Will Talbot on a "strictly confidential" basis that bankers declined to make further loans to Long-Bell and that the firm had been forced to return some of its timber to Weyerhaeuser. R. A. Long and his associates had to spend much of their time fending off angry investors and searching for solutions to their speculation-induced dilemma.[23]

The plight of the Charles R. McCormick Lumber Company also attracted considerable attention. During 1926, McCormick spent large sums rebuilding the Port Gamble sawmill, extending logging operations and integrating manufacturing facilities with its shipping interests. "They are spending money like drunken sailors," observed Alex Polson. "Maybe they can make it out of it, but like the Scotchman, I have my doots." By the end of the year, the company had been driven to the brink of bankruptcy. Flabbergasted by this development, Edwin G. Ames confessed that he had been unable to sleep after receiving the bad news: "I heard the clock chime every quarter of an hour all night long." Pope & Talbot had accepted McCormick stock in payment for its mills and timber and had to assume a leading management role in the McCormick Company in hope of salvaging its own financial position, destroying Will Talbot's plans for a problem-free retirement.[24]

Manufacturers were quick to detect reasons for the end of the boom. The national infatuation with the automobile, some believed, had led consumers to divert funds normally spent on purchase or improvement of homes. Industry leaders also pointed to Robert M. La Follette's independent campaign for the presidency in 1924 as a significant factor in the situation. What one lumberman termed "the La Follette menace" supposedly had caused national business uncertainty and a fall-off in demand. "The United States," complained the editor of the *West Coast Lumberman*, "has been burdened with loud-mouthed prophets, demanding the mantle of Moses to lead God's chosen people out of the wilderness into some dream-land of hokum and bunk."[25] The failure of prosperity to return with the electoral triumph of Calvin Coolidge, however, forced abandonment of this analysis in favor of rational explanations.

Lumbermen blamed their misfortune on a combination of factors. Large new firms like Long-Bell, for one thing, tended to cut prices in order to secure business. The shipment of unsold lumber cargoes—known as "transits"—to the Atlantic coast was a serious problem. Beginning in 1927, a sharp drop in the rate of new construction further demoralized the market. Problems also developed with respect to sales in Japan. Only "an earthquake that will move the Japanese current a thousand or more miles off the coast," Alex Polson had claimed, would prevent continued growth of that trade. There was no such disaster, but excessive shipments, rising competition from Siberia and, finally, imposition of increased Japanese tariff rates reversed the fortunes of exporting mills.[26] Reduced demand in vital markets was a major blow, exacerbating the problem inherent in the postwar expansion of the industry.

A chronic state of overproduction was central to all of the difficulties experienced by owners of logging camps and sawmills. Even in the most profitable years, William Greeley calculated, the industry of Washington and Oregon operated at thirty percent excess capacity. In part, this resulted from the building of new mills, either by investors from the South or by loggers like Simpson and Polson. Additionally, the booming demand of the early 1920s and the eight-hour day had caused older mills to add extra shifts, greatly increasing industry-wide production. "The loggers are putting in too many logs and the mills are sawing too much lumber," George Long summarized the problem, "and as usual, not until they bump into something that really hurts is there apt to be a very positive movement to correct the main evil which is the Pacific Coast production of rather more lumber than the market will absorb at a profitable price."[27] A hindrance when demand was good, overproduction threatened ruin in periods of declining markets.

The pressures leading to overproduction were nicely demonstrated by the experience of Weyerhaeuser. To generate cash flow for the payment of taxes, the firm greatly expanded its manufacturing capacity in the late 1920s. Six hundred acres of waterfront land were acquired at Longview in 1926, and by mid-1929 three sawmills were turning out Weyerhaeuser lumber on the Columbia River. Later in the same year, the company opened a large mill at Klamath Falls to cut its southern Oregon timber. At the same time, the White River Lumber Company operation at Enumclaw, east of Tacoma, was purchased, and negotiations commenced for a merger with three manufacturers on Willapa Bay. In 1929, the Weyerhaeuser plants at Everett, Snoqualmie Falls, and Longview produced 460 million feet of lumber, far and away the largest figure in the region. Somewhat embarrassed by his addition to industry capacity, George Long conceded that "it certainly comes at an unfortunate time for market conditions."[28] Individual firms, though, had to look out for their own interests, even when such behavior added to the general problems of the industry.

Shoring up the Market

Lumbermen turned to something old and something new in the search for solutions to their problems. As in the past, efforts to curtail production failed because of mistrust and conflicting interests within the industry. Reducing competition through a largescale merger had become a standard alternative to individual curtailment. In 1926, John W. Blodgett, Charles Keith, and C. D. Johnson, nationally prominent lumbermen who had invested heavily in Oregon, presented a plan for the merger of at least 40 Northwest firms, including the Long-Bell, St. Paul & Tacoma, and Charles R. McCormick operations. Eastern investors, acting through the National City Bank of New York, were involved to the extent of $180 million in financing the venture. The refusal of Weyerhaeuser to transfer a portion of its timber holdings as a longterm supply of raw material resulted in failure of the merger by mid-1928.[29]

Merger schemes had long been the companions of hard times, and curtailment had been practiced ever since the first collapse of the San Francisco market in the 1850s. Other remedies devised after 1924, however, were of more contemporary vintage. Speaking to the U.S. chamber of commerce in mid-1926, the St. Paul & Tacoma Lumber Company's Everett Griggs, a longtime leader of the trade association movement, called for the setting aside of the anti-trust laws and the authorization of industrial self-government. "The lumbermen can, among themselves," he argued, "correct the faults in their own business, if left to themselves with that determination." By eliminating the fear of prosecution, lumbermen could engage in open and effective control of production and prices. It would thus be possible, pointed out the *West Coast Lumberman,* to create "a logical orderly plan . . . whereby there will be no more lumber cut than can be readily disposed of."[30]

A variation of the Griggs suggestion developed from recollection of the supposed wartime accomplishments of Colonel Brice Disque. In 1929, J. D. Tennant, resident manager of Long-Bell's Pacific Northwest operations, proposed the appointment of William Greeley as "Director General" of the industry. Greeley, who had become secretary of the West Coast Lumbermen's Association in the previous year, would allocate production among the various firms. The idea had superficial appeal and the *West Coast Lumberman* had already advised that "the industry needs a Mussolini." While attractive in concept, however, there were practical obstacles to the creation of a forested police apparatus. "A Director General," George Long noted in deflating the Tennant plan, "could not be given authority, under our legal restrictions, to do anything other than recommend."[31] And recommendations, as decades of experience had affirmed, were unlikely to deter uncooperative manufacturers.

The idea of placing William Greeley in charge of a legion of production policemen was at least in keeping with the trend of events in the world. The growing attention paid to the marketing of lumber was

an additional instance of up-to-date action. By 1928, transit shipments
and declining construction had resulted in demoralization of the Atlantic coast market. "We hardly see any immediate prospect for improvement," reported Corydon Wagner, New York representative of
the St. Paul & Tacoma Lumber Company. In response, St. Paul &
Tacoma, Charles R. McCormick, the Charles Nelson mill at Port Angeles, and the Sudden & Christensen interests on Grays Harbor merged
their eastern sales outlets. Their Pacific-Atlantic Lumber Corporation—known in the industry as "Palco"—controlled a fourth of the
trade between the two coasts and was a relatively effective counter to
the increasingly chaotic state of the market.[32]

Establishment of joint agencies was only one example of the new
emphasis placed on sales. Weyerhaeuser took the lead in stressing the
importance of advertising and in catering to the needs of customers.
"We are confronted with the necessity of selling," George Long pointed
out in early 1926, "and while we may regret the large expense that
will have to be absorbed, yet we should not shrink from the expenditure." Developed during 1927, Weyerhaeuser's "4-Square" campaign represented "a bold challenge to time-honored lumber-selling
methods" in the opinion of the *Timberman*. The program involved the
sawing of high quality timber to exact customer specifications and the
wrapping of lumber in colorful packaging for attractive presentation
and protection during shipment. An advanced instance of the new attention devoted to sales, "4-Square" was an immediate success.[33]

George Long was also closely involved in national and regional trade
association efforts to stimulate demand for lumber and combat the
growing use of wood substitutes in construction. Here, he contended,
was one of those "questions that come up from time to time where,
if they were not handled by an industry, they are very apt to be neglected and forgotten, because no one individual has enough pep or
time or money to accomplish a great deal by himself." Working with
Wilson Compton of the National Lumber Manufacturers Association,
Long developed a trade extension campaign to be mounted by that
organization. Commenced in early 1927, the campaign involved the
expenditure of a million dollars a year on market and laboratory research and on advertising in national magazines. Because the N.L.M.A.
encompassed competing lumber regions within its membership, the
purpose was to arouse general interest in the use of forest products.[34]

To promote specific interest in Douglas-fir, Long led the way in
founding the West Coast Lumber Trade Extension Bureau. Assessing
members five cents per thousand feet of production, the bureau purchased advertising in the *Saturday Evening Post,* the *Literary Digest*
and what Long termed "a lot of other high-brow papers." Despite the
hopes of its sponsors, however, the bureau failed to generate enthu-

siasm among lumbermen. Many operators were hard-pressed by the market collapse and declined to join because of the necessary financial contributions. Disliking the entire concept of advertising, J. H. Bloedel refused to sign up and withdrew from the West Coast Lumbermen's Association for good measure.[35] Combined with hard times, the lukewarm response to market extension actually undermined the trade association movement in the Pacific Northwest.

Responding to the challenge, Long saw to the reorganization of the West Coast Lumbermen's Association in mid-1928. The new format incorporated the trade extension bureau and included departments for the gathering of statistics and the inspection of lumber cargoes. "There is no question at all," Long had earlier pointed out, "but what we are scattering our shots too much for effective bagging of the game on an economic basis." The hiring of William Greeley was meant to improve the management as well as the image of the association. Nevertheless, the same problems of financial hardship and disinterest that had hampered the market extension campaign limited the effectiveness of the W.C.L.A.[36]

Lumbermen also turned to the government for assistance. Federal officials, for example, were sympathetic to the plight of loggers who had acquired cutting rights on the Quinault Indian Reservation in the high-priced atmosphere of the early 1920s. Beginning in 1925, the Bureau of Indian Affairs authorized the reduction of contract prices to reflect market conditions. The "only unsatisfactory condition which will result," advised the reservation superintendent, ". . . is that revenue will not come in for the individual Indians as fast as they expected under contract provisions."[37] Apparently, the financial needs of the agency's Indian clients were of scant concern when compared to the difficulties white loggers had imposed on themselves through imprudent business decisions.

When Herbert Hoover, recently installed in the White House, called a special session of congress to revise the tariff in March 1929, many lumbermen saw another opportunity to secure aid from the government. Under the prevailing system of duties, lumber and shingles were on the free list and only a modest rate was assessed on logs. Benefitting from this situation, British Columbia operators had become strong competitors in the Atlantic coast market. With Mark Reed directing the strategy and George Long arranging the financing, Washington and Oregon lumbermen sought protection from this competition. Significant divisions within the industry, however, hampered the tariff effort. J. H. Bloedel and a number of other important Northwest lumbermen with investments in British Columbia timber resisted any restriction on the shipment of logs south of the line. Mills dependent on open market purchases of logs also opposed the imposition of high rates. The Smoot-

Hawley Tariff, finally passed in the spring of 1930, admitted logs and shingles free of duty and applied a tariff of only one dollar per thousand feet on lumber.[38]

Conflicting interests within the industry also marked the battle fought out over the future of the Olympic Peninsula in 1929. With the Polson, Bloedel-Donovan, Merrill & Ring, and Crescent logging railroads controlling access to private and public timber, small loggers and mill owners faced a bleak future. To save themselves, timberless Grays Harbor firms petitioned the Forest Service for an allocation of timber. At the same time, the Northern Pacific Railway announced plans to build north from Grays Harbor and open an estimated twenty-eight billion feet of timber to logging. Combined with the decision of the Bureau of Indian Affairs to offer the remaining Quinault timber for bid, these actions provoked an angry response from large operators. Competition would be restored to the peninsula and an increased supply of logs would exacerbate the industry's problems. "The tendency of government officials," complained Everett Griggs, "to throw on the market large tracts of timber at a time when the industry is suffering from overproduction is certainly a crime."[39]

Opponents of railroad building and timber sales succeeded in quashing the threat by the fall of 1929. Although submitting a bid on Quinault timber in order to protect his dominant position on Grays Harbor, Alex Polson pressed the B.I.A. and the Forest Service to withhold timber from sale. The Quinault sales were, in fact, withdrawn and the Forest Service announced that it would not sell Olympic timber until justified by market conditions. The government's policy, noted a forestry official, was "to avoid premature exploitation and the aggravation of overproduction." When the Northern Pacific abandoned its construction plans, the threat to the private logging railroads came to an end.[40] For the government, preserving the existence of timberless operators was secondary to the shoring up of the market standing of the industry.

The slumping demand for lumber combined with rapid reforestation in the yellow pine region to halt in mid-shift the Southern move to the Pacific coast. The optimism of the early decade was snuffed out and lumbermen tried in vain to find a solution to the problem of overproduction. Nevertheless, manufacturers were aware that past hard times had always given way to prosperity. By 1929, the feeling was widespread that the time-tested process was about to occur once again. "In view of the handicaps which the industry has successfully surmounted," J. H. Bloedel explained to a newspaper reporter in July 1929, "there is little need to fear that any circumstances can develop in the years to come to impede its progress."[41] The development of progress-impeding circumstances, however, would come not in a matter of years but in a matter of months.

14

Old Man Depression

It isn't necessary for me to enlarge on what Old Man Depression has done, not only to the greatest country in the world, but to all countries and to millions of people who were progressive, but not to blame. There are . . . few in the world who have escaped Old Man Depression's oppressive methods.—Alex Polson[1]

It seems to be the survival of the fittest today.—Everett Griggs[2]

Relaxing with his newspaper in a Klamath Falls hotel room in early August 1930, George Long suffered a fatal stroke. His health suffering from what he termed "an occasional touch of something akin to . . . ptomaine poison," Long had retired from day-to-day management of the Weyerhaeuser Timber Company in the previous year. His successor as general manager was F. Rodman Titcomb, son-in-law of J. P. Weyerhaeuser and a graduate of the "kindergarten" conducted by Long in the early 1920s for members of the family's younger generation. Although in a state of semi-retirement, Long had continued to take an active part in the firm's affairs and was, at the time of his death, in southern Oregon to inspect Weyerhaeuser mills and timber.[3] His position at Weyerhaeuser, his reputation for integrity, and his talent for industrial diplomacy made him the one figure able to exert substantial influence over the direction of Northwest lumbering. Long's death in the early stages of the crisis of the Great Depression was therefore a severe blow, as no other person was able to fill the role he had occupied for so many years.

Collapse on Wall Street

The stock market collapse of October 1929 commenced a period of national despair. Neither the efforts of the great Wall Street bankers to shore up the market nor the confident pronouncements of President Herbert Hoover could stem the rapid decline into the most severe depression in American history. Lumbermen and their spokesmen responded to the crash with public expressions of whistling-through-the-graveyard economics. The *American Lumberman*, for instance, dismissed the Wall Street panic as "certain performances that recently have taken place in a rather limited area of the country and that have been participated in by a relatively small part of the population of the United States." Anticipating the transfer of funds from speculation to investment in construction, industry observers expected a revival in the demand for lumber. When these predictions proved to be unfounded, the *West Coast Lumberman* refused to concede that the economic slump was anything more than a "period of adjustment."[4]

Privately, lumbermen were a good deal less than sanguine about their prospects. The industry, T. D. Merrill informed a friend in early December 1929, "has lapsed into a state of progressive atrophy which has been coming on like old age for a year or two and the break in the stock market has not helped matters." Pessimism became more pronounced when, contrary to public expectation, construction failed to revive in the first half of 1930. There was "not a dollar in sight in the lumber business anywhere," reported Alex Polson. "Everything seems to have gone to Halifax," observed Merrill & Ring's Timothy Jerome after attending a gloomy industry meeting in midsummer, "and what has not already gone seems to be on the way." Mill owners and logging operators were forced to admit that the economic situation involved more than a readjustment.[5]

These private quiverings of the stiff upper lip reflected the statistical performance of the industry. With the national economic collapse producing an equivalent slump in the demand for forest products, national per capita consumption of lumber dropped from 223 feet in 1929 to seventy-nine feet in 1932. Washington lumber production fell from 7.3 billion feet in 1929 to 2.2 billion feet in 1932 and the number of operating sawmills declined by more than half. The industry was reported to be running at thirty-five percent of capacity at the end of 1931 and at nineteen percent in the following spring. By the bottoming-out point of the Depression, lumber prices had fallen to their lowest levels since before the First World War. "In the lumber district liquidation is complete," summed up William C. Butler.[6]

Among individuals associated with the industry, the economic blow fell most heavily on workers. The national lumbering and logging employment index, with 1927 equalling 100, fell to thirty-seven by 1932. Washington logging payrolls dropped from $41 million in 1929 to $13 million in 1931, and the average daily wage paid in the camps declined from $5.80 to $3.11. Although employers were better able to withstand the collapse, they also felt the emotional and monetary strain. "I am not only a physical wreck," Timothy Jerome informed a friend in 1932, "but I am a mental wreck and because of business conditions I am also a financial wreck."[7] From bunkhouse to boardroom, there was no hiding from the debilitating impact of the Depression.

According to the state forester, Washington's forest-based economy was "afflicted with the proverbial 'Black Eye.'" This economic blemish reflected the sorry performance of many of the industry's leading firms. The Charles R. McCormick Lumber Company avoided bank-

ruptcy in 1930 only because Pope & Talbot agreed to forego payment
of principal and interest due from the 1925 sale. Long-Bell lost a reported $2.6 million in 1930 and had to deal with lawsuits by bond-
holders desiring to force it into receivership. In the case of the Polson
Logging Company, the Depression brought to the surface the results
of years of mismanagement. Alex Polson's nervous collapse in 1930
was followed by the revelation that he had borrowed large sums from
the company and from local banks to finance personal speculations.
"It is really pathetic," Timothy Jerome reported after an examination
of the books in mid-1931 led to the conclusion that "Merrill & Ring
have certainly suffered from the unbusinesslike ventures of the Polson
Logging Co." By the end of the year, Polson had been forced out of
management, "to guard against," noted Jerome, "any possibility of his
doing anything he should not do at a time when he is not himself
mentally."[8] In the years following the crash, many a reputation for
business sagacity was put to the test and found wanting.

Crisis in Lumber

Mill owners relied for survival on their own initiative, on aid from the
government, and on cooperation with their erstwhile competitors. The
most spectacular instance of self-help developed from an unlikely source:
the nearly bankrupt pulp operations of Edward M. Mills. During the
Depression, pulp and paper manufacturers were hard-pressed by slumping
demand, by increased capacity—typified by the building of a Weyerhaeuser mill at Longview in 1930—and by heavy imports resulting
from depreciated foreign currencies. "Grasping at straws" in the effort
to save his plants, as an associate later recalled, Mills formed an alliance with the rapidly expanding rayon industry. World production of
rayon tripled in the ten years following 1927 and the so-called "artificial silk" attained second rank to cotton among textiles. The fluc-
tuating supply and uncertain price of cotton, from which rayon was
manufactured in these early years, greatly retarded the new industry's
potential.[9]

The need of rayon manufacturers for a dependable source of raw
material complemented the desperate need of Mills for new market
outlets. In mid-1930, the Rainier mill at Shelton contracted with the
DuPont chemical company for joint development of wood pulp capable
of use in the production of rayon. Scientists turned out the first dis-
solving pulp early the following year and gave it the trade name of
Rayonier. By 1935, production at Shelton, Hoquiam, and Port Angeles
exceeded 100,000 tons and the three mills were responsible for a fifth
of the world's output.[10] Holding the dominant position in a new and

fast-growing industry, Mills had by a stroke of good fortune rescued his firms from Depression-induced oblivion.

Forced to rely on more traditional methods, most operators looked to the federal government for assistance. To reduce the supply of logs and lumber, President Hoover ordered a sharp restriction of Forest Service timber sales in mid-1931. That agency and the Bureau of Indian Affairs were also authorized to grant reductions in contract prices and extensions of the time required for removal of timber. Officials in the nation's capital also responded to demands for protection against competition from British Columbia. A lengthy lobbying effort by Oregon and Washington mills produced success in June 1932 when Hoover signed a measure increasing the duty on lumber to $3 per thousand feet. Thereafter, British Columbia exports to the United States were sharply curtailed, to the benefit of American manufacturers. These actions justified the *West Coast Lumberman*'s faith that the industry had "a sympathetic friend" in the White House.[11]

Other efforts to restrict competition were not as successful. During 1930 and 1931, American lumbermen, arguing that the use of convict labor allowed sales at artificially low prices, attempted to secure a ban on importation of lumber from the Soviet Union. Although Soviet sales in the United States were on the increase, this effort ended in failure. An indignant Everett Griggs joined with other industry leaders to protest this refusal to discriminate against communists who "do not pay their debts, [and] ignore every law of God and man." Mill owners also pressed for equalization legislation to compensate for the depreciated currencies of nations abandoning the gold standard. Depreciated currencies, it was argued, made for cheap imports and negated the protection extended to American industries by the tariff. Here too Congress failed to respond to the pleas for assistance.[12]

President Hoover's principal response to the lumber crisis was appointment, at the urging of the National Lumber Manufacturers Association, of the Timber Conservation Board in January 1931. Assigned the task of devising means for matching production to demand and for preservation of the forests, the board consisted of the secretaries of commerce, agriculture, and interior and of representatives of industry, farmers, and conservation groups. William Greeley, E. T. Allen of the Western Forestry & Conservation Association, David T. Mason, and other experts provided professional assistance and the real energy of the panel. "The plan is a sort of dual board," explained Allen, "a titular board containing men who can give wisdom . . . without much work, and an advisory committee of lesser lights who will carry the work." As suggested by the fact that the N.L.M.A. and other trade groups paid the entire cost of its activities, the board was heavily-oriented toward the views and problems of industry. "The immediate purpose [of the Timber Conservation Board]," noted the N.L.M.A.'s Wilson Compton, "is to make the forest industries prof-itable."[13]

The board's final report, submitted to Hoover in July 1932, came down on the side of practical and time-honored measures. Among the major recommendations, the report called for increased federal timber purchases and expenditures for fire prevention, as well as for modi-fications of property tax systems at the state level. On the manufac-turing front, the board encouraged mergers and supported the "control of production under competent Federal supervision" as the best means of restoring prosperity to the industry.[14] Although no action was taken in the closing months of the Hoover Administration, the proposals sug-gested that lumbermen were receptive to drastic changes in their stan-dard methods of operation, changes that would greatly reduce the free-dom of action of individual operators and greatly increase the ability of industry to regulate itself.

In their advocacy of the Timber Conservation Board proposals, lum-bermen looked to the future of industrial self-government. For the mo-ment, however, they continued to apply the methods of the past. In-dustry meetings in 1930 resulted in voluntary agreement to reduce total output through elimination of night shifts and weekend work. "The curtailment (with the powerful support of Mother Necessity) is the most drastic and general of any time in the last two years," William Greeley reported in midsummer, "and we have at least practically brought cur-rent production into balance with current orders." Nevertheless, the reduction failed to reduce inventories or to keep pace with sagging demand. Voluntarism also proved the undoing of Greeley's "firm price policy," announced in the late summer of 1930. Because efficient op-eration of its vast manufacturing complex required the constant move-ment of lumber, Weyerhaeuser declined to take the pledge to maintain price levels. Overlooking the inherent weaknesses of the voluntary ap-proach, many manufacturers blamed the failure of Greeley's initiative on Weyerhaeuser.[15]

The economic crisis undermined the spirit of cooperation and almost drove the West Coast Lumbermen's Association out of business in the spring of 1932. The organization's trustees voted to disband if mem-bership could not be increased, especially among the larger firms. "There are a good many, such as Mr. J. H. Bloedel," Everett Griggs had complained, "who do not in any way assist in associated activities and those of us who are attempting . . . to cooperate get pretty sick at times." A concerted membership drive, spurred by reduced dues and recognition that the collapse of the W.C.L.A. would be, as the *Tim-berman* put it, "nothing short of a calamity" resulted in the salvation of the association. New members included Bloedel-Donovan, the Charles

R. McCormick Lumber Company, and other prominent operators, sparking a modest revival of optimism for the future of the trade association movement.[16]

Market extension had become a recognized cooperative device during the 1920s. Under the pressure of the Depression, however, expenditures on advertising and research were sharply reduced, to the consternation of advocates of these modern tools. On the other hand, the crisis accelerated the trend toward joint sales agencies. Led by Mark Reed, the Puget Sound Associated Mills was founded in mid-1931 as Atlantic coast agent for Simpson and two dozen other mill operators. Earlier in the year, the Douglas Fir Exploitation & Export Company had been reorganized, bringing into the fold firms previously unaffiliated with the combine.[17] Mill owners were more responsive to those measures that promised to increase efficiency and to reduce their individual financial burden than they were to those that involved the expenditure of scarce funds.

Concern for efficiency and for control of production brought negotiations for a largescale merger close to success. Leading the way in the effort to consolidate sawmill properties was J. D. Tennant of the financially-troubled Long-Bell Lumber Company. "Their only hope of surviving," F. R. Titcomb observed of the Longview firm, "is working out some arrangement of this kind." In the spring of 1930, Tennant proposed the merger of mills with a total annual capacity of five billion feet of lumber. Prior to going into business, the new company would conclude a longterm agreement with Weyerhaeuser and other timber owners for a supply of logs. While agreeable to the scheme in theory, many operators doubted that it could be implemented in the practical world of everyday business life. Holders of stumpage opposed the terms of the draft logging contract and concern was expressed over financing and over the failure to include a sufficient number of mills.[18] Depression severity, though, meant that discussion of the project continued in spite of such reservations.

By the spring of 1931, the original Tennant plan had evolved into a larger and more ambitious consolidation. The general outline grew out of a conference held in February under the auspices of the Weyerhaeuser Timber Company. On the Olympic Peninsula, all of the major logging operations would be involved, including the Clemons Logging Company, Weyerhaeuser's Grays Harbor affiliate. Among the major mill properties to be combined were Bloedel-Donovan, St. Paul & Tacoma, and—at least as far as most industry leaders were concerned—the Weyerhaeuser plants at Everett. Serious obstacles remained, including the need to overcome traditional mistrust. Alex Polson, for one, declined to "fall for a combination without the entire holdings of the Weyerhaeusers in the state of Washington" and opposed association with the untrustworthy J. H. Bloedel. Interest and need, however, were sufficient for work on valuations to be commenced and for forester Carl Stevens to be retained as supervisor of the negotiations. If successful, the merger would result in domination of the state by two giant firms: Weyerhaeuser and the new Washington Timber Company.[19]

Weyerhaeuser's interest in mergers, quipped Mark Reed, resulted from "a state of affairs which they have not heretofore realized in their operations; that is, one of actually losing money." This unhappy development had in fact accelerated the company's desire to improve cost-cutting efficiency in its operations. Weyerhaeuser had finalized details for the merger of various properties on Willapa Bay at the end of 1930. The following spring, formation of Potlatch Forests through consolidation of the Idaho investments of syndicate members reflected acceptance of the merger concept. Although Weyerhaeuser disappointed advocates of the Washington scheme by refusing to contribute its Everett mills, the company encouraged the plan by making available sizable tracts of timber and by assigning Laird Bell, the son of company president F. S. Bell, to work with Stevens on the details.[20]

At the end of 1931, the younger Bell began a series of trips to the Pacific Northwest that continued throughout the subsequent year and on into the early months of 1933. Attention focused on the need to bring in recalcitrant operations like Polson, on the necessity of working out complementary mergers in Oregon and British Columbia, and on resolving the potential threat of antitrust prosecution. As the plan moved slowly ahead, it became evident that mergers involved reconciliation of so many individual interests and personalities that they were not a viable emergency response to depressed conditions. "It is, . . . very doubtful in my judgment," William Greeley had pointed out at the beginning of serious negotiations in 1930, "if such developments will come fast enough or go far enough to substantially remedy the problems of excess capacity and overproduction."[21]

Crisis in Labor

Curtailed production, whether planned or not, meant inevitable wage reductions and unemployment for the industry's workforce. In the lack of unemployment insurance and federal relief programs, workers faced disaster. By the end of 1931, the best estimates indicated that half of those normally employed in the mills and forests of Washington and Oregon were out of work. The situation was so bleak that holders of public and private timberland announced a freeze on the hiring of extra personnel to fight forest fires. "Hundreds and hundreds of men are in need of and are seeking employment," N. O. Nicholson, superintendent of the Quinault Indian Reservation, explained in mid-1932, "and would . . . start fires in a minute if they thought it would make employment available to them or others in fighting these forest fires."[22] Here was a direct Depression-related threat to both conservation and the value of timber investments.

Some operators opposed sharp wage reductions, the traditional response to hard times, as unfair to employees. "I sincerely trust and hope," wrote Alex Polson, "that we can and will keep going without depressing the man who works to that condition to which some of our competitors are doing." Polson's associate T. D. Merrill, though, expressed the prevailing sentiment. "We should make it a condition of our operating," he advised, ". . . that our employees must meet the situation in order for us to continue logging." Meeting the situation involved acceptance of an industry-wide ten percent reduction in the summer of 1930, the first in what observers anticipated would be a series of cuts.[23]

Sensitive to the possible repercussions of wage reductions, lumbermen decried any attempt to portray themselves as other than interested in the welfare of their employees. They insisted that workers were grateful for continued employment and not opposed to cuts in pay. "The rank and file of the employees understand the situation clearly," observed William C. Butler at the end of 1931, "and are ready and willing to accept wage reductions." Alex Polson reported that his men had "thanked me for our efforts to do something to keep them from starving." Employers also claimed that they were primarily interested in providing employment and in maintaining the stability of their local communities. At Shelton, for instance, Mark Reed allowed unemployed workers to shop on credit at the company store after being forced to close the Simpson sawmill in 1932.[24] Such efforts, however, could not prevent the rapid development of discontent among the region's working men and women.

Spreading layoffs and slumping wage rates encouraged the spirit of radical protest in the Pacific Northwest. Beginning in 1930, wildcat strikes closed mills and demonstrators vented their wrath at employers. "The whole audience appeared fired with a determination to sweep out present-day unemployment with a new deal," wrote Loyal Legion president William Ruegnitz after attending a communist rally in Portland. Pressed on the one side by market conditions and on the other by worker discontent, lumbermen worried that a conflict with labor would bring on a revolutionary situation. J. H. Bloedel feared, according to the report of a 4L official, "the possible effect of further wage reductions causing an increase in the Communist movement."[25] Whether to take action or to stand back and hope for the best was the great dilemma facing employers.

Rising labor unrest caused a revival of interest in the Loyal Legion. Although 4L membership slumped to an alltime low under the pressure of the Depression, lumbermen were alive to the tension-defusing potential of the organization. This awareness was encouraged by William Ruegnitz in an effort to rebuild the Legion. "The economic, political, social, and religious revolutions that prevail in the world today," he informed one employer, "point very clearly to the need for a closer co-operation of men and management." Ruegnitz suggested to another mill operator that the 4L "is a positive method of personnel administration which gets valuable results."[26] As always, the Legion faced the problem of its inherent contradiction: how to serve at the same time the interests of its employer sponsors and its laboring members.

For Ruegnitz, the key point was to demonstrate the validity of the 4L's cooperative approach by halting the downward wage spiral. "The history of labor depressions," he reminded a Weyerhaeuser executive in mid-1931, "shows that resistance, disturbances and riots follow when the pressure of reductions pinch men into action." Despite pressure from mill owners, the 4L board refused to reduce the minimum daily rate below $3 in the summer of 1931. "Without the 4L," boasted the Legion journal as the nation entered the third year of the Depression, "it is safe to say, there would at present be a lower scale and a more chaotic situation than there is." Ruegnitz and his subordinates believed that they deserved the credit for maintaining a semblance of labor peace.[27]

Ultimately, the 4L front crumbled under the weight of the Depression. The refusal of the Weyerhaeuser and Long-Bell mills at Longview to join the organization crippled the drive for new members and the final collapse of the wage line added new difficulties. Weyerhaeuser's F. R. Titcomb, claiming that "no one feels more concerned about low wages than do I," nevertheless insisted that the scale "be determined by the economics of the situation." The 4L minimum was reduced to $2.60 in August 1932 and at the end of the year was suspended altogether. By then, the imperatives of cost-cutting were paramount to sensitivity over the feelings of workers.[28] The 4L's failure to ameliorate worker grievances demonstrated that, in the midst of the Depression, voluntarism was as weak an instrument for the handling of labor problems as it was for the control of production.

Political Disaster

Like many Americans with unexpected time for contemplation, lumbermen devised remedies for the Depression. To Alex Polson, the old free-silver advocate of 1896, salvation lay in inflation through coinage of the white metal. "Civilization," he contended, "never advanced in a scarce money market." Most lumbermen, though, advocated more conventional approaches. They welcomed the loans made to railroads and financial institutions by the Reconstruction Finance Corporation as sure to result in increased demand for lumber. Considerable support was expressed for the guarantee of bank deposits and for the repeal of prohibition as morale-boosting devices. Above all, restoration of faith in the future was necessary for an end to the crisis. "Much of our trouble," reflected R. D. Merrill in the spring of 1932, "is due to lack of confidence in ourselves."[29]

Lumbermen differed from most of their fellow citizens when it came to assessing responsibility for the Depression. More than anything else, they believed, the reelection of Herbert Hoover in 1932 was essential to the recovery of confidence. "Hoover is the right man in the right place at this time," asserted Alex Polson, "and I think what he is doing is a wonderful thing, not only for the American people, but for the entire world." As far as mill owners and logging operators were concerned, the president was not to blame for the economic collapse. Rather, his policies had saved the nation from total disaster. A perusal of the history books, insisted Mark Reed, "does not record any administration during times of depression when more constructive leadership came from the administration in power than has originated during President Hoover's term."[30]

In their effort to secure a second term for Hoover, lumbermen were also driven by fear of the Democratic alternative, Franklin D. Roosevelt. Apprehension was enhanced by recognition that the Depression had provided the Democrats with a powerful and winning issue. "The hour has struck for these opportunists," wrote a worried William C. Butler in May 1932. Roosevelt's public support for the "stabilization of the forest products industry" failed to down suspicion of the Democratic standard-bearer. Lumbermen were especially concerned over the apparent behind-the-scenes role of longtime industry critic William Randolph Hearst in securing the nomination of the Roosevelt-Garner ticket. "It is Hearst or Hoover," R. D. Merrill informed a friend, continuing with a reference to the publisher's well-known companion: "how would you like Marion Davies as 'First Lady of the Land' of the unmarried set?"[31]

Industry leaders threw themselves into the campaign on behalf of Hoover. Mark Reed, benefitting from years of political experience, assumed direction of the president's reelection effort in Washington state. A modest late summer revival of the nation's economy offered hope of an upset, but unemployment was simply too large an obstacle to overcome. "It is only natural," observed a philosophical Merrill, "that the laboring man should be dissastified with present conditions and look about for some change which it would seem might benefit his position." This sentiment, lumbermen concluded, doomed Hoover to defeat and the nation to further economic troubles.[32]

The election-day tally confirmed the first of these expectations. Nationally, Roosevelt buried the discredited Hoover by a landslide margin. In Washington state, Democrats were victorious in all but one statewide contest, won control of the legislature, and swept the congressional races. "It is too bad the general public were not able to appreciate the effort which President Hoover made in their behalf," mourned Merrill of the repudiation of his political hero. Should the Democrats elected to congress elsewhere in the country prove to be of the same stripe as those chosen in Washington, Merrill favored "giving up in despair and moving over to British Columbia."[33] Here was another in the string of disasters to come crashing down since 1929: first the depression, then the rising militance of labor, and now the displacement of a safe-and-sane administration by one of uncertain loyalty to industry.

With the rest of the nation, Pacific Northwest lumbermen waited through the long and somber winter for the inauguration of Roosevelt in March 1933. During this period, the victorious Democrats lacked power and the Republicans held office without a mandate to govern, adding political demoralization to the depressed condition of the economy as cause for worry. "More than for any other reason," wrote R. D. Merrill, "business is now suffering on account of the uncertainty of what is going to happen in Washington." Within the industry, there was agreement, however repugnant, on one certainty. "Radical changes in the industry's methods," contended the *American Lumberman* in January 1933, "will have to be made if lumber is to recover its place in modern commerce."[34] United on this point, owners of sawmills and logging camps stood with faint hearts and damp palms on the brink of the New Deal.

15

On the Road to Recovery

It has been as though one were looking into absolute darkness. Now,
however, it is possible to see ahead. . . . We are still beset with troubles
and difficulties of various kinds, but I do believe sincerely that we are
now well along on the road to recovery.—J. H. Bloedel[1]

I do not agree with everything Roosevelt is doing and really don't know
just what his plans for the future are—in fact, I don't believe he has any
definite plans, but I have to give him credit for trying anyway. If he does
not make Socialists out of everybody, I will be happy.—Timothy Jerome[2]

A small group of lumbermen and government officials gathered about
the desk of the president of the United States on the nineteenth day of
August in the Depression year of 1933. With a flourish of his pen,
Franklin D. Roosevelt signed a document known, without hint of irony,
as a "code of fair competition" for the nation's forest industries. Providing, as the first article stated, for "an undertaking in industrial self-
government," the code set aside the antitrust laws and authorized the
control of production and prices to bring recovery to the sawmills and
logging camps. Additional clauses provided for the implementation of
conservation measures, for the right of labor to organize, and for the
setting of maximum hours of work and minimum rates of pay. Supervised by the National Recovery Administration, the code empow-
ered lumbermen to set their house in order by eliminating the chaos
and overproduction that had contributed to the economic collapse. The
industry, in the opinion of various trade and forestry publications, had
opened "a new frontier," started down "a new and untried path" and—
the favorite metaphor—set out on "the uncharted seas of the new deal."[3]

Renewed Optimism

At the beginning of 1932, lumbermen, while relieved that they had
survived the suffering of the past, joined together in apprehension over
the future. The year-end figures revealed that in 1932 Washington had
produced the smallest amount of lumber since 1904. The state's mills,
moreover, operated at less than a fifth of normal capacity in January
of the new year. Concern mounted in the final days of February and
the first days of March, as panic-stricken depositors forced the closure
of banks in many states and Roosevelt, in his initial dramatic act, declared a nation-wide Bank Holiday. For the moment, trade in lumber

came to a halt because of the inability to conclude financial transactions. Adding to the worry, the closing of the banks seemed to place in peril the social and economic fabric of the nation.[4]

During this time of distress, attention within the industry focused on changes in the management and policy of the Weyerhaeuser Timber Company. By the end of 1932, the firm's directors had become disenchanted with the work of F. R. Titcomb, feeling that he had failed to measure up to the standards set by George Long. Titcomb, complained company president F. S. Bell, had "lost initiative to a certain extent." In January 1933, this situation was resolved by the appointment of J. P. Weyerhaeuser, Jr. as executive vice-president. Known as "Phil" to avoid confusion with his father, the thirty-four-year-old Weyerhaeuser had for several years been a key figure in the syndicate's Idaho operations. He was well-respected within the industry, although the combination of youth and shyness often made his communication with older lumbermen difficult. And, to the consternation of operators used to the widespread vision of George Long, he viewed the company's interest from a narrow perspective.[5]

This go-it-alone attitude was reflected in the collapse of the merger negotiations. The company's representative, Laird Bell, continued in the early weeks of 1933 to work on the Oregon and British Columbia combinations believed necessary to complement the Washington merger. A sharp reduction in the price of logs by the Weyerhaeuser-operated Clemons Logging Company in mid-January, however, was taken by the industry as a signal that the firm was no longer interested in cooperation. Mark Reed, for one, assailed the "lack of a disposition to coordinate among those whom we should look up to for leadership." Bell tried for a time to salvage the project, but eventually had to admit defeat.[6] With Weyerhaeuser apparently losing interest, other members of the industry concluded that there was no point in continuing work on the scheme.

Concern over the merger's failure faded before an unexpected increase in the demand for lumber in the spring of 1933. "We are feeling a little burst of business," reported Phil Weyerhaeuser at the end of March, "and are trying to keep from being stampeded into added mill production." For the first six months of the year, as one instance of the upswing, Washington cargo shipments increased by fourteen percent over the corresponding period in 1932. Wary industry leaders feared that the boomlet was the work of buyers eager to stock up in anticipation of the higher prices likely to result from the Roosevelt recovery program. The passage to sustained prosperity, knowledgable lumbermen warned, would be neither rapid nor comfortable.[7]

A reviving lumber market, however uncertain its composition, was but one source of optimism in the first weeks of the New Deal. Timber owners anticipated that they would soon be able to sell their surplus land. In the Copeland Report—officially titled *A National Plan for American Forestry*—the Forest Service presented a blueprint for federal acquisition of timber and other conservation measures. Holders of timber would be able to reduce the cost of taxes and fire protection by disposing of land that was either logged-off or covered with noncommercial species. "The sooner the private timber owner can divest himself of the timber burden," observed the *Timberman,* "the better it is for the industry and the nation." Most industry leaders endorsed the report and looked forward to rapid implementation of its major recommendation.[8]

Other aspects of the early New Deal program also received the support of the industry. Lumbermen approved Roosevelt's initial effort to reduce government spending as the proper course for restoring business confidence. The administration's farm program was expected to result in new trade for sawmills. And the Civilian Conservation Corps meant increased attention to the campaign against fire. "Mr. Roosevelt is doing the very best he can," wrote an approving Timothy Jerome in mid-April, "and many of the things he has done and is trying to do should help us on our way back to better times." If anything, the New Deal was regarded as too restrained in the reach of its objectives. "Nothing has as yet been initiated," complained the *West Coast Lumberman,* "looking toward a solution of the forest industries' problems." The nation's lumbermen, continued the trade journal, were "clamoring for federal help."[9]

New Deal for Lumber

Passed by congress and signed by the president in mid-June, the National Industrial Recovery Act was the New Deal's principal response to the plight of industry. The legislation represented a compromise between the demand of labor for the setting of minimum wages and the demand of industry for suspension of the antitrust laws. To remain in effect for two years, the recovery act authorized the drawing up of industrial codes and the creation of a federal administrative apparatus, the National Recovery Administration. The labor provisions mandated inclusion in the codes of wage and hour standards and the right of workers to organize and bargain with management. Although businessmen were not overjoyed by these latter requirements, the way had been opened, as the *American Lumberman* put it, "for each industry to do its own house-cleaning, and thus avoid any oppressive Federal regulation."[10]

Industrial self-government, as the N.R.A. approach was known, had long been advocated by supporters of the trade association approach to efficient business management. Since the stock market crash, for instance, lumber industry figures had revived earlier plans for government-approved regulation of output and prices. In 1930, William Greeley of the West Coast Lumbermen's Association called for the establishment of a federal natural resources board. Under the panel's supervision, regional trade organizations would draw up plans for "a rational control of production by collective or cooperative action within the industry." A similar scheme had been proposed by Hoover's Timber Conservation Board.[11] Provided that the role of government was carefully delineated, there was widespread acceptance of the collective approach.

Reflecting this general support, the nation's lumbermen approved the first draft of a lumber code prior to the signing of the recovery act by Roosevelt. Meeting in Chicago in late May, the National Lumber Manufacturers Association reached agreement on the outline of a code. Mid-June conferences held in Tacoma, Portland, and Eugene by the West Coast Lumbermen's Association resulted in endorsement of the document by mill owners from the Douglas-fir regions of Washington and Oregon. These actions were taken, noted an amazed R. D. Merrill, even though "we have not yet seen the bill in its completed form." Anticipating federal approval of the code, a special W.C.L.A. committee set to work in the third week of June on the control of production.[12]

Work on the code necessitated further advances on the trade association front. Unwilling to entrust their interests to the mill-dominated lumbermen's association, Northwest loggers formed their own group to deal with the code. The Pacific Northwest Loggers Association, with Mark Reed as president, was established in late June. "Unless we govern ourselves," Reed informed the loggers, "we are going to be governed and, in all likelihood, in a way we do not like." An immediate problem faced by the new organization was the widespread demand for inclusion in the code of an embargo on the importation of logs from British Columbia, a demand vigorously opposed by American owners of Canadian timber. The dispute was temporarily resolved when negotiations with representatives of the B.C. industry resulted in imposition of an unofficial import quota to complement south-of-the-line efforts to control output.[13]

At the end of June, a delegation of Northwest lumbermen traveled to Chicago to take part in the final drafting process. Because of the complex nature of the American lumber industry, it was decided that the code would be administered through regional trade associations and that the wage and hour differentials between the Southern and Pacific coast branches of the industry would not be disturbed. Another problem was the desire of President Roosevelt that the code contain conservation provisions. "Undoubtedly," noted F. E. Weyerhaeuser, youngest son of Frederick Weyerhaeuser and now the leading member of his family, "there will be a tremendous drive made by our emotional forester friends who think there is rare opportunity to force upon the lumber industry all of their theories and imaginings." To please the president and to counter the revival of the regulatory impulse, lumbermen added a conservation clause to their draft. Eventually known as Article X, this clause promised increased effort in the area of practical forestry and extended theoretical support to the concept of sustained yield.[14]

Draft in hand, the Emergency National Committee appointed to act on behalf of the industry departed for the nation's capital on the seventh of July. With its members went the belief that speedy N.R.A. acceptance of the code was essential to recovery. "It is unfortunate that the code cannot go into effect at once," observed R. D. Merrill. "Many of the operators are now increasing their production and taking unfair advantage of the present situation." Running three shifts a day, Bloedel-Donovan was "again the greatest offender" according to Merrill. The West Coast Lumbermen's Association reported that the industry was operating at forty-eight percent of capacity, compared to thirty-two percent in early June.[15] In their eagerness to dispose of low-priced orders before the code took effect, mill owners threatened to produce a market glut.

Northwest lumbermen faced in the increasing militance of labor an additional reason for rapid approval of the code. Unemployment, low wages, and the recovery act's Section 7a sparked formation of American Federation of Labor locals in Bellingham, Everett, and Longview and strikes in the mills and camps of the Olympic Peninsula. Lumbermen recognized that the code's labor provisions could be used to their advantage in facing this developing crisis. When negotiations with the N.R.A. stretched into August, industry spokesmen announced that the eventual code wage scale would be made retroactive to the first of the month. "In localities where the [labor] pressure is growing embarrassing," pointed out an official of the Pacific Northwest Loggers Association, "the retroactive position might be a talking point that would compose the situation."[16]

Despite the urgency of things, several problems necessitated a lengthy process of public hearings and private meetings in the nation's capital. The owners of large sawmills, for instance, worried over the method to be used for allocation of production. Under the "soup bowl" approach devised by William Greeley, quarterly quotas would be allocated according to mill capacity in a base period of 1925–29. Greeley also argued for "birth control," the denial of quotas to plants built after implementation of the code. The big mills contended that the "soup bowl" would force efficient two- and three-shift operations to cut back, but allow small and inefficient one-shift mills to run almost fulltime. The Weyerhaeuser Timber Company also opposed "birth control" because it would be prevented from adding to existing plants or constructing new mills. "I was damned if I would sit by and see him get away with both 'soup-bowl' and 'birth control' and joined issue," wrote Weyerhaeuser attorney A. W. Clapp of his response to Greeley's proposals. The result was a tradeoff in which the former—with some technical modifications to allow mills built since 1929 to secure quotas—was retained and the latter eliminated.[17]

Under pressure from the Roosevelt Administration, the industry also resolved the question of wages and hours. Upon its arrival in Washington, the Emergency National Committee was informed by N.R.A. director Hugh Johnson that "temporarily and only because of the emergency, wages and hours had assumed an immediate importance." Final action on these points required compromise between the South and the Pacific coast. The forty-hour workweek was made mandatory throughout the country. On the wage front, however, the Pacific Northwest hourly minimum was set at 42.5¢ and the Southern rate at 23¢ for loggers and 24¢ for mill workers.[18] By continuing the historical tradition in which the Northwest paid higher wages than any other lumbering section, this solution enabled the South to retain its competitive advantage in the rail markets east of the Mississippi.

Settlement of the production allocation and wage and hour issues took place amidst hectic weeks of formal hearings and private consultations between industry representatives and N.R.A. officials. The code, William Greeley informed his wife in the second week of August, "is running the gauntlet of 15 different varieties of Hebrew economists and numerous other technical and legal 'advisors' of the Administration." Final approval by N.R.A. lawyers and a little last-minute tinkering with the forestry article by President Roosevelt himself culminated in the signing of the code on the nineteenth of August. Although dissatisfied with some code provisions, industry leaders expressed satisfaction with the outcome. The document, David Mason recorded in his diary, was "generally recognized . . . as [the] best code worked out so far."[19]

There had been little time, in the intense atmosphere of the drafting process, for serious reflection. "Now that it is all over," admitted F. E. Weyerhaeuser in late August, "I don't know what the Code means any more than I did when we introduced it for public hearing at Washington." As the initial exhilaration wore off, owners of large sawmills worried that the price and production controls favored inefficient small operators. Many lumbermen pointed out that the code provided no mechanism for stimulating demand. "It seems a shame," wrote Timothy Jerome, "that the President could not have enacted a law that would compel every man, woman and child in the United States to buy a few logs." There was for all concerned a sense of uncertainty about the changes sure to result from the code. "Every one," observed R. D. Merrill, "seems to be up in the air as to how the NRA is going to work out."[20]

Confusion and doubt aside, most lumbermen were ecstatic over their principal achievement: the code's sanction to control production and prices. R. W. Wetmore, a prominent investor in Northwest timber, looked forward to "the supreme satisfaction of hearing the squeals, howls and yells from a certain section of the industry which has always indulged in anything but constructive practices. These boys are going to be good now and toe the mark." The N.R.A. was, according to this view, sure to achieve success where eight decades of curtailment and merger schemes had failed.[21] The historical fight for restrictions on output had combined with the Depression emergency to produce the reconciliation of traditional divisions between large and small operators, between mill owners and loggers, and between geographical divisions of the industry. In the actual working of the code would be found the proof of that reconciliation.

Administering the Lumber Code

Whether the code would succeed or fail depended in the end on the ability of lumbermen to maintain the cooperative spirit. President Roosevelt "can lead the gang to water," commented the Loyal Legion's William Ruegnitz, "but can't make them lie down aside each other and be happy about it." Chaired by J. D. Tennant of the Long-Bell Lumber Company, the Lumber Code Authority administered the code under the jurisdiction of the N.R.A. On behalf of the Authority, the West Coast Lumbermen's Association and the Pacific Northwest Loggers Association supervised the code in the Douglas-fir regions of the Northwest. The logging group was hampered by the death in late August of Mark Reed, for years the leading influence among Washington loggers. Association members worried in a prophetic vein that without Reed's "judicial minded" presence, things would "go haywire" in the difficult work of implementing the code.[22]

Nationally and regionally, the administrative apparatus focused on the control of production and prices. An emergency production allocation went into effect in September 1933, followed by the first official quota for the final three months of the year. Complaint over the quotas was immediate and widespread, focusing on the apparent lack of fairness. The use of the late 1920s as a base period meant that Bloedel-Donovan, a nonparticipant in the curtailment programs of those years, received a higher allotment than companies that had restricted output. Loggers cutting Forest Service or Bureau of Indian Affairs timber found that the quotas forced them to choose between violating the code or breaking their government contracts. Weyerhaeuser and other large mill operators charged that trade association voting procedures gave inordinate weight to small mills, resulting in a quota formula that ignored their timber holdings, tax burdens, and up-to-date machinery.[23]

Together, the means selected to control production and prices prevented effective administration of the code. Determining the proper level of prices—"cost protection" in official terminology—was a difficult undertaking. "I should like to be around," F. E. Weyerhaeuser quipped, ". . . when the Code Authority attempts to fix minimum prices on all the items of lumber of all the various species manufactured in the United States." A procedure was adopted whereby prices were determined by dividing the total cost of production of all manufacturers by the amount of lumber produced. Because most producers were small and inefficient and had high production costs, the resultant prices were above prevailing market levels and canceled the competitive advantages of low-cost operations. The artificial rates, many lumbermen feared, would stifle demand, attract renewed competition from British Columbia, and make it impossible for American lumber to be sold in export markets.[24]

Pegging prices at a high level set up what Phil Weyerhaeuser termed "a tent over the little fellow." Under the protection of this tent, an estimated 5,000 mills were either built or resumed operation in the United States during the life of the N.R.A. In the absence of "birth control," these mills had to be dealt an allocation of production from the common "soup bowl." The Lumber Code Authority's national quota declined in the face of stagnant demand, from 5 billion feet in the first quarter of 1934 to 4.5 billion in the second and 3.8 billion in the third. Thus, a decreasing amount of production had to be divided among an increasing number of production units. The owners of large sawmills reacted with especial and understandable outrage to the necessary limitation on their business.[25]

By early 1934, six months into the great experiment, lumbermen concluded that their code was in need of drastic alteration. Aside from the malfunctioning controls on production and prices, there were constant irritants. The refusal of mill companies to place their satellite logging camps under the jurisdiction of the loggers association caused considerable antagonism. Dealings with the New Deal bureaucracy—described by David Mason as "like a kaleidoscope in four dimensions"—were sure to end in frustration. The suspicion was growing that the N.R.A. could not enforce the code's provisions, a suspicion that would likely tempt those of an uncooperative bent into violations.[26] Finally, there was the need to deal with conservation and labor, issues given new immediacy by the requirements of the code.

Article X Conservation

The Great Depression was a period of considerable tumult in the area of conservation. At the state level, the lumber industry won a major victory on the practical forestry front. In 1931, Washington legislators authorized a special property tax classification for logged-off land. West of the Cascades, such land would be assessed at an annual rate of a dollar an acre during the period of regrowth, to be followed by imposition of a yield tax at the time of harvest. By early 1932, applications for inclusion under the classification were filed on 209,000 acres, with half of this acreage owned by Weyerhaeuser, Long-Bell, and the St. Paul & Tacoma Lumber Company. Thereafter, the process was stalled as the result of lawsuits protesting the special treatment extended to owners of timberland. Lumbermen, moreover, shifted their attention to a campaign for extending the yield tax from cutover land to old growth timber.[27]

Although victorious on the reforestation tax, lumbermen came under increasing pressure from conservation advocates, as the general unpopularity of business focused public attention on the exploitation of natural resources by private interests. Sustained yield, argued those with a forestry bent, would lead to enlightened management of the nation's timber supply. "Some appear to see in it a veritable panacea," observed the *Journal of Forestry* in December 1933, "a quick and certain solution of many of the country's forest ills." Washington timber owners agreed in theory and supported passage of a bill in 1933 mandating sustained yield management of state timber on the Olympic Peninsula. They continued to maintain, however, that taxes, fire danger, and the stagnant demand for lumber prevented application of the concept to private land. David Mason, the most enthusiastic proponent of sustained yield, conceded that the Depression would have to be overcome "before there can be anything like the full development that there ought to be along this line."[28]

Spurred by the availability of new technology, the concept of selective logging generated considerable controversy in the Pacific Northwest. Introduced into the region's forests in the early 1930s, caterpillar tractors made possible the abandonment of traditional logging methods that required the removal of all trees. Building on this advance, University of Washington forestry professors Burt Kirkland and Axel Brandstrom argued that only the best quality timber should be logged and the remainder left standing for future use and for encouragement of regeneration. "The loggers' ax," they wrote, "should work with Nature rather than against her, and guide and speed her productive processes rather than destroy them." Tacoma-based logger L. Tom Murray became well-known for his adoption of the theory, impressing such personages as President Roosevelt and Secretary of the Interior Harold Ickes with his apparent devotion to environmental protection.[29]

Most lumbermen joined with practical foresters to denounce selective logging as impractical and as little more than a new gloss on an old approach to forest exploitation. In the midst of the Depression, loggers argued that they could not absorb the financial cost of abandoning old equipment in favor of new. Selective logging opponents also pointed to the fact that there was no evidence the technique would encourage regrowth. Early-day loggers, moreover, had cut only the trees for which there was a market and, behind the theory, critics contended that selective logging amounted to nothing more than a return to oldtime methods. "There is nothing novel or unique in Tom Murray's selection operation," contended John B. Woods of the National Lumber Manufacturers Association.[30] The reputation earned by Murray in the east, his colleagues in the industry believed, was one more example of the inability of outsiders to understand the forest situation in the Northwest.

Fending off the proponents of sustained yield and selective logging, lumbermen also worried about the implications of the appointment of Ferdinand Silcox as chief of the Forest Service in 1933. Although a trained forester, Silcox had spent much of his career as a labor mediator and represented a throwback to the progressive belief in conservation as social reform. Concern over his apparent devotion to the views of Gifford Pinchot, however, did not mean that the industry was opposed to a greater forest role for the government. In the fall of 1933, Northwest lumbermen, encouraged by the Copeland Report, drafted a formal request for federal acquisition of their marginal timberland. The clash of these forces—the popularity of new forestry concepts, the return of the forester militant to the nation's capital, and the problems facing timber owners—would determine the final form of the lumber code's mandate that "the applicant industries undertake, in cooperation with public and other agencies, to carry out such practicable measures as may be necessary . . . in respect of conservation and sustained production of forest resources."[31]

To reach agreement on the practical meaning of Article X's rather vague pronouncement, conferences were held in Washington, D.C. in October 1933 and January 1934. These meetings resulted in agreement on an accelerated private and public effort against the fire menace and in recommendations for federal financing of sustained yield and application of the yield tax to old growth timber. The industry promised that selective logging "shall be practiced wherever possible" and that sustained yield "shall be adopted as rapidly as possible." Commitment on these latter points was less than binding and Article X merely gave renewed impetus to the concepts long advocated in the Northwest. In the aftermath, a joint conservation committee formed by the West Coast Lumbermen's and Pacific Northwest Loggers associations urged upon logging operators the adoption of measures for improved fire protection and encouragement of regrowth. Under the moderate terms of the code, the argument that only the federal government could insure responsible management of timber resources was deflected.[32]

Despite the benign nature of Article X, lumbermen grew increasingly concerned over the conservation views of the Roosevelt Administration. "The Government is going to tell us what to do," wrote a concerned George Jewett, grandson of Frederick Weyerhaeuser, "so why go through the motions of figuring out cooperative measures. It is like the victim of a holdup cooperating by handing over his purse!" Timber owners complained that the Forest Service refused, in illogical juxtaposition, to recognize the progress being made by industry in the Northwest or to do its own part by spending the funds requested for fire protection, research, and purchase of forest land.[33] Held to what they considered a one-sided standard of conduct, lumbermen added lack of fair treatment on conservation matters to their list of complaints about the code.

Section 7a in the Northwest

To employers, the most controversial provisions of the code were those relating to labor and its right to organize. Splitting logical hairs of exceptional fineness, some lumbermen actually denied that Section 7a required them to deal with unions. "Employes have the right to bargain collectively with an employer," conceded J. D. Tennant. Employers, though, "may bargain with employes, or may not, at the discretion of management." With the 4L in the field as a time-tested representative of workers, industry leaders argued that there was no need for additional labor organizations in the Pacific Northwest. Wishful thinking, however, could not prevent Section 7a from providing the stimulus for a major organizing drive by the American Federation of Labor. National union membership increased by a third in the first year of the N.R.A., to the consternation of those who saw in the codes the cause of this development. "The N.R.A. is intended to increase jobs," spluttered one logging operator, "and not to impose A.F. and L. regulations."[34]

Lumbermen looked to the Loyal Legion for protection against the forces set loose by Section 7a. Under the N.R.A., William Ruegnitz warned employers, the organization offered the only alternative to radicalism. Mill owners strengthened the 4L by giving preference in hiring to members and by setting up a joint Legion-trade association committee to handle labor disputes. These efforts, however, failed to generate enthusiasm among employees. When the Legion postponed a conference scheduled to raise wages in late 1933, additional force was given to the argument that only independent unions could defend the interests of labor. The average worker, admitted one 4L organizer, wanted "something beside wage-cuts to come his way once in a while" and regarded the Legion as little more than "a shell game."[35]

The American Federation of Labor moved to fill the gap caused by the 4L's inability to appeal to workers, stepping up the formation of locals in the fall of 1933. This was a menacing development to employers, one that portended all sorts of future difficulties. "The American Federation of Labor, . . . is grasping for causes to fight and win," wrote Phil Weyerhaeuser, "and there is no measuring the reasonableness of the causes which they will seize." The A.F. of L., though, made only slight progress in the first year of the lumber code. Beset at the national level by internal disputes and by the problems inherent in rapid expansion, federation leaders were unable to supply assistance to Northwest timberworkers.[36]

In sharp contrast, the union movement fastened itself upon the region's pulp and paper industry. Locals of the International Brotherhood of Pulp, Sulphite, and Paper Mill Workers were formed in Port Angeles, Everett, Longview, and other manufacturing centers in the summer of 1933. By the end of the year, representatives of a dozen locals had established the Pacific Northwest Council of Pulp and Paper Unions to bargain with management. Pulpworker leaders exuded an uncertainty about their goals and about the proper relations between management and organized labor. The head of the Port Angeles local, for instance, requested of international president John Burke advice on "how to proceed. Would you please send me any information you have on the proper manner in which to take the next steps." Applying for a job with the international, an officer of the Longview local gave as his reference the manager of the town's Weyerhaeuser pulp mill.[37]

While modest in their ambitions, the new pulp unions were militant in their demands for higher wages. Deluged with demands for strikes, the cautious Burke warned against any action that would antagonize management. The union, he advised, must "proceed carefully as would a mariner sailing through unchartered waters." Pulp and paper mill owners also favored the avoidance of trouble, as the heavy financial requirements of the industry made constant operation and cash flow a necessity. In the summer of 1934, Washington and Oregon employers agreed to a general contract recognizing the locals as bargaining agents and creating machinery for the determination of wages and the handling of complaints.[38] Here was a rare instance in which management and labor were able to achieve a relative harmony of interests.

Mutual antagonism, rather than any semblance of harmony, marked the situation in the lumber industry by early 1934. Labor relations became increasingly bitter and the industry seemed to be nearing a major crisis. A dramatic preview of that crisis came in May, June, and July with closure of the ports of the Pacific coast by striking longshoremen. The strike made it impossible to load vessels and forced the closure of waterfront mills on Puget Sound, Grays Harbor, and the Columbia River. At Longview in late June, sawmill workers abandoned their jobs for several days in sympathy with the longshoremen. The longshore strike and its culmination in a general strike in San Francisco raised the prospect of a united labor campaign against the traditional prerogatives of management.[39]

The complexity of the labor situation was reflected in the fact that by mid-1934 each faction had concluded that the N.R.A. favored the opposition. The A.F. of L. expressed outrage over the industry's use of the 4L as a vehicle for worker complaints. The Legion's William Ruegnitz, in stark contrast, complained that the A.F. of L. had been placed "in the pile-driver seat" by the Roosevelt Administration. Employers insisted that the National Labor Board, set up to monitor compliance with N.R.A. labor provisions, was biased against private enterprise. Northwest board officials, charged Phil Weyerhaeuser in March 1934, were helping the A.F. of L. "to organize our industry" through "liberal interpretation of the scant provisions of the National Recovery Act with reference to labor's right to bargain collectively."[40] The code failed to satisfy any of the participants in the struggle that would eventually result, in 1935, in a major convulsion.

Collapse of the Code

When not fretting over the developing labor crisis, lumbermen spent their time contemplating the mixed performance of the recovery program. Held down by controls and sluggish demand, Washington lumber production stood at 3.1 billion feet in both 1933 and 1934, a rate that was forty-two percent of output in 1929. In 1934, waterborne shipments from the state decreased from the level of the previous year, in large part because of the longshoremen's strike. Although the government ordered a reduction in the official minimums as part of an effort to stimulate construction, there was a sharp increase in the price of lumber. Douglas-fir prices mounted from $10.73 per thousand feet in 1932 to $14 in 1933 and $16.55 in 1934.[41] These rates, however, were insufficient to down a growing suspicion that the lumber code was retarding recovery and that it was working against the interest of many firms.

By the summer of 1934, a year of frustrated expectations forced a number of lumbermen to the conclusion that they had taken on an impossible task. "We are playing with a problem in economics that is so far flung," pointed out C. H. Kreienbaum of the Simpson Logging Company's sawmill affiliate, "covering so many operating units, and such a diversity of distribution, I feel the foundation is crumbling by sheer weight of the situation." Large manufacturers were especially fed up with the impact of production controls and artificially high prices. "There are more in the industry," reported Phil Weyerhaeuser, "who question the advisability of setting up an umbrella [over inefficient operators] through minimum prices and guaranteed hours of operation."[42] Led by Weyerhaeuser, a campaign got underway to eliminate the code's production and price provisions.

Focusing at first on price controls, the "insurgents"—the description was David Mason's—moved against the code in the fall of 1934. A meeting of the West Coast Lumbermen's Asociation in late September, likened in its tumult by J. H. Bloedel to a longshoremen's convention, voted by only a narrow margin in favor of continuing the controls. Representatives from the Pacific coast disrupted a national conference of the Lumber Code Authority in Chicago in October with demands for repeal of the minimum prices. Finally, at Tacoma on the next-to-last day of October, the W.C.L.A.'s trustees voted 14-7 to place the association's resources behind the fight for repeal. The industry had come apart over the conflicting interests of big and small firms and was, according to a *Timberman* editorial on the subject, "A House Divided Against Itself."[43]

The inability of the N.R.A. to prevent violations of the code was an additional cause of the general denunciation of controls in the Northwest. As dissatisfaction with the code grew during 1934, so did the willingness to ignore its provisions. An industry leader estimated in mid-September that only a small portion of the lumber shipped by water to California and the Atlantic coast was sold at code prices. The government filed suit against the Simpson Logging Company mill and a half-dozen other manufacturers in October, but otherwise lacked the manpower and the funds to prosecute violators.[44] Recognizing these all-too-apparent facts, those breaking the code felt triumphant and those standing by it regarded themselves as softheaded idealists.

Wholesale abandonment by those following their own individual interests made the code something of an open joke. When allocations for the final quarter of 1934 were announced, for instance, loggers association president Joseph Irving stated that he would run fulltime in defiance of the code. His friends in the organization had moved to secure an injunction when Irving, well known as a heavy drinker, showed up at an early November meeting of the trustees. "He went out many times," wrote Timothy Jerome of the affair, "and when he came in right after luncheon, he was fuller than a boiled owl." Irving, continued Jerome, "had great difficulty in talking, so that he could be understood. But he did not have any difficulty in trying to talk almost every minute, making it very hard to transact any business." The meeting ended, in graphic demonstration of the belief that there was little reason to abide by the code, with the granting of a special allocation to Irving and the withdrawal of the legal action.[45]

With the code disintegrating as if written on yellowed newsprint, the N.R.A. held urgent hearings in the nation's capital in mid-December. The testimony revealed intense opposition to price controls on the Pacific coast and suggested that continued agitation would place the entire code apparatus in danger of collapse. Following brief consideration of these points, the Lumber Code Authority announced the suspension of the minimum price provisions. Although a relief to everyone, the action aroused fears for the remainder of the code. "In spite of all our difficulties under the code," noted J. D. Tennant, "we still have made great progress toward industry stabilization and self-government, and we do not want to lose ground."[46] Few lumbermen advocated a return to the old days of competitive chaos and attention focused on strengthening the code and on reaching agreement on long-term revisions to be sought when the N.R.A. came up for congressional renewal in June 1935.

For all concerned, the key to preservation of the code was the government's willingness to prosecute violators. When the Justice Department announced the withdrawal of a heavily publicized suit against an Alabama lumberman in March 1935, therefore, the code was abandoned as a worthwhile instrument. The government's action, wrote F. R. Titcomb, who had remained at Weyerhaeuser in a subordinate capacity, came "more or less as a bolt out of the sky" and meant "the complete break-down of the lumber code." The document, according to the rather inelegant mixed metaphor of one trade publication, was "thoroughly emasculated and impotent" and "a dead cock in the pit." Working to contain the damage, David Mason, since mid-1934 the executive director of the Lumber Code Authority, and other N.R.A. officials met with industry leaders in an effort to restore faith in the Roosevelt Administration's commitment to the integrity of the code. Ignoring their pleas, trade associations throughout the country demanded the scrapping of the N.R.A.[47]

While the lumber situation was still in flux, the Supreme Court ruled in the Schecter poultry case that the N.R.A. was unconstitutional, forcing the dismantling of all codes. Lumbermen greeted the decision with relief, worrying only that the collapse of the N.R.A. might lead to runaway production. The *Timberman* contrasted the relaxed response with the state of things in the summer of 1933, "when business executives by the thousands were scurrying to Washington, from every corner of the land, pockets stuffed full of codes and theories guaranteed to change human nature overnight."[48] In the intervening two years, enthusiasm had given way to discontent and the pleas for federal assistance to a demand for the return of free enterprise.

Industrial self-government, while attractive in concept, had become a frustrating illusion long before the demise of the N.R.A. The economic crisis of 1933 had produced agreement among competing regions, among lumbermen and loggers, and among large and small operators. The very compromises needed to reach agreement, however, made the limitations on prices and production unworkable and demonstrated, as a trade journal put it, that "new deals may not always be the best."[49] The failure of code enforcement meant that compliance depended on the weak instrument of voluntarism. In the final analysis, the lumber code failed for the same fundamental reason as past efforts at cooperation. The nation's lumber industry was too large and composed of too many diverse elements for there to be, except in occasional brief instances, a single point-of-view on any problem or issue.

At the beginning of the New Deal, lumbermen had applauded the actions taken by the Roosevelt Administration to restore confidence and to free industry from the outmoded shackles of the antitrust laws. But disappointment over the N.R.A. and opposition to the leftward trend of the New Deal caused approval to give way to denunciation. However illogical the combination, lumbermen charged that the president was "intoxicated" with power and the puppet of "college professors" who were "strongly socialistic in their leanings." Roosevelt, reflected R. D. Merrill, was "the worst man we ever had in the White House." By mid-1934, Timothy Jerome was convinced that those who had voted the Democratic ticket in 1932 should "be taken out and shot at sunrise."[50] In the aftermath of the N.R.A., vituperation would increase as issues of labor and conservation came to the fore in the lumbering regions of the Northwest.

16

Like a Moving Film

The outlook for West Coast lumber runs across the screen like a moving film. The story is often tragic; now and then lively with action; and frequently throws new problems into the plot. But it is always moving. You can't get a 'still shot' of the fir lumber business.—William B. Greeley[1]

I was told the other day that people were beginning to talk about Mr. Roosevelt's being out of the picture sometime early this fall. Just what they mean by this I do not know, but I would not be surprised if some disgruntled person might think it his duty to eliminate him from the picture. If this is done, it would be an awful thing, but it might make a difference in the picture all over the country.—Timothy Jerome[2]

Rioting disrupted the city of Tacoma in the final week of June 1935. The national guard, backed up by state patrolmen, blocked access to the tideflat sawmills, preventing pickets from interfering with lumber production. Turning their attention to the office of the Loyal Legion of Loggers and Lumbermen, frustrated strikers surrounded the building and threatened to assault its occupants. "For several hours," wrote a frightened 4L official, "we were not sure what was going to happen to us." In early evening, the guard and the patrol finally arrived to disperse the demonstrators, tossing tear gas bombs and spewing gas from the exhaust pipe of a police vehicle. "It sure scattered the crowd hell-west and crocked," reported a relieved J. A. Ziemer following his rescue from the Legion office. That night and the next day, soldiers occupied the streets of Tacoma to prevent pickets from gathering and patrolled the city's residential areas to intimidate workers.[3] After two years of the New Deal, labor was at war with management and, as dramatized by the attack on the 4L, with itself.

The Great Strike

Hostility mounted on the Northwest labor front in the final months of the National Recovery Administration. Attention centered on Longview and on the likelihood that unionization of the Weyerhaeuser and Long-Bell mills would stimulate American Federation of Labor efforts in other localities. In early 1935, the National Labor Relations Board ruled, on the basis of elections held in Longview, that the A.F. of L. timberworkers must be recognized as the bargaining agent for employees of the two firms. The response of management was immediate and hostile. "We cannt grant the principle of majority rule to the exclusion of minorities," insisted Phil Weyerhaeuser. To do so would be to accept the inevitability of the closed shop and dictation by organized labor. F. R. Titcomb vowed that the closed shop "is the last thing we are going to allow to take place in our plants."[4]

This dispute was soon absorbed into a region-wide conflict between management and labor. At its October 1934 convention, the American Federation of Labor, responding to pressure from John L. Lewis and other advocates of industrial unions, approved a plan to organize unskilled and semi-skilled workers. In furtherance of this effort, the Washington and Oregon timber locals were placed under the jurisdiction of the United Brotherhood of Carpenters and Joiners, a rather bizarre linkage considering the growing militance of the former and the craft-oriented conservatism of the latter. Assigned to direct the organizing campaign in the Pacific Northwest, carpenters official A. W. Muir experienced immediate difficulty controlling his new followers. Over his objections, the Northwest Council of Sawmill and Timber Workers called for an industrywide strike in May 1935 to secure a thirty-hour workweek, a seventy-five-cent hourly minimum wage and union recognition.[5]

Faced with this crisis, lumbermen turned for assistance to their old ally, the Loyal Legion of Loggers and Lumbermen. The Legion denounced the pending strike, but otherwise was unable to reconcile the demands of both employers and employees. Hoping to stop the loss of members, 4L officials, in a moment of independent action, called for an increase in the minimum wage from 45 to 50¢ an hour. "There is no industry justification for the raise," complained Phil Weyerhaeuser when he heard the news, "but the 4-L seems to think its fighting for its existence and helping itself by such action." Employers believed that the wage demand undermined their bargaining position and demonstrated the unreliability of the organization. Contending that "the 4-L has done us no good whatsoever," Weyerhaeuser withdrew his firm's financial support.[6] Although the Loyal Legion limped on until the spring of 1937, when the supreme court's decision validating the Wagner Act's ban on company unions forced its final collapse, the loss of industry backing meant its immediate disappearance as a significant factor.

Workers began abandoning their jobs a week before the official strike deadline of May 6. Within days, most of the sawmills in Portland and on Puget Sound and Grays Harbor were closed by employers hoping to avoid violence. Labor and management alike looked to Longview, where Weyerhaeuser and Long-Bell negotiated directly with A. W. Muir, for clues as to the length of the walkout. Without some form of agreement at Longview, Phil Weyerhaeuser was "afraid the union leaders were going to lose control of the situation." On May 9, a compromise settlement was agreed upon, providing a 50¢ minimum wage, a forty-hour workweek, and restriction of the union's bargaining rights to representation of its membership. Rumors of these terms caused workers at the two firms to disavow Muir's leadership and go on strike. Although the Longview mills briefly reopened in early June, the collapse of the negotiations portended a long and paralyzing strike. By the middle of May, 90 percent of the Northwest industry's capacity was shut down and 30,000 workers walked the picket lines.[7]

Although the Charles R. McCormick and Bloedel-Donovan firms soon settled with the union, the remainder of the industry was determined to stand fast until protection could be secured from the state. In late June, Tacoma operators developed a plan to reopen and pay workers a 50¢ minimum for a forty-hour week. To protect those accepting the terms from pickets, employers urged Governor Clarence Martin, a conservative Democrat, to send the national guard and the state patrol to Tacoma. "The great majority of the men want to go back to work," F. R. Titcomb informed the governor. These workers, though, were being intimidated "by men who have an ulterior purpose in mind—that of overthrowing our present system of government." Despite opposition from Tacoma city officials, the governor acceded to the request. "I insist on respect for collective bargaining, for peaceful picketing, for the right to strike," stated Martin in a curious pronouncement, ". . . but also I shall continue to demand respect for the right to work."[8]

Five hundred guardsmen and seventy state patrolmen arrived in Tacoma on June 24 to prevent strikers from interfering with the reopening of the mills. The guard, unused to such harrowing duty, received crucial backing from the no-nonsense patrol. "These state patrolmen are sure some men," wrote an admiring observer, "and they sure mean business. They will not stand for any back talk." Through intimidation and the use of gas, the streets were cleared and those wishing to work—men who were for the most part young and new to the industry—were able to enter and leave the mills without fear of violence. Governor Martin, firm in his self-styled defense of the rights of labor, turned a deaf ear to complaints that he was breaking the strike.[9]

Defying a guard-imposed ban on picketing, strikers marched on the mills on the 25th of June. The demonstrators were met by tear-gas and by a line of determined guardsmen. "The guards followed up the patrol with set bayonets," reported the 4L's J. A. Ziemer, "and they began to push the pickets up the hill." Within minutes, the strikers were "scattered all over town." Confrontations between workers and soldiers continued, but mill production was soon back to seventy percent of normal. The Tacoma strike came to an end in early August with approval of an agreement for a 50¢ hourly minimum, a forty-hour week, union collective bargaining on behalf of its membership, and the re-hiring of strikers.[10] Lumbermen, disgusted with the attitude of the national Democratic administration, had found a valuable ally in the state's Democratic governor.

A similar course of events marked the final stages of the strike in other localities. In early July, mills at Longview, Everett, Aberdeen, and Hoquiam reopened under the protection of the patrol and the guard. Final settlement, though, was delayed by occasional outbursts of violence. On Grays Harbor, the wives and children of strikers stoned guardsmen, houses and stores were bombed, and employers hired teamsters to escort strikebreakers through the picket lines. By the third week of August, exhausted lumbermen and workers reached agreement in each of the mill towns on the same terms as those in Tacoma.[11]

In the aftermath, lumbermen could for a moment congratulate themselves on winning a battle. They had, after all, surrendered nothing in the settlement with the union. The forty-hour week had been the standard since 1918, the 50¢ minimum wage had already been instituted by the 4L, and the closed shop was denied. These terms, however, failed to disguise the fact that employers had in reality lost the war with labor. Considering the prevailing sentiment among workers, Phil Weyerhaeuser pointed out, "I do not think that we can refuse recognition of the union in some way in the future." With the demise of the Loyal Legion as a legitimate force, he continued, there was "no vehicle other than the sawmill union for any sort of bargaining . . . between our employees and ourselves." Some operators tried to avoid acceptance of this difficult fact by setting up new company unions, but the clear result of the strike was the fastening of organized labor on the lumber industry.[12]

Rise of the Woodworkers

For employers, the constant pressure to increase wages brought home with especial force the new power of labor. Despite slumping domestic and foreign trade, the average hourly wage paid in Washington sawmills mounted from 57¢ in 1935 to 62¢ in 1936 to 70¢ in 1937 and to 73¢ in 1938. Over the same period, the logging camp rate increased from 66¢ to 84¢. "The wage scale in the state of Washington," observed one industry leader in early 1939, "stands up so-to-speak like Mount Rainier." Mill owners complained that high labor costs made it impossible to sell lumber in competition with other regions and would eventually force them to close their operations. The industry, they mourned, had fallen under the dictation of the union movement.[13]

Much of the energy of organized labor, however, was expended on internal squabbling. Rank-and-file hostility toward A. W. Muir's lack of militance was exacerbated by craft union raids on the various timberworker locals. In issue after issue, the union newspaper, the *Timberworker,* attacked "the old die-hard labor leaders" and their craft philosophy, "which in this day and age, still belongs to the dark ages of many years ago." Responding to the threat to his leadership, Muir replaced the Northwest Council of Sawmill and Timber Workers with district councils in hope of retarding the spread of anticarpenter influence. Under these circumstances, considerable interest was expressed by Muir's opponents in the Committee for Industrial Organization, formed in November 1935 by John L. Lewis, Sidney Hillman and other advocates of the industrial union approach.[14]

This development posed a quandary for employers. On the one hand, the relative conservatism of the A.F. of L. encouraged cooperation with that organization. On the other, though, it was more efficient to negotiate with one party than with several unions. "It would be almost impossible for us to deal with the many crafts which could be developed on our plants," pointed out Phil Weyerhaeuser. "If we are to have unions," agreed the Weyerhaeuser syndicate's Laird Bell, "we are better off with industry unions than with craft unions." Whatever their sympathies, lumbermen were caught between the fast-developing rival movements. Dealing with one group resulted in trouble with the other, and both based membership drives on demands for higher wages. "The whole thing, so far as I can see," observed one perplexed mill company executive, "is just one sweet mess."[15]

Worker opposition to the carpenters and to the A.F. of L. increased during 1936. Craft union leaders, insisted the *Timberworker,* had "become fat, fatuous and entirely futile." The dissidents were reinforced in March 1936 when the Lumber Workers Industrial Union of British Columbia voted to affiliate with south-of-the-line timberworkers. In September, anti-carpenter forces formed the Federation of Woodworkers as a rival organization within the A.F. of L. Finally, in July 1937, the federation transformed itself into a new union, the International Woodworkers of America, and moved into the C.I.O. camp.[16]

The I.W.A.—the spelling was changed from the original International Wood Workers of America to prevent opponents from using the provocative initials I.W.W.A.—was one of the most militant and controversial labor organizations of the Depression era. Under the leadership of such avowed leftists as Harold Pritchett, who became the first president, and O. M. Orton, the union claimed that businessmen were deliberately ruining the nation's economy and that the A.F. of L. was a "stooge" of industry. I.W.A. leaders supported the programs of the Roosevelt Administration and worked for the expansion of the

New Deal. "We must correct the weaknesses of the economic and social level of the people of this country," insisted Orton.[17] From the beginning, the organization was burdened with charges of communist penetration and by the efforts of conservative members to dispose of the leadership of the Pritchett-Orton group.

Unwilling to concede defeat to the I.W.A., the carpenters formed the Oregon-Washington Council of Lumber & Sawmill Workers from the remnant of the old timberworkers organization in August 1937. The council secured a court order allowing retention of union funds and insisted that employers honor contracts providing for recognition of A.F. of L. locals. Between mid-1937 and late 1941, the Northwest industry was disrupted by a constant series of jurisdictional strikes. On Grays Harbor, for instance, most mills were closed throughout 1938 by the dispute. Relations between management and labor were greatly complicated by the fact that efforts to negotiate with one of the unions meant inevitable boycotts and picketing by members of the other. "The problem of what will satisfy the CIO and not put the A.F. of L.'s nose out of joint (and vice versa)," admitted Phil Weyerhaeuser, "is completely baffling."[18]

Within the I.W.A. itself, the issue of communism added to the complexity of the labor situation. Adolph Germer, a C.I.O. official sent to the Northwest to aid the organizing effort, charged that Pritchett and Orton were using the union as a "cloak to conceal their real purpose which is, namely, to carry out the often changed line of the Communist Party." In the summer of 1940, Pritchett, a Canadian citizen, was denied an extension of his visa by U.S. immigration authorities and had to resign the presidency, to be succeeded by Orton. At the I.W.A.'s international convention in October 1940, the radical forces defeated an attempt to condemn communism and began a campaign to purge their opponents from the union. The following year, however, strong C.I.O. assistance enabled the anti-radical faction to wrest control of the organization.[19]

Despite its divisiveness, labor's internal struggle failed to undermine the overall triumph of the union movement. By 1940, the I.W.A. claimed 100,000 members in the Northwest, north and south of the international boundary, and the carpenters boasted of 35,000 adherents. That employers had conceded a measure of control over the industry's destiny was demonstrated by the abortive effort to reduce wages in the late 1930s. In late 1938, Weyerhaeuser joined with other operators in a movement to cut wages, but only as the result of industry-wide bargaining with the unions. Negotiations continued until a revival of demand in 1940 transformed the talks into a forum for increased rates of pay.[20] The days when lumbermen could meet hard times by arbitrarily slashing payrolls had passed into history. Here, in the new status of labor, was the real New Deal in the Northwest lumber industry.

Failure of Recovery

The war with labor was fought amidst deteriorating economic conditions for Northwest lumbermen. Released from the fetters of the N.R.A. in the spring of 1935, the region's industry leaped forward toward apparent prosperity. Water shipments for 1935 were the heaviest in four years. Production and prices, boosted by a reduction in railroad freight charges, made further advances in 1936. "We are on the threshold of better times," enthused Long-Bell's J. D. Tennant. Close analysis of the figures, however, revealed a faltering performance. Washington's lumber output in 1936 was more than twice the 1932 level, but still well below the production of the 1920s. Symbolizing the decline from those boom years, the state surrendered its position as the nation's leading lumber producer to Oregon in 1938.[21]

Increased labor costs, as far as mill owners were concerned, forced corresponding price hikes and the exclusion of American lumber from the foreign market. In 1936, for instance, British Columbia shipped twice as much lumber abroad as the combined total of Washington and Oregon mills. By 1939, the United States had fallen from first to fifth place among lumber exporting nations and was about to drop another notch below, of all countries, Poland. The federal government, complained William Greeley, had created "ghost towns" in the Pacific Northwest "by permitting our lumber to be excluded from many foreign markets through discriminatory tariffs and other trade barriers." Providing a graphic example of this development, the Charles R. McCormick Lumber Company dismantled its Port Ludlow mill in 1935 and, in a bankrupt state, reverted to the ownership of Pope & Talbot in 1938.[22]

What was left of the overseas trade fell victim to the mounting international tensions of the late 1930s. Economic controls and the danger to vessels reduced lumber shipments to Asia following the outbreak of war between Japan and China in 1937. The European political crises leading up to and following upon the Munich agreement of September 1937 had a dampening effect on demand. Together, noted Timothy Jerome, the situations in Asia and Europe "make it harder than ever to sell all of the lumber that is cut." Northwest lumbermen, for the first time in the history of their industry, were forced into what amounted to total reliance on the domestic market.[23]

That market, unfortunately, was undercut by the combined impact of the Roosevelt Administration's foreign policy and economic stagnation. To stimulate American exports, the administration negotiated reciprocal trade agreements with Canada and Great Britain in 1935 and 1938. Canadian mill owners, in return for accepting a relaxation of the British imperial preference system, demanded access to California and the Atlantic coast. The result was a reduction in the lumber tariff to 50¢ per thousand feet. This, with the lower labor costs and taxation north of the line, supposedly gave British Columbia lumbermen a competitive advantage and represented a needless sacrifice of the American industry.[24] Once again, the New Deal had treated lumbermen in a thoughtless and unfair manner.

Coming at the same time as reciprocity, the failure of economic recovery meant that there was increased competition to service a decreasing demand for lumber. The "Roosevelt recession" in the winter of 1937–1938 wiped out most of the gains of the New Deal and produced pessimism among investors and a sharp fall-off in construction. "Our business activity index," reported Phil Weyerhaeuser in late November 1937, "is going down so fast it makes my head swim." The collapse of the market was accompanied by a loss of confidence among lumbermen. "As our friend the President said in 1932," observed a depressed Timothy Jerome, "things cannot be possibly any worse, but as a matter of fact I think they are worse now than they were then."[25]

Only one Northwest forest products firm managed to prosper in spite of the circumstances. Pulp output continued to increase at the Shelton, Hoquiam, and Port Angeles mills producing Rayonier, the special pulp for use in the manufacture of rayon. These plants benefited from the rapid expansion of the Japanese rayon industry during the 1930s and from the dependence of that industry on imported pulp. Even with the dislocations of the war in China, sales to Japan in 1937 accounted for over half of the Shelton mill's business. Enhanced by the Japanese military's uniform requirements, the trade produced a close and profitable economic linkage between western Washington and the island empire. To increase efficient handling of the business, the three mills were merged in late 1937 to form Rayonier Incorporated.[26] Here, thanks to modern chemistry, was a lone beacon shining through the gloom of economic stagnation.

Those less fortunately situated could do little more than blame the New Deal for the failure of recovery. New Deal spending and taxation policies had administered, according to a favored charge, a confidence-sapping dose of what one lumber salesman called "Roosevelt-phobia" to investors and businessmen. The administration's apparent assault on the constitution, especially in the court-packing fight of 1937, supposedly placed private property itself in danger. "The whole current cheapening and vulgarizing of our national institutions," mourned E. T. Allen, ". . . is disgustingly suggestive of a horde of marauding and obscene apes defiling our dedicated temples." The president, claimed R. D. Merrill, was "determined to make this country a second Russia." The return of prosperity, all in the industry agreed, would be delayed until what Timothy Jerome termed "our damned Democratic Administration" was succeeded by a government devoted to the principles of Herbert Hoover.[27]

Nevertheless, industry leaders had to admit that the New Deal appealed to most Americans. "The great number of people that were misled into voting for the President," observed Jerome after Roosevelt's landslide reelection in 1936, "shows conclusively that they were not all rif-raf, because there are not as many of the rif-raf in the world as the number of votes Roosevelt got." This recognition, of course, provided little comfort for worried lumbermen. The mounting public debt, the taxes levied on the income of corporations and the wealthy, and the alarming rate of federal spending were all cause for concern.[28] Even worse was the apparent subservience of the Roosevelt Administration to those who would arbitrarily preserve the environment from exploitation.

Forestry after Article X

Following the abandonment of the lumber code, timber owners moved quickly to resuscitate the Article X forestry program. David Mason and other industry leaders met with government officials to propose continued cooperation on fire prevention and tax reform. Under a program adopted in November 1935, the joint forestry committee of the West Coast Lumbermen's and Pacific Northwest Loggers associations spent nearly a million dollars promoting improved logging practices in the region. These efforts preserved valuable stands of timber and demonstrated the good faith of voluntary conservation measures. Owners of timber, insisted Phil Weyerhaeuser, had to counter "the efforts of . . . the administration to force government control of private forest lands by asserting the very real progress which has been made by the industry . . . in spite of the government's complete lack of cooperation."[29]

Fear of the regulators mounted as the New Deal moved away from an accommodation with business in the aftermath of the N.R.A. experiment. Beginning in 1935, Forest Service chief Ferdinand Silcox stepped up his campaign for mandatory adoption of forestry by private timber owners. "A large portion of the lumber industry is still operating on a basis of quick liquidation," he argued, "draining off the few remaining reserves of virgin timber." The "entire program" of the Forest Service, wrote Silcox, was based on the concept that "when private initiative fails to meet the needs of the people . . . the government must step in to protect the public interests." In the absence of improved industry practices, the chief forester threatened to establish public logging and lumbering operations. Although Silcox explained that his proposals would be implemented only as a last resort, his protestations of benign intent failed to convince suspicious lumbermen.[30]

Despite the emotions generated by Silcox, cooperation continued to mark relations between the Forest Service and industry. Dependent on the government for technical and financial assistance, lumbermen believed that a public battle with Silcox would be counterproductive. For its part, the Forest Service relied on industry lobbying support, especially after the conservative resurgence in the 1938 elections. Indicative of the conciliatory spirit, a federal-private forestry conference in the spring of 1937 approved resolutions in favor of increased government expenditures, state tax reform and adoption of sustained yield "as rapidly as conditions justify." The following year, lumbermen endorsed creation of a joint congressional committee on forestry, expecting it to produce public appreciation of industry problems.[31]

More than anything else, the struggle over government reorganization in 1937 and 1938 revealed the mutual dependence of the Forest Service and timbermen. Early in the New Deal, preservationists called for the transfer of the Forest Service from the Department of Agriculture to the Department of the Interior and the creation from the latter of a new Department of Conservation. Transfer became a distinct possibility with introduction of legislation authorizing the president to reorganize the executive branch. Foresters and commercial groups, joined in opposition, pointed out that Interior's supervision of the nation's forests in the nineteenth century had been marked by corruption and mismanagement. "The fact is," insisted Gifford Pinchot, the leading spokesman for the antitransfer forces, "that the Interior Department is Uncle Sam's real estate agent."[32]

Northwest lumbermen took part in the struggle to save the Forest Service from falling under the control of Secretary of the Interior Harold Ickes, at one point even meeting with Pinchot, their ancient antagonist, to coordinate strategy. If carried out, they believed that the transfer would mean the "hoarding of timber" through eventual inclusion of the forests in the national parks. The defeat of the reorganization bill in 1938 reaffirmed the importance of the Forest Service to the industry. Although there had been no explicit agreement in which abandonment of federal regulation was exchanged for support on the transfer issue, the assistance provided by timber owners encouraged government foresters to back away from the controversial subject.[33]

Even when working in harmony with the Forest Service, lumbermen were unable to down their basic mistrust of the Roosevelt Administration's conservation course. Timber owners were quick to abandon their early support of administration forestry measures. Initial admiration for the Civilian Conservation Corps gave way to a conviction that agency staff appointments were made on the basis of politics rather than forestry qualifications. Endorsement of the Copeland Report's proposal for federal purchase of marginal and cutover timberland was withdrawn by lumbermen, "because," wrote William Greeley, "they read into it a further extension of Government control."[34] Above all else, the dramatic battle over creation of Olympic National Park posed serious problems in western Washington.

The Olympic Park Fight

Beginning with Yosemite and Yellowstone in the late nineteenth century, the nation had protected scenic wonders west of the Mississippi from despoilment by private interests. Located for the most part in remote and rugged regions and thus effectively barred from commercial exploitation, national parks generated little opposition. The creation of Mount Rainier National Park in 1899, for instance, was approved by Washington lumbermen, many of whom became frequent and enthusiastic visitors. During the 1930s, environmentalists hoped to take advantage of the conservation sensibilities of the Roosevelt Administration to advance the cause of preservation. "I don't know what the New Deal will do for the nation generally," wrote journalist Irving Brant, a leader of the environmental movement, "but in conservation it is certainly a time to apply the motto of the Indiana lady, 'while you're gittin,' git a plenty.'"[35] In the Pacific Northwest, park advocates focused on the majestic Olympic Peninsula, where timber resources were neither worthless nor inaccessible.

The jumbled mountain peaks, alpine meadows and lush rain forest valleys of the peninsula had been the subject of a series of presidential actions. Originally, Grover Cleveland had proclaimed a forest reserve of 1.5 million acres. Theodore Roosevelt withdrew 615,000 of these acres in 1909 to create the Mount Olympus National Monument. Responding to pressure from logging interests and ignoring the protests of environmentalists, Woodrow Wilson halved the size of the monument by returning valuable timber stands to the Olympic National Forest.[36] The plan in the 1930s was to create a park by combining the monument with as much acreage as possible from the forest.

This proposed transfer of land was opposed by the Forest Service, in part because many participants in the fight linked the issue with the struggle over government reorganization. Of more importance, the addition of its timber to the monument would force the Forest Service to abandon longrange plans for the peninsula. Concentrated on the western slopes of the Olympics, the predominant commercial species

in the region was hemlock. Government foresters focused on this timber as the basis for the future expansion of pulp and paper manufacturing on the Strait of Juan de Fuca and on Grays Harbor. Removing
this timber from exploitation would seriously retard the orderly development of the industry.[37]

The debate, then, was not so much over the wisdom of creating a
national park as over the boundaries of the preserve. No one objected
to basing a park on the existing and isolated monument, but pulp and
paper manufacturers and allied timber owners contended that inclusion
of the hemlock between the mountains and the ocean would ruin the
economy of the Olympic Peninsula. A large park, argued R. D. Merrill, "would sound the death knell to any possible increase in the pulp
industry." That industry became the principal opponent of the park
plan, while lumbermen not directly interested in production of pulp
and paper on the peninsula stood aloof. For instance, Weyerhaeuser,
always sensitive to legislation effecting its interests, ignored the issue
altogether.[38]

In March 1935, Representative Monrad C. Wallgren of Washington
introduced a bill for the creation of Mount Olympus National Park.
Drawn up in consultation with environmentalists, the measure would
add 400,000 acres of national forest to the existing monument. Opposition to the Wallgren boundaries was intense and prevented any
chance of passage. Failure to take into consideration the economy of
the Olympic Peninsula, the bill's critics warned, would prevent any
chance of establishing a sensible park. Over the next three years, the
future of the Olympics would be embroiled in what one participant in
the struggle termed "a seething mass of controversy."[39]

Those favoring a large park looked to the Emergency Conservation
Committee, an organization of eastern environmentalists, for leadership. Chaired by Rosalie Edge and relying on the promotional talents
and political connections of Irving Brant, the committee concentrated
on two tasks. Fearing that the compromising tendency of the Roosevelt
Administration would result in the public interest being "double-crossed,"
it kept up a steady pressure on federal officials. The committee also
dispatched a representative, William Schulz, to Washington state to
organize local supporters of the park. This, wrote Brant, "would avert
the charge that easterners who don't know anything about western conditions are butting in and telling them what to do."[40]

Opponents of the Wallgren bill, while not as well organized, relied
on assistance from the Washington state planning council and the University of Washington forestry school. The anti-park forces charged
that the bill was the work of misguided idealists who couldn't tell the
difference between spruce and hemlock and who would gladly sacrifice the jobs and investments of Olympic Peninsula residents to pro-

vide scenic attractions for effete tourists. The typical park supporter, contended C. S. Cowan of the Washington Forest Fire Association, "regrets that you have but one piece of property he can take for his individual pleasure." Opposition leaders, though, feared that neither invective nor reasoned analysis were likely to overcome public sympathy for environmental preservation.[41]

Hoping to win support for a compromise, Wallgren introduced a new bill in February 1937, reducing the park to 648,000 acres through elimination of the Hoh, Bogachiel and Queets valleys and other regions. "That part which was included in the former bill but not in the present," noted Seattle photographer and outdoorsman Asahel Curtis, a leading advocate of compromise, "is in the lower rain belt and is largely pulpwood species." This went far to satisfy opponents, although they pressed for exclusion of additional hemlock from the park. Preservationists, though, were outraged by the removal of the rain forest valleys. The new bill was "a shocking retreat," contended Irving Brant, and "a capitulation by Congressman Wallgren to the lumbering interests of the peninsula."[42] With this vocal opposition, the issue remained stalled on the big park–small park question.

As in so many squabbles of the New Deal years, the participants looked to President Roosevelt as the final arbiter. The first step in this process was a personal inspection of the Olympic Peninsula by the president in the fall of 1937. News of the impending visit set off a scramble to influence Roosevelt's perception of the situation. Arranging the details of the tour, Forest Service officials removed national forest signs from offensive cutover areas and attempted to bar park supporters from the presidential party. In response, Irving Brant made use of his White House connections to inform Roosevelt of these manipulations and of "the plain purpose" of the Forest Service "to turn government timber over to the Grays Harbor mills."[43] Here, all believed, was the climactic moment of the park fight.

Roosevelt toured the peninsula on September 30 and October 1, traveling by car from Port Angeles along the coast to Grays Harbor amidst heavy rain and confusion over lost baggage. Passing through the Quinault Indian Reservation, just south of the national forest, the president was horrified by what he later recalled as "criminal devastation" on the part of the "lumber interests." With this impression imbedded in his mind, Roosevelt declared himself in favor of a large park. Reading the news, Brant was overjoyed. "With Roosevelt behind the park," he wrote, "there is no danger of the Forest Service running things at Washington, no reason to accept a compromise for fear of not getting anything."[44] The first dispatches, however, failed to convey the complexity of the president's views.

During his tour, Roosevelt had offered a typical off-the-cuff proposal calculated to win over all groups. A large park, he suggested, could be jointly administered by the Forest Service and the Park Service and logging could be allowed. "The President compared the Olympic forests to the Black Forest of Germany," William Schulz informed Brant, "where logging has been carried on, and felt perhaps the same thing could be done here." Although the state planning council endorsed the idea, it soon proved to be a defective trial balloon. If timber cutting was approved, argued park opponents, then there was no need to transfer land from the sole control of the Forest Service. Environmentalists were horrified, believing that logging activities were incompatible with the basic national park concept. "If this is permitted in one park," Brant warned Roosevelt, "it will be an entering wedge for commercial encroachments elsewhere." Any compromise, he advised, must leave the park unsullied by exploitation.[45]

Meetings among Brant, Forest Service and Interior officials, and Representatives Wallgren and Martin Smith, the latter from Grays Harbor, resulted in the drawing of new park legislation. Introduced in March 1938, the third Wallgren bill provided for a park of 898,000 acres, excluded some pulpwood areas and added a strip of land along the wild ocean coast of the peninsula. The measure was too extreme for commercial interests and provoked immediate criticism. The park boundaries have "been changing with the different phases of the moon," complained one opponent, "and God knows what the brilliance of the Eastern Super-Conservationist will try next."[46]

At this point, opponents of the large park played their final card, sending Governor Clarence Martin to the nation's capital to plead for boundary reductions. Meeting with Roosevelt in late April, Martin—who, reported William Schulz, "is considered pretty much of a nitwit"—secured agreement for creation of a smaller park with presidential authority to enlarge the boundaries after "consultation" with state officials. Martin returned to the Northwest convinced that he had produced "a satisfactory solution" to the Olympic question. Upon reflection, though, the opposition began to worry about the exact meaning of the word "consultation" and about the implications of giving Roosevelt the power to determine the final outlines of the park.[47]

With congress moving toward adjournment and Washington's Democratic senators, Homer Bone and Lewis Schwellenbach, hesitant to antagonize powerful constituents, Roosevelt believed that a compromise was necessary to secure passage of park legislation. Following house approval of the Wallgren bill in May, the senate passed a measure creating a park of 648,000 acres and authorizing the president to increase the park to a maximum size of 898,000 acres. Inclusion of these terms in the conference report and elimination of drafting errors—one would have allowed the president to *add* 898,000 acres to the preserve—concluded with final passage of the bill creating Olympic National Park on June 16, 1938. Roosevelt's signature on June 29 ended the first and most emotional stage of the battle over the Olympics.[48]

Over the next year-and-a-half, attention focused on the expected additions to the park. Opponents were willing to extend the boundaries to the north, east, and south as long as the hemlock on the western slopes remained available for exploitation. Environmentalists fought for the inclusion of those pulpwood areas as vital to the integrity of the park. "After all," reflected William Schulz, "the whole fight was really over timber in the west, and to lose that is to accept defeat." On January 2, 1940, President Roosevelt proclaimed the addition of 187,000 acres to the park, including the disputed valleys of the Hoh and Bogachiel rivers. He was convinced, the president informed Governor Martin, that preservation of the westside timber "would not harm the established pulp industries of the peninsula." This contention, needless to say, was not shared by development groups.[49] For those who had opposed the large park, the New Deal had delivered one more blow to the forest products industry of western Washington.

The beginning of the Second World War in September 1939 distracted attention from the final stages of the park fight and brought a calculated response from lumbermen. Speaking for the industry, the *Timberman* called for harsh measures against Japan in protest of the cutoff in trade with China. "The Japanese military powers," argued the trade journal, "must be made to understand the United States . . . will not abjectly permit their rights to be trampled upon." Despite this militance, the publication's editor strenuously opposed American involvement in either the European or the Asian war, pointing out that the military preoccupations of the other great powers would result in "an extended period of prosperity, occasioned by the flow of European gold into American trade channels." United States participation would necessitate the termination of normal civilian business activity and therefore be "a national catastrophe."[50] The lumber industry had suffered enough disruption during the preceding years of depression, labor conflict, and environmental dispute. Against their will, however, lumbermen had to confront a final transformation in their traditional means of operation and, as well, in their importance to the economic scheme of things in the Pacific Northwest.

Epilogue

On the quiet Sunday morning of December 7, 1941, Japanese planes swooped in upon the Pearl Harbor anchorage of the United States Pacific fleet. As in so many areas of American life, the sudden propelling of the nation into the Second World War marked a watershed in the history of western Washington lumbering. For nearly a century—beginning with the first extensive American settlement on Puget Sound—lumbermen had stood at the center of life in the land west of the Cascades, employing most of the workers and supplying the crucial stimulant to the region's economy. Before Pearl Harbor, the economic structure of western Washington was uncomplicated, depending almost entirely on the time-honored exploitation of the forest. During and after the war, the economy expanded and altered into a state of complexity, the rise of new industries shattering the dominance of forest products and, within the forest sector, the further expansion of pulp and paper manufacturing challenging the old-fashioned production of lumber. With America's entrance into the war, the story of Pacific Northwest lumbering passed from the realm of history into the realm of modern times, from the province of the historian to that of the contemporary observer of events.

Despite shortages of logs and labor, forest-related enterprise benefited from the war. During 1941, as the Roosevelt Administration mobilized the nation's economy for defense, lumbermen turned out great quantities of lumber for cantonments, shipyards and aircraft plants. By the time of Pearl Harbor, the camps and mills of the Pacific Northwest were "at battle strength" according to the *Timberman* and three-fifths of production was directed toward the defense effort. Thereafter, national lumber consumption mounted under the pressure of military demand to the highest point since the early 1920s and the lumber price index—with 1926 equalling 100—increased from 93 in 1939 to 122 in 1941 and 155 in 1945. "The war has so thoroughly absorbed all the energies of the West Coast lumber industry," wrote William Greeley in early 1943, ". . . that for practical purposes we are part of the Armed Services."[1]

Money flowing into western Washington from defense contracts, however, brought to an end the economic domination of the region by lumbermen. The growth rate of their industry lagged far behind that of enterprises deemed more essential to national security. Under the stimulus of war mobilization, manufacturing employment in Washington state increased from 115,000 in 1939 to 302,000 in 1944. Between those years, though, the percentage of wage-earners employed in lumbering declined from forty-six percent to seventeen percent. In contrast, shipyard employment increased from one percent of the total to thirty-two percent and aircraft plant employment from four percent to fourteen percent.[2]

Although lumbering briefly regained its dominant position in 1946, the impact of Cold War defense spending and the rapid development of new industries soon reduced lumber manufacturing to the permanent status of a major, rather than the sole, industrial endeavor in western Washington. Between 1947 and 1953, lumbering employment fell by eight percent, compared to a 154 percent increase in aircraft, a forty percent increase in chemicals and a thirty-one percent increase in primary metals. Of every six wage-earners, one worked in an airplane factory. Economic analysts calculated that prior to the war Washington, with its reliance on relatively crude lumber production, was not an industrial state. In the postwar years, however, the economy became more diversified and more oriented toward high technology enterprise.[3]

Within the forest products industry, significant alterations took place in the postwar era. Washington had surrendered first place in lumber production to Oregon in the late 1930s and continued to lose ground to its southern neighbor in the years following the war. Washington turned out sixteen percent of the lumber produced in the eleven western states and British Columbia in 1939, but only ten percent in 1954. High stumpage, labor and shipping costs, moreover, altered traditional market patterns. In the first forty years of the twentieth century, American lumber imports exceeded exports only twice. Between 1941 and 1957, though, imports were greater than exports in all but one year. Dependent on a highly competitive and housing-oriented domestic market, Northwest lumbermen experienced repeated recessions in the postwar decades.[4]

The greatest transformation came in the relative importance of lumber and pulp and paper manufacturing. The trend toward improved utilization of timber resources through production of pulp accelerated after the war. Ninety percent of the logs hauled from the woods in 1925 had gone into lumber and only three percent into pulp. A quarter century later, the lumber share had declined to seventy-one percent and the pulp share increased to sixteen percent. By 1960, the value of pulp produced in Washington actually exceeded the value of lumber.[5] The rising cost of timber and the restricted market for lumber placed pulp at the center of the forest products industry, a placement dramatized by the passing of many oldtime firms into the control of the

new order. At war's end, Rayonier Incorporated purchased the Bloedel-Donovan logging operation on the Strait of Juan de Fuca and the
Polson Logging Company. In subsequent years, such well-known sawmill companies as Bloedel-Donovan, St. Paul & Tacoma, and Long-
Bell were taken over by pulp and paper concerns.

Timber became an increasingly precious commodity in the postwar
years, reflecting the increasing demand for pulp and paper and the
declining stock of privately held old growth. Stumpage prices increased from $5.20 per thousand feet in 1944 to $16.40 in 1950 and
$28.90 in 1955. The rising cost of raw material was, of course, re-
flected in the price of lumber. Douglas-fir lumber sold for $66 per
thousand feet by 1965 and then mounted in breathtaking fashion to
$152 in 1973 and to $206 in early 1977. With the war, the long historical period of relatively cheap timber and lumber came to an end,
to be succeeded by a new epoch of rapidly escalating values.[6]

As the value of timber increased, so did the interest of lumbermen
in forestry. Under the auspices of the West Coast Lumbermen's Association, two million acres of logged-off land were organized into
private timber farms during the Second World War. Founded in 1943,
for instance, the South Olympic Tree Farm Company combined 176,000
acres of Simpson and Weyerhaeuser land for purposes of intensive fire
protection and regrowth. A cooperative industry nursery, set up near
Nisqually on Puget Sound, was growing six million seedlings by the
end of the war. "Many of our hardheaded lumbermen," observed William Greeley in mid-1946, "have become convinced that it pays to
grow trees."[7] Increased prices and decreased supplies of privately owned
timber meant that there was a considerable financial incentive for reforestation.

Depleted private holdings and the waiting period for regrowth forced
increased reliance on Forest Service timber. Concentrated in the Pacific Northwest, federal timber sales doubled during the war and continued to grow in subsequent years, reaching 6.2 billion feet in 1955.
Except for Weyerhaeuser, all of the firms in the industry were to some
extent dependent on purchases from the Forest Service. This development was accompanied by an end to the feud over regulation. "The
Forest Service is recovering from its 'regulation fever,'" wrote Greeley in December 1947, "and is headed back toward a more cooperative
attitude toward forest industry." Pointing out that "we just have too
much at stake in our dependence upon the federal timber," Greeley
urged the maintenance of close relations between the industry and government foresters.[8]

Both the dependence of industry and the renewed spirit of cooperation were reflected in passage of the Sustained-Yield Forest Management Act in early 1944. The legislation authorized creation and

joint Forest Service–private management of sustained yield units as a means of insuring the economic stability of small towns dependent on adjacent timber resources. Federal timber within each unit would be reserved for sale to the participating firm, eliminating the competitive factor and its attendant uncertainties. Under the act, the Shelton Sustained-Yield Unit agreement, concluded in December 1946, provided for cooperative administration of 270,000 acres of Simpson Logging Company land, most of it cutover, and Forest Service timber.[9]

Although opposition from small timber companies and from organized labor precluded additional agreements, in the Pacific Northwest or elsewhere in the country, other actions testified to the value of industry-government cooperation. Earnings from timber sales were made subject to capital gains taxation, rather than the income tax, by congressional action in 1944. Developed during the postwar era, the Forest Service's multiple use approach to land management reduced the likelihood that public timber would be withheld from exploitation. At the state level, decades of industry action culminated in success in the early 1970s when the yield tax was extended to all private timber holdings, bringing substantial property tax savings in the name of practical forestry.

To environmentalists, an increasingly vocal and influential element in the postwar years, the close relationship between business and government was inimical to the public interest. The Forest Service, charged Ralph Nader in 1974, had long since abandoned conservation in favor of turning the nation's forests into "timber factories." The profit-oriented efforts of timber owners to apply forestry received only faint praise, with Weyerhaeuser being immortalized as the "Best of the S.O.B.s." That company, observed one wilderness advocate, was "managed by men who believe they can grow trees better than nature can." Analysts pointed with alarm to the fact that forest reproduction was lagging far behind timber removal and to the increasing rate of Pacific Northwest log exports.[10] In the postwar age of timber scarcity, the fate of the forested land became a matter of intense public debate. The timber of the modern Northwest was too important a resource to be left to lumbermen and foresters.

The modern lumbermen of western Washington operate in a more complex economic, political and legal setting than that of their forebears. Moving beyond the traditional production of lumber, the region's forest concerns turn out pulp and plywood and other byproducts of the age of timber utilization. Gone are the old-fashioned mill towns, the symbols of the era of wasteful exploitation of natural resources. At Port Blakely, all signs of what had once been one of the largest sawmill complexes in the world have long since disintegrated and only a few houses line the shore of the deserted bay. Port Ludlow has become a resort, complete with condominiums, golf course and yacht harbor. Only at Port Gamble, where Pope & Talbot began the production of lumber in 1853, does the modern resident of Washington find industrial activity.

From the earliest Indians through the colorful lumber barons of the late nineteenth century to the modern corporate executive and unionized mill worker, human inhabitants of the land west of the Cascades have depended on the forest. The magnificent Douglas-fir, the decay-resistant cedar, the light and supple spruce, and the once-despised hemlock have provided shelter and sustenance from the time of whaling and barter to the time of pulp and computers. Indians took to the forest with hatchet and adze, the early white loggers with axes, oxen, and rickety logging railroads, and the modern men and women of the woods with chainsaws, tractors, and powerful trucks. In more recent times, a growing segment of the population has found in the forested wilderness a necessary respite from the pressures of life in urbanized America. Northwesterners have always lived in close economic and psychic dependence on their natural surroundings.

Late-twentieth-century residents of the Northwest are fortunate that largescale lumbering came late to their region. Concern for protection of the environment, including the self-interested concern exhibited by timber owners, became of paramount importance before the forests of Washington and Oregon had been devastated and the Northwest thereby avoided the fate of the Great Lakes states. The historian Bruce Catton, growing up in the cutover region of northern Michigan, recalled that "the bigger part of a state had been treated, not as a region where people might happily live but as an expendable resource."[11] Washington west of the Cascades would be treated, as even lumbermen at the end of their trek across the continent conceded, not as an expendable resource but as a part of the country with an environmental future. Scenic and wilderness areas were preserved in Rainier, Olympic, and North Cascades national parks, the last created in 1968. Logging practices, it was true, continued to produce depressing vistas of stumps and slash, offending the sensibilities of even those who recognized the economic necessity of timber exploitation. The environment of modern Washington produced profit for investors, wages for workers, and aesthetic satisfaction for lovers of nature. With recognition that it paid to grow trees, there was reason for guarded hope for the future of the forest.

Notes

In order to prevent the notes from overwhelming the text, citations are limited to quotations and to immediate sources of information and interpretation. The bibliography contains a complete list of sources used in the writing of this book.

1: The Best Ever Planted

1. *Journal Kept by David Douglas during His Travels in North America, 1823–1827* (London, 1914), 103.
2. John Muir, "The American Forests," *Atlantic Monthly* 80(August 1897):145.
3. C. M. Scammon, "Lumbering in Washington Territory," *Overland Monthly* 5(July 1870):55.
4. James Stevens, *Paul Bunyan* (Garden City, N.Y., 1925), 170–87.
5. John Muir to James Davie Butler, 1 September 1889, in William Frederic Badé, *The Life and Letters of John Muir*, 2 vols. (Boston, 1924), 2:232.
6. John Muir, *Travels in Alaska* (Boston, 1915), 8; J. W. Boddam-Whetham, *Western Wanderings* (London, 1874), 283–84.
7. Sinclair Lewis, *Free Air* (New York, 1919), 251.
8. George M. Colvocoresses, *Four Years in a Government Exploring Expedition* (New York, 1852), 252. Willapa Bay guide quoted in Jean Hazeltine, *The Historical and Regional Geography of the Willapa Bay Area* (South Bend, Wa., 1956), 11.
9. George H. Emerson to C. H. Jones, 27 November 1906, George H. Emerson Letterbooks, Manuscripts Collection, University of Washington Library.
10. Ernest Ingersoll, "From the Fraser to the Columbia—Part II," *Harper's New Monthly Magazine* 68(May 1884):872; F. I. Vassault, "Lumbering in Washington," *Overland Monthly* 20(July 1892):24.
11. Scammon, "Lumbering in Washington Territory," 55; Michael F. Luark Diary, 31 August 1853, Manuscripts Collection, University of Washington Library; Address to Pacific Coast Lumber Manufacturers Association, 28 February 1908, Emerson Letterbooks.
12. Archie Binns, *The Laurels Are Cut Down* (New York, 1937), 142–43.
13. Quoted in Athelston George Harvey, *Douglas of the Fir* (Cambridge, Mass., 1947), 57–58.
14. Information on tree species derived from Donald Culross Peattie, *A Natural History of Western Trees* (Boston, 1953), 147–54, 157–61, 169–81, 217–24; Edmond S. Meany, Jr., "The History of the Lumber Industry in the Pacific Northwest to 1917" (Ph.D. diss., Harvard University, 1935), 23–36; Washington State Division of Forestry, *Forest Resources of Washington* (Olympia, 1940), 7, 10.
15. Alexis de Tocqueville, *Democracy in America* (New York, 1838), 8–9.

16. James G. Swan, "The Indians of Cape Flattery," *Smithsonian Contributions to Knowledge* (Washington, D.C., 1870), 36.

17. Ibid., 4–5.

18. Myron Eells, "The Twana, Chemakum, and Klallam Indians, of Washington Territory," *Annual Report of the Board of Regents of the Smithsonian Institution, 1887* (Washington, D.C., 1889), 614–15, 641–42; George Gibbs, *Tribes of Western Washington and Northwestern Oregon* (Washington, D.C., 1877), 218.

19. Richard White, *Land Use, Environment, and Social Change: The Shaping of Island County, Washington* (Seattle, 1980), 9–11, 19–25. George Vancouver, *A Voyage of Discovery to the North Pacific Ocean, and Round the World*, 3 vols. (London, 1798), 1:254.

20. J. B. Tyrrell, ed., *David Thompson's Narrative of His Explorations in Western America, 1784–1812* (Toronto, 1916), 502, 507; Josiah Keller to Charles Foster, 29 September 1854, Josiah Keller Papers, Western Americana Collection, Beinecke Library, Yale University.

2: So Delightful a Prospect

1. George Vancouver, *A Voyage of Discovery to the North Pacific Ocean, and Round the World*, 3 vols. (London, 1798), 1:210.

2. Frederick Merk, ed., *Fur Trade and Empire: George Simpson's Journal*, rev. ed. (Cambridge, Mass., 1968), 123–24.

3. J. C. Beaglehole, ed., *The Journals of Captain James Cook on His Voyages of Discovery* (Cambridge, Eng., 1967), Vol. 3, *The Voyage of the Resolution and Discovery, 1776–1780*, 293–94.

4. Quoted in W. A. Carrothers, "Forest Industries of British Columbia," in A. R. M. Lower, *The North American Assault on the Canadian Forest* (Toronto, 1938), 254.

5. John Meares, *Voyages Made in the Years 1788 and 1789, from China to the North West Coast of America*, 2 vols. (London, 1791), 1:253–54.

6. Vancouver, *Voyage of Discovery*, 1:227–28; C. F. Newcombe, ed., *Menzies' Journal of Vancouver's Voyage, April to October 1792* (Victoria, 1923), 48.

7. Vancouver, *Voyage of Discovery*, 1:258–59; Bern Anderson, ed., "The Vancouver Expedition: Peter Puget's Journal of the Exploration of Puget Sound, May 7–June 11, 1792," *Pacific Northwest Quarterly* 30(April 1939):191.

8. F. W. Howay, ed., *Voyages of the 'Columbia' to the Northwest Coast, 1787–1790 and 1790–1793*, (Boston: Massachusetts Historical Society Collections, 1941), 435–38.

9. Vancouver, *Voyage of Discovery*, 1:420.

10. F. W. Howay, "Early Days of the Maritime Fur Trade on the Northwest Coast," *Canadian Historical Review* 4(March 1923):34–35; Archibald Campbell, *A Voyage round the World, From 1806 to 1812* (Edinburgh, 1816), 152–3.

11. Reuben Gold Thwaites, ed., *Original Journals of the Lewis and Clark Expedition, 1804–1806*, 8 vols. (New York, 1905), 3:279, 285–86.

12. Kenneth A. Spaulding, ed., *On the Oregon Trail: Robert Stuart's Journal of Discovery* (Norman, Okla., 1953), 29.

13. Reuben Gold Thwaites, ed., *Ross's Adventures of the First Settlers on the Oregon or Columbia River, 1810–1813* (Cleveland, 1904), 90–92.

14. F. W. Howay, "Brig Owhyee in the Columbia, 1827," *Oregon Historical Quar-*

terly 34(December 1923):328; Peter Corney, *Early Voyages in the North Pacific, 1813–1818* (1896, reprint, Fairfield, Wa., 1965), 118.

15. *Annals of Congress,* 16th cong., 2nd sess., 955; 17th cong., 2nd sess., 46.

16. *Reg. of Debates in Congress,* 18th cong., 2nd sess., 695.

17. Merk, ed., *Fur Trade and Empire,* 123–24.

18. E. E. Rich, ed., *Part of a Dispatch from George Simpson . . . to the Governor and Committee of the Hudson's Bay Company, London, March 1, 1829* (Toronto, 1947), 84–89; John McLoughlin to Gov. and Com., 5 August 1829, in E. E. Rich, ed., *The Letters of John McLoughlin from Fort Vancouver, First Series, 1825–38* (Toronto, 1941), 77.

19. George Simpson to Aemelius Simpson, ca. October 1828, in Merk, ed., *Fur Trade and Empire,* 298; Rich, ed., *Part of a Dispatch,* 86–89.

20. Herman Leader, ed., "A Voyage from the Columbia to California in 1840: From the Journals of Sir James Douglas," *California Historical Society Quarterly* 8(June 1929):112.

21. Richard Henry Dana, Jr., *Two Years before the Mast* (New York, 1842), 98; Thomas R. Cox, *Mills and Markets: A History of the Pacific Coast Lumber Industry to 1900* (Seattle, 1974), 13–20.

22. Simpson to McLoughlin, 10 July 1830, in Merk, ed., *Fur Trade and Empire,* 327; McLoughlin to Richard Charlton, 18 November 1830, in Bert Brown Barker, ed., *Letters of Dr. John McLoughlin, Written at Fort Vancouver, 1829–1832* (Portland, 1948), 157.

23. A company accountant later calculated that trade with Hawaii earned an average annual profit of $10,000 down to 1846. Simpson to Gov. and Com., 1 March 1842, in Glyndwar Williams, ed., *London Correspondence Inward from Sir George Simpson, 1841–42* (London, 1973), 127; Memoranda for Chief Factor Tolmie, 17 April 1866, Hudson's Bay Co. Misc. Records, Oregon Historical Society.

24. McLoughlin to Gov. and Com., 30 November 1835 and 31 October 1837, in Rich, ed., *McLoughlin Letters, 1825–38,* 139, 204, 206.

25. James Douglas to Simpson, 18 March 1838; to Gov. and Com., 18 October 1838, in Ibid., 259–60, 285; Thomas J. Farnham, *Travels in the Great Western Prairies, the Anahuac and Rocky Mountains, and in the Oregon Territory* (New York, 1843), 103.

26. Samuel Parker, *Journal of an Exploring Tour Beyond the Rocky Mountains,* 2nd ed. (Ithaca, N.Y., 1840), 144; Farnham, *Travels,* 91; John Charles Frémont, *Memoirs of My Life* (Chicago, 1887), 278–79.

27. Simpson to H. U. Addington, 5 January 1826; to Gov. and Com., 5 March 1830, in Merk, ed., *Fur Trade and Empire,* 264, 322.

28. William F. Tolmie Journal, 16, 26 and 28 August 1833, Bancroft Library, University of California, Berkeley; Fort Nisqually Journal of Occurences, 27 December 1833, Manuscripts Collection, University of Washington Library; Simpson to McLoughlin, 25 June 1836, in "Fort Langley Correspondence," *British Columbia Historical Quarterly* 1(July 1937):188–89.

29. Douglas to Gov. and Com., 14 October 1839, in E. E. Rich, ed., *The Letters of John McLoughlin from Fort Vancouver to the Governor and Committee, Second Series, 1839–44* (Toronto, 1943), 217; Simpson to John Henry Pelly and Andrew Colvile, 25 November 1841, in Williams, ed., *London Correspondence,* 99–100.

30. Douglas to Simpson, 5 March 1845, in E. E. Rich, ed., *The Letters of John McLoughlin from Fort Vancouver to the Governor and Committee, Third Series, 1844–46* (Toronto, 1944), 185–86.

31. Douglas to William F. Tolmie, 4 November 1846 and 19 April 1847, William F. Tolmie Papers, Manuscripts Collection, University of Washington Library. British

foreign secretary quoted in James O. McCabe, "Arbitration and the Oregon Question," *Canadian Historical Review* 41(December 1960):308.

32. Douglas to Simpson, 4 April 1845, in Rich, ed., *McLoughlin Letters, 1844–46*, 190.

33. Twentieth-century Northwest lumbermen recognized John McLoughlin as the founder of their industry. See, for example, *Timberman* 22(May 1921):30–31, 33–34.

3:   Gold and Lumber

1. Alfred Hall to brother and sister, 1 January 1853, Alfred Hall Correspondence, Bancroft Library, University of California, Berkeley.

2. Ezra Meeker, *Pioneer Reminiscences of Puget Sound* (Seattle, 1905), 34.

3. Arthur A. Denny, *Pioneer Days on Puget Sound* (Seattle, 1888), 13, 16–17.

4. Thomas R. Cox, *Mills and Markets: A History of the Pacific Coast Lumber Industry to 1900* (Seattle, 1974), 23–25; Edmond S. Meany, Jr., "The History of the Lumber Industry in the Pacific Northwest to 1917" (Ph.D. diss., Harvard University, 1935), 80–82, 87.

5. Cox, *Mills and Markets*, 26–28; Meany, "History of the Lumber Industry," 79–80; Michael F. Luark Diary, 24 October 1855, Manuscripts Collection, University of Washington Library.

6. Gustavus Hines, *Oregon: Its History, Condition and Prospects* (Buffalo, 1851), 340–41; "Diary of an Emigrant of 1845," *Washington Historical Quarterly* 1(April 1907):156.

7. Later residents adopted the name Tumwater for the settlement. William F. Tolmie Reminiscences, Bancroft Library, University of California, Berkeley, 10; Antonio B. Rabbeson Reminiscences, ibid., 10; Notes Copied From the Hudson's Bay Company's Account Books at Fort Nisqually, Northwest Collection, University of Washington Library, 17, 19–22.

8. William F. Tolmie to James Douglas, 14 November 1848, in Fort Nisqually Journal of Occurrences, Manuscripts Collection, University of Washington Library; Rabbeson Reminiscences, 10–11; Samuel Hancock, *The Narrative of Samuel Hancock, 1845–1860* (New York, 1927), 58; "Edmund Sylvester's Narrative of the Founding of Olympia," *Pacific Northwest Quarterly* 36(October 1945):334–35.

9. "Sylvester's Narrative," 335, 337; Fort Nisqually Journal, 1 January 1850.

10. James Fenimore Cooper to Mrs. Cooper, 27 February 1849, in James Franklin Beard, ed., *The Letters and Journals of James Fenimore Cooper*, 6 vols. (Cambridge, Mass., 1960–1968), 6:8.

11. Brian H. Smalley, "Some Aspects of the Maine to San Francisco Trade, 1849–1850," *Journal of the West* 5(October 1967):594–96.

12. Walter Colton, *Three Years in California* (New York, 1850), 414; Marco G. Thorne, ed., "Bound for the Land of Canaan Ho!: The Diary of Levi Stowell," *California Historical Society Quarterly* 27(March 1948):44.

13. Andrew J. Pope to Edwin Pope, 13 December 1850, Andrew J. Pope Papers, Pope & Talbot Archives.

14. Miss A. J. Allen, comp., *Ten Years in Oregon: Travels and Adventures of Doctor E. White and Lady, West of the Rocky Mountains* (Ithaca, N.Y., 1850), 51; William Penn Abrams Diary, 11–14 April 1851, Bancroft Library, University of California, Berkeley.

15. A. S. Mercer, *Washington Territory* (Utica, N.Y., 1865), 8; George Pope to Wm. Pope and Sons, 4 March 1861, Pope Papers.

16. Quoted in Robert W. Vinnedge, *The Pacific Northwest Lumber Industry and Its Development* (New Haven, 1923), 9.

17. DeLancey Floyd-Jones to Sarah Floyd-Jones, 31 April 1853, Floyd-Jones Family Papers, Bancroft Library, University of California, Berkeley; Meeker, *Pioneer Reminiscences*, 51–53, 73–74.

18. John R. Finger, "Henry L. Yesler's Seattle Years, 1852–1892" (Ph.D. diss., University of Washington, 1968), 13–17. Olympia newspaper quoted in Clarence B. Bagley, *History of Seattle, From the Earliest Settlement to the Present Time*, 3 vols. (Chicago, 1916), 1:30.

19. Finger, "Yesler's Seattle Years," 24–26, 28, 40, 55–56, 61–63; Kate P. Blaine to Father and Mother Blaine, 30 October 1854, David E. and Kate P. Blaine Letters, Northwest Collection, University of Washington Library.

20. Henry Roder Reminiscences, Bancroft Library, University of California, Berkeley, 8–12; Edward Eldridge Reminiscences, ibid., 2–4.

21. Meeker, *Pioneer Reminiscences*, 53; Charles Prosch, *Reminiscences of Washington Territory* (Seattle, 1904), 13–14.

22. Phoebe Goodell Judson, *A Pioneer's Search for an Ideal Home* (Bellingham, 1925), 97; Meeker, *Pioneer Reminiscences*, 35.

23. Meeker, *Pioneer Reminiscences*, 34–35.

24. Luark Diary, 4 and 5 September, 8–26 October, and 3 and 19 December 1853; 5 April–15 June 1854.

4: The Live Yankee

1. "Business Broadside of 1853," *Washington Historical Quarterly* 20(July 1929): 229–30.

2. Quoted in Robert W. Johannsen, *Frontier Politics on the Eve of the Civil War* (Seattle, 1955), 7.

3. William C. Talbot to Charles Foster, 4 September 1853, Josiah P. Keller Papers, Western Americana Collection, Beinecke Library, Yale University; Edwin T. Coman, Jr. and Helen M. Gibbs, *Time, Tide and Timber: A Century of Pope & Talbot* (Stanford, Cal., 1949), Chap. 5.

4. W. T. Sayward Reminiscences, Bancroft Library, University of California, Berkeley, 34; H. H. Bancroft, *Chronicles of the Builders of the Commonwealth*, 7 vols. (San Francisco, 1892), 4:621–24; Iva L. Buchanan, "An Economic History of Kitsap County, Washington, to 1889" (Ph.D. diss., University of Washington, 1930), 69–70.

5. Thomas R. Cox, *Mills and Markets: A History of the Pacific Coat Lumber Industry to 1900* (Seattle, 1974), 63, 108, 110; Thomas F. Gedosch, "Seabeck, 1857–1886: The History of a Company Town" (M.A. thesis, University of Washington, 1967), 1–7; J. Ross Browne to James Guthrie, 4 September 1854, in David Michael Goodman, *A Western Panorama, 1849–1875: The Travels, Writings and Influence of J. Ross Browne* (Glendale, Cal., 1966), 58.

6. Charles Foster remained in Maine and did not become active in the company. Frederic Talbot, Andrew Pope's original partner, had by this time returned to New England. Between 1855 and 1862, the California outlet was known as W. C. Talbot & Company. Pope & Talbot is used in the text to minimize confusion. Puget Mill Co.

Agreement, 20 December 1852, Keller Papers; Coman and Gibbs, *Time, Tide and Timber*, 31–47, 355–57.

7. Josiah P. Keller to Foster, 7 September and 9 October 1853; 4 March 1854; 15 January 1855, Keller Papers; James S. Lawson Reminiscences, Bancroft Library, University of California, Berkeley, 76–77.

8. Keller to Foster, 25 November 1853; 4 March 1854; 15 January 1855; 2 January 1856, Keller Papers; Andrew J. Pope to Wm. Pope & Sons, ? April 1855; 4 and 19 February 1856, Andrew J. Pope Papers, Pope & Talbot Archives; Iva L. Buchanan, "Lumbering and Logging in the Puget Sound Region in Territorial Days," *Pacific Northwest Quarterly* 27 (January 1936):38–39.

9. Pope to Pope & Sons, 2 July, ? September and 18 November 1857, Pope Papers; Keller to Foster, 2 January 1856, Keller Papers; Coman and Gibbs, *Time, Tide and Timber*, 65–66.

10. R. G. Dun & Co. Credit Reports, 15:117; 16:156, Harvard Business School Library.

11. Pope to Pope & Sons, 19 January 1861, Pope Papers; Edwin G. Ames to W. H. Talbot and Frederick Talbot, 2 May 1912, Pope & Talbot Records, Ames Coll., Manuscripts Collection, University of Washington Library; Coman and Gibbs, *Time, Tide and Timber*, 93–94, 116.

12. Arthur A. Denny, *Pioneer Days on Puget Sound* (Seattle, 1888), 40; R. G. Dun & Co. Credit Reports, 14:155.

13. Edith Sanderson Redfield, *Seattle Memories* (Boston, 1930), 31; Caroline C. Leighton, *Life at Puget Sound* (Boston, 1884), 26.

14. J. Murphy, "Summer Ramblings in Washington Territory," *Appletons' Journal* 3(November 1877):393; Edward Clayson, Sr., *Historical Narratives of Puget Sound: Hoods Canal, 1865–1885* (Seattle, 1911), 5; Archie Binns, *The Timber Beast* (New York, 1944), 221–22.

15. Clayson, *Historical Narratives*, 11; Leighton, *Life at Puget Sound*, 27–30.

16. Ames to F. Talbot, 2 October 1913, Pope & Talbot Records; Clayson, *Historical Narratives*, 6–7; Keller to Foster, 5 August 1855, Keller Papers.

17. Keller to Foster, 9 October 1853, Keller Papers; Log of the Mayflower, 9 September 1859, New York Public Library; Frederick J. Yonce, "Public Land Disposal in Washington" (Ph.D. diss., University of Washington, 1969), 229.

18. Adams, Blinn & Co. to Washington Mill Co., 9 November 1857 and 5 April 1858, Washington Mill Co. Records, Manuscripts Collection, University of Washington Library; Pope to Pope & Sons, 21 February 1859, Pope Papers; Gedosch, "Seabeck," 55–56, 59, 63–65; Buchanan, "Economic History of Kitsap County," 117–20.

19. N. A. Batchelder to Glidden & Williams, 1 April 1858, Pope Papers; Log of the Mayflower, 12–30 September 1859; W. B. Seymore, "Port Orchard Fifty Years Ago," *Washington Historical Quarterly* 8(October 1917):259.

20. Pope to Pope & Sons, 2 and 18 July 1857, Pope Papers; Keller to Foster, 3 May 1861, Keller Papers; Coman and Gibbs, *Time, Tide and Timber*, 86–89; John S. Hittell, *The Commerce and Industries of the Pacific Coast*, 2nd ed. (San Francisco, 1882), 592–93.

21. Quoted in Edmond S. Meany, Jr., "The History of the Lumber Industry in the Pacific Northwest to 1917" (Ph.D. diss., Harvard University, 1935), 110–12.

22. Keller to Foster, 24 April 1854, Keller Papers; Gedosch, "Seabeck," 83–95; Buchanan, "Economic History of Kitsap County," 128–31.

23. *McCormick's Almanac, 1870* (Portland, 1870), 30–34; *Ninth Census* (Washington, D.C., 1872), 1:71, 283; Buchanan, "Economic History of Kitsap County," 2, 108–12, 133–39.

24. *Report of the Commissioner of the General Land Office, 1868* (Washington, D.C., 1868), 344; *Ninth Census*, 3:580.

25. Thomas Lane to James Birnie, 23 May 1853, James Birnie Papers, Oregon Historical Society; Keller to Foster, 29 September, 29 October, and 21 November 1854; 11 April 1855, Keller Papers.

26. Cox, *Mills and Markets,* 78–81; Pope to Pope & Sons, 31 March, 30 April and ? 1855; 19 February 1856; 19 June 1860, Pope Papers; Keller to Foster, 24 April 1854 and 5 June 1856, Keller Papers; Port of Honolulu Records, 1855, Hawaii State Archives.

27. Pope to Pope & Sons, 30 September 1852; ? February 1855; 21 July 1860, Pope Papers; Keller to Foster, 29 October 1854; 26 August 1855; 1 March 1858; 4 January 1862, Keller Papers; Talbot & Co. to Pope & Sons, 19 March 1859, William C. Talbot Papers, Pope & Talbot Archives; Adams, Blinn & Co. to Washington Mill Co., 5 November 1858, Washington Mill Co. Records; Cox, *Mills and Markets,* 95–96.

28. In demonstration of the volatility of the Latin trade, five cargoes were shipped to Peru from Puget Sound in both 1870 and 1874. A total of 59 were dispatched in the intervening three years, however, as boom followed upon bust and bust upon boom. Cox, *Mills and Markets,* 82–87, 97–98; Coman and Gibbs, *Time, Tide and Timber,* 80; Puget Sound Customs District Records, RG 36, Federal Records Center, Seattle.

29. Pope to Pope & Sons, 15 September and 4 November 1857; 19 October 1860, Pope Papers; Adams, Blinn & Co. to Washington Mill Co., 16 October 1857; 12 March, 5 April, 8 September, 20 October and 5 November 1858, Washington Mill Co. Records; Cox, *Mills and Markets,* 87–88, 118–20.

30. Keller to Foster, 1 and 11 November 1855, Keller Papers; Pope to Pope & Sons, 19 March 1856, Pope Papers.

31. Keller to Foster, 11 November and 31 December 1855; 17 March and 9 May 1856, Keller Papers; Charles M. Gates, ed., *Messages of the Governors of the Territory of Washington to the Legislative Assembly, 1854–1889* (Seattle, 1940), 26–27; Pope to Pope & Sons, 4 March 1856, Pope Papers.

32. Keller to Foster, 17 March, 9 May, and 5 June 1856, Keller Papers.

33. Keller to Foster, 11 November 1855 and 6 July 1856, Keller Papers.

34. Gates, *Messages of the Governors,* 28; William N. Bell Reminiscences, Bancroft Library, University of California, Berkeley, 19; Keller to Foster, 6 July 1856 and 14 April 1857, Keller Papers.

35. *Ninth Census,* 3:612–13; Meany, "History of the Lumber Industry," 121–29; Cloice R. Howd, "Development of the Lumber Industry of West Coast," *Timberman* 25(August 1924):194; Adams, Blinn & Co. to Washington Mill Co., 9 March 1858, Washington Mill Co. Records.

36. W. Kaye Lamb, "Early Lumbering on Vancouver Island, 1844–1855," *British Columbia Historical Quarterly* 2(January 1938):43–46, and "Early Lumbering on Vancouver Island, 1855–1866," *British Columbia Historical Quarterly* 2(April 1938), 101–10; F. W. Howay, "Early Shipping in Burrard Inlet, 1863–1870," *British Columbia Historical Quarterly* 1(January 1937):4–16.

37. W. A. Carrothers, "Forest Industries of British Columbia," in A. R. M. Lower, *The North American Assault on the Canadian Forest* (Toronto, 1938), 266–67; Robert B. Cranston, "The Forests and Forest Industries of British Columbia" (M.F. thesis, University of Washington, 1952), 75–76; Lamb, "Early Lumbering on Vancouver Island, 1855–1866," 118–21.

38. Meany, "History of the Lumber Industry," 121–22.

5: The Loose Laws

1. Henry George, Jr., ed., *The Complete Works of Henry George,* 10 vols. (New York, 1904), 8:89.

2. George S. Long to E. T. Allen, 29 September 1922, Weyerhaeuser Timber Co. Records, Weyerhaeuser Co. Archives.

3. Hazard Stevens to J. W. Sprague, 23 August 1871, Hazard Stevens Papers, University of Oregon Library.

4. In 1865, Burrard Inlet lumberman Edward Stamp contended that "the best trees on the American side have already been cut." Andrew J. Pope to Wm. Pope & Sons, 20 June 1858, Andrew J. Pope Papers, Pope & Talbot Archives; Edward Stamp to Governor Seymour, 30 August 1865, Hastings Saw Mill Records, University of British Columbia Library.

5. Pope to Pope & Sons, 30 November 1860, Pope Papers; Roy Robbins, *Our Landed Heritage* (Princeton, N.J., 1942), Chaps. 1–7.

6. Josiah Keller to Charles Foster, 3 May and 26 November 1861; 22 April 1862, Josiah Keller Papers, Western Americana Collection, Beinecke Library, Yale University; Pope to Pope & Sons, 30 April 1861, Pope Papers; Daniel Bagley to Marshall Blinn, 5 November 1862; 19 May and 26 June 1863; University Land Sale Record Book, Daniel Bagley Papers, Manuscripts Collection, University of Washington Library; Charles M. Gates, "Daniel Bagley and the University of Washington Land Grant, 1861–1868," *Pacific Northwest Quarterly* 52(April 1961):58–9, 62.

7. J. M. Edmunds to John McGilvra, 16 February 1863, General Land Office Records, RG 49, National Archives; Cyrus Walker to Blinn, 17 January 1863; Arthur A. Denny to Blinn, 26 May 1863, Washington Mill Co. Records, Manuscripts Collection, University of Washington Library; Gates, "Bagley," 60–64; Frederick J. Yonce, "Public Land Disposal in Washington" (Ph.D. diss., University of Washington, 1969), 175–79. Seattle newspaper quoted in Roy M. Robbins, "The Federal Land System in an Embryo State," *Pacific Historical Review* 4(December 1935):371.

8. Military bounty warrants and agricultural college scrip were also used to some extent for land purchases. Pope to Pope & Sons, 4 April 1863, Pope Papers; R. G. Dun & Co. Credit Reports, 4:108, Harvard Business School Library; Edwin G. Ames to W. H. Talbot, 21 December 1926, Pope & Talbot Records, Ames Coll., Manuscripts Collection, University of Washington Library; Edmond S. Meany, Jr., "The History of the Lumber Industry in the Pacific Northwest to 1917" (Ph.D. diss., Harvard University, 1935), 182–92.

9. Meany, "History of the Lumber Industry," 196–201; Paul W. Gates, "The Homestead Law in an Incongruous Land System," *American Historical Review* 41(July 1936):667–68.

10. Ivan C. Doig, "John J. McGilvra: The Life and Times of an Urban Frontiersman, 1827–1903" (Ph.D. diss., University of Washington, 1969), 40–41, 49, 54; Keller to Foster, 28 August 1861, Keller Papers.

11. Doig, "McGilvra," 54–55; Walker to Washington Mill Co., 17 January 1863, Washington Mill Co. Records.

12. Ames to W. Talbot and Frederick Talbot, 20 March 1912, Pope & Talbot Records; Meany, "History of the Lumber Industry," 169–70.

13. Doig, "McGilvra," 54–60; Edmunds to Reg. and Rec., Olympia, 17 January 1863; to McGilvra, 16 February and 19 May 1863; 15 August 1865, General Land Office Records.

14. Edmunds to McGilvra, 19 January, 16 February, and 19 May 1863; 15 August and 20 November 1865; to Reg. and Rec., Olympia, 17 January 1863, General Land Office Records; Doig, "McGilvra," 60–62; Blinn to McGilvra, 7 June and 16 July 1863, John J. McGilvra Papers, Manuscripts Collection, University of Washington Library; McGilvra to Blinn, 20 June 1863, Washington Mill Co. Records.

15. Edmunds to McGilvra, 15 August 1865; to Denny, 11 January 1866; to Leander

Homes, 3 February 1866; Joseph S. Wilson to Holmes, 6 October 1866 and 6 November 1868, General Land Office Records; McGilvra to Blinn, 20 October 1865; Denny to Blinn, 15 April 1866, Washington Mill Co. Records.

16. Eldridge Morse Notebooks, 23:27–28, Bancroft Library, University of California, Berkeley; Meany, "History of the Lumber Industry," 170–72; *Annual Report of the Commissioner of the General Land Office, 1875* (Washington, D.C., 1875), 10.

17. Ross R. Cotroneo, "The History of the Northern Pacific Land Grant, 1900–1952" (Ph.D. diss., University of Idaho, 1966), 47.

18. Sprague to Stevens, 1 April 1871, Stevens Papers; Meany, "History of the Lumber Industry," 217–18.

19. Ironically, the Northern Pacific itself was guilty of stealing timber from public land for use in construction. Sprague to Stevens, 1 April and 1 August 1871; 26 February 1872, Stevens Papers; Morse Notebooks, 23:31–34; Harold K. Steen, "Forestry in Washington to 1925" (Ph.D. diss., University of Washington, 1969), 46–47.

20. Sprague to Stevens, 1 and 16 August 1871; Stevens to Sprague, 9 August 1871, Stevens Papers; Willis Drummond to C. Delano, 27 April 1871; to Holmes, 7 April and 20 June 1871, General Land Office Records.

21. Stevens to Sprague, 9 and 26 August and 21 October 1871; 5 January 1872; Sprague to Stevens, 9 August and 16 September 1871, Stevens Papers.

22. Stevens to Sprague, 19 and 23 August and 12 September 1871, Stevens Papers.

23. Drummond to Reg. and Rec., Olympia, 2 May 1872; to Delano, 22 March 1873; W. M. Curtis to Reg. and Rec., Olympia, 2 November 1872, General Land Office Records; Stevens to Sprague, 16 March and 19 April 1872; Sprague to Stevens, 20 July and 20 August 1872; 19 April 1873, Stevens Papers.

24. Sprague to Stevens, 16 June 1873; N. P. Jacobs to Stevens, 8 July and 29 August 1873, Stevens Papers.

25. Jenks Cameron, *The Development of Governmental Forest Control in the United States* (Baltimore, 1928), 172–3; Morse Notebooks, 23:34–35.

26. U.S. Department of Agriculture, *Report Upon Forestry,* 4 vols. (Washington, D.C., 1878–84), 1:193–96; Paul W. Gates, *History of Public Land Law Development* (Washington, D.C., 1968), 546–55; Bernard DeVoto, "The West: A Plundered Province," *Harpers Magazine* 169(August 1934):358.

27. *Cong. Globe,* 42nd cong., 1st sess., 178; Samuel Trask Dana, *Forest and Range Policy: Its Development in the United States* (New York, 1956), 31–3; Robbins, *Landed Heritage,* 247–49; Yonce, "Public Land Disposal," 154–57.

28. The act was extended to all public land states in 1892. *Cong. Rec.,* 44th cong., 1st sess., 1145; Robbins, *Landed Heritage,* 287–88; Gates, *History of Public Land Law,* 485.

29. Cameron, *Development of Governmental Forest Control,* 113–14; Robbins, *Landed Heritage,* 288–91; Meany, "History of the Lumber Industry," 202–5.

30. Frederick J. Yonce, "Lumbering and the Public Timberlands in Washington: The Era of Disposal," *Journal of Forest History* 22(January 1978):14–15; Renton, Holmes & Co. to Port Blakely Mill Co., 23 March 1882; Port Blakely Mill Co. to Renton, Holmes & Co., 26 May 1882, Port Blakely Mill Co. Records, Manuscripts Collection, University of Washington Library.

31. An industry legend claims that the holdings of the Puget Mill Company tended to be close to tidewater because Cyrus Walker used visiting sailors to enter claims and feared they would become lost away from saltwater. Port Blakely Mill Co. to Renton, Holmes & Co., 7 April 1882; to J. B. Forbes, 10 April 1882; to J. P. Sullivan, 1 August 1882, Port Blakely Mill Co. Records; G. W. Andrews to Commissioner, General Land Office, 26 September 1895, General Land Office Records, RG 48, National

Archives; Edwin T. Coman, Jr. and Helen M. Gibbs, *Time, Tide and Timber: A Century of Pope & Talbot* (Stanford, 1949), 113. Seattle newspaper quoted in Iva L. Buchanan, "Lumbering and Logging in the Puget Sound Region in Territorial Days," *Pacific Northwest Quarterly* 27(January 1936):47.

32. Port Blakely Mill Co. to Renton, Holmes & Co., 7 April 1882; Memorandum of Payments made for Timber Land for Blackman Bros., 26 June 1882, Port Blakely Mill Co. Records; Steen, "Forestry in Washington," 44.

33. Port Blakely Mill Co. to Renton, Holmes & Co., 17 May 1887; to Forbes, 30 June 1883, Port Blakely Mill Co. Records; W. Talbot to Walker, 28 May and 15 June 1887; 24 January 1888; C. F. A. Talbot to Walker, 19 and 23 January 1888, Pope & Talbot Records.

34. Renton to Walker, 15 June 1883; Port Blakely Mill Co. to Renton, Holmes & Co., 3 August 1883, Port Blakely Mill Co. Records; W. Talbot to Walker, 3 and 15 August 1888, Pope & Talbot Records; Richard C. Berner, "The Port Blakely Mill Company, 1876–89," *Pacific Northwest Quarterly* 57(October 1966):161–62.

35. Walker's letter is the last before a gap appears in surviving Pope & Talbot records. A search of the Vilas papers at the State Historical Society of Wisconsin failed to discover evidence of the scheme, but did reveal an interest in Oregon politics during 1888. Walker to W. Talbot, 26 May 1888, Pope & Talbot Records, Pope & Talbot Archives; W. Talbot to Walker, 29 May 1888, Pope & Talbot Records, University of Washington Library.

36. Renton, Holmes & Co. to Port Blakely Mill Co., 9 March and 10 November 1888; 22 September and 2 November 1889; Port Blakely Mill Co. to Sol G. Simpson, 24 May 1889, Port Blakely Mill Co. Records; W. Talbot to Walker, 10 November 1888; Ames to W. Talbot and F. Talbot, 20 March 1912, Pope & Talbot Records, University of Washington Library.

37. Coman and Gibbs, *Time, Tide and Timber*, 227; John S. Hittell, *The Commerce and Industries of the Pacific Coast*, 2nd ed. (San Francisco, 1882), 588–89, 592–93; Port Blakely Mill Co. to Renton, Holmes & Co., 17 January 1884 and 15 February 1886, Port Blakely Mill Co. Records; Berner, "Port Blakely Mill Company," 161–62; R. G. Dun & Co. Credit Reports, 14:148; 16:156.

38. *Forestry & Irrigation* 9(September 1903):419; U.S. Bureau of Corporations, *The Lumber Industry*, Part 1, *Standing Timber* (Washington, D.C., 1913), xiii; Gifford Pinchot, "How Conservation Began in the United States," *Agricultural History* 11(October 1937):260.

39. Ames to W. Talbot and F. Talbot, 20 March 1912, Pope & Talbot Records, University of Washington Library; *Mark Twain's Speeches* (New York, 1923), 236.

40. Ray Allen Billington, "The Origin of the Land Speculator as a Frontier Type," *Agricultural History*, 19(October 1945):204–12; Roy E. Appleman, "Timber Empire from the Public Domain," *Mississippi Valley Historical Review*, 26(September 1939):193.

41. Yonce, "Lumbering and the Public Timberlands," 16–17; William J. Trimble, "The Influence of the Passing of the Public Lands," *Atlantic Monthly* 113(June 1914): 756–57.

42. *Lumber Workers Bulletin*, 1 March 1923.

43. *General Land Office Annual Report, 1875*, 349; J. M. Harrison, *Harrison's Guide and Resources of the Pacific Slope*, 2nd ed. (San Francisco, 1876), 29; Herman M. Johnson, comp., *Production of Lumber, Lath, and Shingles in Washington and Oregon, 1869–1936*, Forest Research Notes No. 24(Portland, 1938), Table 1; *Tenth Census of the United States*, Vol. 9, *Report on the Forests of North America* (Washington, D.C., 1884), 486–87; Meany, "History of the Lumber Industry," 125–26.

44. Thomas R. Cox, *Mills and Markets: A History of the Pacific Coast Lumber Industry to 1900* (Seattle, 1974), 257–59; Buchanan, "Lumbering and Logging," 48–

51; Thomas F. Gedosch, "Seabeck, 1857–1886: The History of a Company Town" (M.A. thesis, University of Washington, 1967), 19–21; Berner, "Port Blakely Mill Company," 164–65.

45. Renton, Holmes & Co. to Port Blakely Mill Co., 29 January, 23 July and 2 and 7 August 1878; Port Blakely Mill Co. to Renton, Holmes & Co., 1 December 1878, Port Blakely Mill Co. Records.

46. Lumbermen had supported approval of the reciprocity treaty as a means of increasing trade between Puget Sound and Honolulu. Renton, Holmes & Co. to Port Blakely Mill Co., 28 February, 29 July and 27 September 1878, Port Blakely Mill Co. Records; *Hawaiian Almanac and Annual for 1876* (Honolulu, 1876):61; *1878* (Honolulu, 1878):53; *1882* (Honolulu, 1882):16–17, 20; *1885* (Honolulu, 1885):16–20.

47. Renton, Holmes & Co. to Port Blakely Mill Co., 8 January, 24 March, 1 and 29 July, 16 August and 5 December 1878; Port Blakely Mill Co. to Renton, Holmes & Co., 23 July and 22 and 24 December 1878, Port Blakely Mill Co. Records.

48. Quoted in W. H. Ruffner, *A Report on Washington Territory* (New York, 1889), 70.

49. *First Report of the Secretary of State, 1890* (Olympia, 1891), 58; Johnson, comp., *Production of Lumber,* Table 1; "A British Report on Washington Territory, 1885," *Pacific Northwest Quarterly* 35(April 1944):147–48; *Report Upon Forestry,* 4:389.

6: A Change of Front

1. Cyrus Walker to W. H. Talbot, 10 May 1888, Pope & Talbot Records, Pope & Talbot Archives.

2. F. I. Vassault, "Lumbering in Washington," *Overland Monthly* 20(July 1892):24.

3. J. Murphy, "Summer Ramblings in Washington Territory," *Appletons' Journal* 3(November 1877):390; Charles Sweet Diary, 16 October 1882, Washington State Historical Society; Septima Collins, *A Woman's Trip to Alaska* (New York, 1890), 34–36; Carrie Adell Strahorn, *Fifteen Thousand Miles by Stage* (New York, 1911), 353.

4. Rudyard Kipling, *From Sea to Sea and Other Sketches,* 2 vols. (Garden City, N.Y., 1925), 2:90–93; *Graphic* 5(3 October 1891):12; Collins, *Woman's Trip,* 37–38; William M. Thayer, *Marvels of the New West* (Norwich, Conn., 1887), 337–41. Tacoma newspaperman quoted in Thomas Emerson Ripley, *Green Timber: On the Flood Tide of Fortune in the Great Northwest* (New York, 1972 ed.), 60.

5. George H. Emerson to E. J. Holt, 18 March 1897, George H. Emerson Letterbooks, Manuscripts Collection, University of Washington Library.

6. James S. Lawson Reminiscences, Bancroft Library, University of California, Berkeley, 93–94; William G. Bek, trans., "From Bethel, Missouri, to Aurora, Oregon: Letters of William Keil, 1855–1870," *Missouri Historical Review* 48(January 1954):143; Caroline Gale Budlong, *Memories of Pioneer Days in Oregon and Washington Territory* (Eugene, 1949), 40; Michael F. Luark Diary, 25 November 1861, Manuscripts Collection, University of Washington Library.

7. H. H. Bancroft, *History of Washington, Idaho, and Montana, 1845–1889* (San Francisco, 1890), 35–36; William Albason, "History of Willapa Harbor Lumbering," *Timberman* 33(June 1932):20.

8. Emerson, who apparently was a distant relative of Ralph Waldo Emerson, believed that "man is a hardened thoughtless brute and wasteful of all things nature has provided." Wallace J. Miller, *South-western Washington* (Olympia, 1890), 188–93; Edward G. Jones, *The Oregonian's Handbook of the Pacific Northwest* (Portland, 1894),

319–20, 333–34; Address to Pacific Coast Lumber Manufacturers Association, 28 February 1908, Emerson Letterbooks.

9. Herbert Hunt and Floyd C. Kaylor, *Washington West of the Cascades*, 2 vols. (Chicago, 1917), 2:92–95, 262, 265, 278–9; Jones, *Oregonian's Handbook*, 29–33; Miller, *South-western Washington*, 173; *Graphic* 5(3 October 1891):52–53; Edwin T. Coman, Jr. and Helen M. Gibbs, *Time, Tide and Timber: A Century of Pope & Talbot* (Stanford, 1949), 399–401.

10. Julian Hawthorne, ed., *History of Washington, the Evergreen State*, 2 vols. (New York, 1893), 2:194, 196; Miller, *South-western Washington*, 112; *Graphic* 5(3 October 1891):48–50; Jones, *Oregonian's Handbook*, 325–26, 333.

11. *Twelfth Census of the United States, 1900* (Washington, D.C., 1902), vol. 9, Part 3, 812, 830; Bureau of the Census, *Manufactures, 1905*, Part 3, *Special Reports on Selected Industries* (Washington, D.C., 1908), 642.

12. *Eleventh Census of the United States, 1890* (Washington, D.C., 1895), 6:597–99, 603–4; James E. Defebaugh, *History of the Lumber Industry of America*, 2 vols. (Chicago, 1907), 1:299–300; Wilson Compton, *The Organization of the Lumber Industry* (Chicago, 1916), 69–71.

13. Minutes, Board of Directors Meetings, 28 October 1887 and 20 March 1888, Northern Pacific Railroad Records, Minnesota Historical Society; Draft agreement, n.d.; unsigned report, n.d., Laird, Norton Co. Records, Minnesota Historical Society; George F. Cornwall, "Fifty Years: History of St. Paul & Tacoma Lumber Company," *Timberman* 39(May 1938):18–19.

14. Emerson to A. M. Simpson, 5 August 1892; to Harry C. Heermans, 17 August 1895, Emerson Letterbooks; Vassault, "Lumbering in Washington," 28–32; *West Coast Lumberman* 4(January 1893):6.

15. *The Puget Sound Catechism* (Seattle, 1891), 1; Walker to W. Talbot, 11 and 17 June 1889; to A. W. Jackson, 1 August 1889, Cyrus Walker Letterbooks, Ames Coll., Manuscripts Collection, University of Washington Library.

16. Alexander N. MacDonald, "Seattle's Economic Development, 1880–1910" (Ph.D. diss., University of Washington, 1959), 176; Thomas D. Stimson, "A Record of the Family, Life and Activities of Charles Douglas Stimson," Manuscripts Collection, University of Washington Library, 9–11; Emerson to Heermans, 17 August 1895, Emerson Letterbooks; *West Coast Lumberman* 3(July 1892):5–6.

17. John H. Cox, "Organizations of the Lumber Industry in the Pacific Northwest, 1889–1914" (Ph.D. diss., University of California, Berkeley, 1937), 31–33; Walker to W. Talbot, 20 May and 11 July 1892, Walker Letterbooks; *West Coast Lumberman* 4(June 1893):9.

18. *West Coast Lumberman* 4(November 1892):9; *Pacific Lumber Trade Journal* 11(January 1906):20–21; Walker to W. Talbot, 20 May, 11 July and 21 September 1892, Walker Letterbooks.

19. Norman H. Clark, *Mill Town* (Seattle, 1970), Chap. 2; Walker to W. Talbot, 14 December 1891; 4 June and 11 and 19 July 1892, Walker Letterbooks; William F. Prosser, *A History of the Puget Sound Country*, 2 vols. (New York, 1903), 1:240.

20. Phoebe Goodell Judson, *A Pioneer's Search for an Ideal Home* (Bellingham, 1925), 278–79; Walker to W. Talbot, 13 May 1888, Pope & Talbot Records, Pope & Talbot Archives; Walker to W. Talbot, 31 March 1890 and 10 March 1891, Walker Letterbooks.

21. Pope & Talbot to Walker, 10 December 1888 and 5 September 1892; W. Talbot to Walker, 23 March 1889, Pope & Talbot Records, Ames Coll., Manuscripts Collection, University of Washington Library; Renton, Holmes & Co. to Port Blakely Mill Co., 1 February 1889, Port Blakely Mill Co. Records, Manuscripts Collection, University of Washington Library.

22. MacDonald, "Seattle's Economic Development," 95–96; *West Coast Lumberman* 3(September 1892):1; Cox, "Organizations of the Lumber Industry," 7.

23. Pope & Talbot to Walker, 5 September 1892, Pope & Talbot Records, University of Washington Library; *West Coast Lumberman* 3(December 1891):12; 4(February 1893):3; (May 1893), 12; Cox, "Organizations of the Lumber Industry," 6–8; MacDonald, "Seattle's Economic Development," 96–97.

24. Edmond S. Meany, Jr., "The History of the Lumber Industry in the Pacific Northwest to 1917" (Ph.D. diss., Harvard University, 1935), 145; MacDonald, "Seattle's Economic Development," 98–99; Cox, "Organizations of the Lumber Industry," 8; *West Coast Lumberman* 4(July 1893):14.

25. Henry B. Steer, comp., *Lumber Production in the United States, 1799–1946*, USDA Misc. Pub. No. 669(Washington, D.C., 1948), 11; *Twelfth Census, 9*, Part 3, 812; Bureau of the Census, *Manufactures, 1905*, Part 3, *States and Territories* (Washington, D.C., 1907), 1148.

26. *First Report of the Secretary of State, 1890* (Olympia, 1891), 58.

27. Charles S. Holmes to Port Blakely Mill Co., 30 August 1889, Port Blakely Mill Co. Records; Walker to W. Talbot, 22 July 1890, Walker Letterbooks.

28. Renton, Holmes & Co. to Port Blakely Mill Co., 17 September 1886; 9 June 1888; 20 May and 28 December 1889; Port Blakely Mill Co. to Renton, Holmes & Co., 6 January and 28 May 1886, Port Blakely Mill Co. Records; *Hawaiian Almanac and Annual for 1890* (Honolulu, 1889), 18, 21; *1892* (Honolulu, 1891), 29; Walker to W. Talbot, 17 September 1889; 17 January and 13 December 1898, Walker Letterbooks.

29. Port Blakely Mill Co. to Renton, Holmes & Co., 20 January, 1 April, and 17 August 1886; Renton, Holmes & Co. to Port Blakely Mill Co., 13 March 1888, Port Blakely Mill Co. Records; Export Registers, Puget Sound Customs District Records, RG 36, Federal Records Center, Seattle; Thomas R. Cox, *Mills and Markets: A History of the Pacific Coast Lumber Industry to 1900* (Seattle, 1974), 216–17.

30. Renton, Holmes & Co. to Port Blakely Mill Co., 11 June 1888; 15 and 19 January, 4 February, and 20 October 1889, Port Blakely Mill Co. Records; Export Registers, Puget Sound Customs District Records; W. Talbot to Walker, 20 March 1889, Pope & Talbot Records, University of Washington Library; Seattle Chamber of Commerce Annual Report, 1891, Northwest Collection, University of Washington Library.

31. Renton, Holmes & Co. to Port Blakely Mill Co., 5 May 1886 and 9 March 1889, Port Blakely Mill Co. Records; W. Talbot to Walker, 14 February 1888; Pope & Talbot to Puget Mill Co., 21 October 1891, Pope & Talbot Records, University of Washington Library; Cox, *Mills and Markets*, 214–16; Coman and Gibbs, *Time, Tide and Timber*, 180–81.

32. Renton, Holmes & Co. to Port Blakely Mill Co., 16 January, 10 and 12 May, 6 November, and 29 December 1886, Port Blakely Mill Co. Records; W. Talbot to Walker, 26 July 1887; 3 January and 29 December 1888; 4 February 1889; Pope & Talbot to Puget Mill Co., 20 February 1889, Pope & Talbot Records, University of Washington Library; Coman and Gibbs, *Time, Tide and Timber*, 201–7; Cox, *Mills and Markets*, 217–19.

33. Walker to W. Talbot, 6 March 1888, Pope & Talbot Records, Pope & Talbot Archives; W. Talbot to Walker, 28 February, 30 June and 22 December 1888; Pope & Talbot to Puget Mill Co., 20 February and 19 March 1889, Pope & Talbot Records, University of Washington Library; Renton, Holmes & Co. to Port Blakely Mill Co., 11 January, 8 March, and 4, 6, and 24 April 1888; 15 and 21 January 1889, Port Blakely Mill Co. Records; Cox, *Mills and Markets*, 219.

34. Pacific Pine could fill export orders, but members were free to participate in the

foreign trade and the combination for the most part limited itself to San Francisco. Renton, Holmes & Co. to Port Blakely Mill Co., 16 and 31 July, 4 August, 17 September, 20 October, and 4, 23 and 29 December 1886; Port Blakely Mill Co. to Renton, Holmes & Co., 5, 7 and 12 August 1886, Port Blakely Mill Co. Records; Cox, *Mills and Markets*, 259–60; Coman and Gibbs, *Time, Tide and Timber*, 409–11.

35. Port Blakely Mill Co. to Renton, Holmes & Co., 12, 17, and 26 August 1886; Renton, Holmes & Co. to Port Blakely Mill Co., 7 and 9 September 1886, Port Blakely Mill Co. Records; W. Talbot to Walker, 19 May, 22 and 25 June, and 15 July 1887; Jackson to Walker, 22 April 1890; Edwin G. Ames to L. H. Pierson, 12 August 1918, Pope & Talbot Records, University of Washington Library; Cox, *Mills and Markets*, 260–61.

36. Holmes to Port Blakely Mill Co., 6 August 1888; Renton, Holmes & Co. to Port Blakely Mill Co., 14 August, 1 September, and 21 December 1888, Port Blakely Mill Co. Records; Pope & Talbot to Puget Mill Co., 20 August 1890, Pope & Talbot Records, University of Washington Library.

37. Renton, Holmes & Co. to Port Blakely Mill Co., 20 August and 15 November 1888; 9 and 10 October and 9 December 1889, Port Blakely Mill Co. Records; Walker to W. Talbot, 2 October 1889, Walker Letterbooks; Pope & Talbot to Puget Mill Co., 23 May and 9 October 1889; 3 and 11 February 1891; W. Talbot to Walker, 4 February 1889, Pope & Talbot Records, University of Washington Library; Cox, *Mills and Markets*, 262; Coman and Gibbs, *Time, Tide and Timber*, 179.

38. Walker to W. Talbot, 9, 16, and 29 January, 5 February and 3 August 1892, Walker Letterbooks; Frederick C. Talbot to Walker 25 February 1892, Pope & Talbot Records, University of Washington Library.

39. Walker to W. Talbot, 14 May 1892, Walker Letterbooks; *West Coast Lumberman* 4(November 1892):10; (October 1893), 8; Cox, *Mills and Markets*, 264.

40. Corydon Wagner Address to West Coast Lumbermen's Association, 31 January 1941, West Coast Lumbermen's Association Records, Oregon Historical Society.

41. Rodney C. Loehr, "Saving the Kerf: The Introduction of the Band Saw Mill," *Agricultural History:* 22(July 1949):169–72; Emerson to Simpson, 5 August 1892 and 6 June 1893, Emerson Letterbooks; *West Coast Lumberman* 4(February 1893):2.

42. Emerson to Simpson, 14 September 1891, Emerson Letterbooks; Renton, Holmes & Co. to Port Blakely Mill Co., 9 November 1887; 8 and 18 November 1889, Port Blakely Mill Co. Records; Walker to W. Talbot, 1 February 1888, Pope & Talbot Records, Pope & Talbot Archives; W. Talbot to Walker, 25 June 1887; Pope & Talbot to Puget Mill Co., 3 and 11 February 1891, Pope & Talbot Records, University of Washington Library.

43. Walker to W. Talbot, 16 February 1888, Pope & Talbot Records, Pope & Talbot Archives; W. Talbot to Walker, 6, 9, and 22 February and 9 April 1888, Pope & Talbot Records, University of Washington Library; Walker to W. Talbot, 8 August 1891, Walker Letterbooks.

44. Ames eventually married the daughter of Cyrus Walker's brother, Will Walker. W. Talbot to Walker, 17 July, 25 August, and 24 October 1888; Pope & Talbot to Puget Mill Co., 21 April 1890, Pope & Talbot Records, University of Washington Library; Coman and Gibbs, *Time, Tide and Timber*, 158–61.

45. "Two Railroad Reports on Northwest Resources," *Pacific Northwest Quarterly* 37(July 1946):187; *Tenth Census of the United States*, vol. 9, *Report on the Forests of North America* (Washington, D.C., 1884), 574.

46. J. C. Moe Reminiscences, Oregon Historical Society, 30; Iva L. Buchanan, "Lumbering and Logging in the Puget Sound Region in Territorial Days," *Pacific Northwest Quarterly* 27(January 1936), 52-53; Port Blakely Mill Co. to Renton, Holmes & Co., 21 March and 12 April 1882; 26 and 31 August 1885; 28 April, 14 July, and 16 December 1887; Renton, Holmes & Co. to Port Blakely Mill Co., 29 August 1888,

Port Blakely Mill Co. Records; Pope & Talbot to Walker, 22 August 1887; W. Talbot to Walker, 28 February 1888, Pope & Talbot Records, University of Washington Library; Walker to W. Talbot, 23 February 1888, Pope & Talbot Records, Pope & Talbot Archives.

47. Stewart H. Holbrook, "Daylight in the Swamp," *American Heritage* 9(October 1958):77–78; H. Brett Melendy, "Two Men and a Mill: John Dolbeer, William Carson, and the Redwood Lumber Industry in California," *California Historical Society Quarterly* 38(March 1959):64–65.

48. George H. Emerson, "Logging on Grays Harbor," *Timberman* 8(September 1907):20; Emerson to Harry Pennel, 11 August 1891; to Holt, 11 December 1891; to Simpson, 2 May 1892, Emerson Letterbooks; *Twelfth Census*, vol. 9, Part 3, 821–22.

49. Meany, "History of the Lumber Industry," 255–59; Cox, *Mills and Markets,* 229–30; Renton, Holmes & Co. to Port Blakely Mill Co., 23 October 1888, Port Blakely Mill Co. Records.

50. Kramer A. Adams, *Logging Railroads of the West* (Seattle, 1961), 12–13; Walker to W. Talbot, 30 September 1885, Pope & Talbot Records, Pope & Talbot Archives; *West Coast Lumberman* 4(February 1893):6; Miller, *South-western Washington,* 110.

51. Renton, Holmes & Co. to Port Blakely Mill Co., 1 January 1885; 5 February 1886; 6 April, 11, and 24 May, 15 August and 9 October 1888; 2 January, 19 March, 26 April, and 16 November 1889; Port Blakely Mill Co. to Renton, Holmes & Co., 27 June and 22 and 24 December 1885; 24 April 1886; 19 April 1887, Port Blakely Mill Co. Records; Richard C. Berner, "The Port Blakely Mill Company, 1876–89," *Pacific Northwest Quarterly* 57(October 1966)1:161–63; Robert E. Ficken, *Lumber and Politics: The Career of Mark E. Reed* (Seattle, 1979), 12–14.

52. Renton, Holmes & Co. to Port Blakely Mill Co., 9 March, 14 April, and 9 May 1888; Sol Simpson to William Renton, 13 January and 23 March 1888, Port Blakely Mill Co. Records; John Campbell Diary, 1 February 1886, Manuscripts Collection, University of Washington Library.

53. Of the 22 women employed in Washington mills, however, only a third received more than $40 a month. *Eleventh Census,* 6:615–16, 626–30, 640; James N. Tattersall, "The Economic Development of the Pacific Northwest to 1920" (Ph.D. diss., University of Washington, 1960), 93–100; Andrew M. Prouty, "Logging with Steam in the Pacific Northwest: The Men, the Camps, and the Accidents, 1885–1918" (M.A. thesis, University of Washington, 1973), 20–27.

54. Pope & Talbot to Puget Mill Co., 12 April 1889, Pope & Talbot Records, University of Washington Library.

55. Ellis Lucia, *The Big Woods* (Garden City, N.Y., 1975), 69.

56. Carlos A. Schwantes, *Radical Heritage: Labor, Socialism, and Reform in Washington and British Columbia, 1885–1917* (Seattle, 1979), 25–27; Walker to W. Talbot, 31 October 1885, Pope & Talbot Records, Pope & Talbot Archives; Iva L. Buchanan, "An Economic History of Kitsap County, Washington, to 1889" (Ph.D. diss., University of Washington, 1930), 257–58; Thomas F. Gedosch, "Seabeck, 1857–1886: The History of a Company Town" (M.A. thesis, University of Washington, 1967), 58.

57. Walker to W. Talbot, 26 and 31 October and 6, 9, and 13 November 1885, Pope & Talbot Records, Pope & Talbot Archives.

58. Walker to W. Talbot, 31 October and 9 November 1885; Ames to W. Talbot, 10 October 1885, Pope & Talbot Records, Pope & Talbot Archives.

59. Port Blakely Mill Co. to Renton, Holmes & Co., 24 June, 16 and 26 July, 6 August, and 6 and 9 December 1886; Renton, Holmes & Co. to Port Blakely Mill Co., 3 May 1886, Port Blakely Mill Co. Records.

60. Renton, Holmes & Co. to Port Blakely Mill Co., 23 and 24 September and 26 October 1886; Port Blakely Mill Co. to Renton, Holmes & Co., 30 September, 28 October, and 4 November 1886, Port Blakely Mill Co. Records.

61. Port Blakely Mill Co. to Renton, Holmes & Co., 12 and 25 November and 3 December 1886; Renton, Holmes & Co. to Port Blakely Mill Co., 19 November and 7 and 10 December 1886, Port Blakely Mill Co. Records.

62. Port Blakely Mill Co. to Renton, Holmes & Co., 6 and 15 December 1886; 14 and 18 May 1887, Port Blakely Mill Co. Records.

63. Walker to W. Talbot, 27 April 1890 and 3 August 1893, Walker Letterbooks.

64. *Special Reports on Selected Industries,* 617; Meany, "History of the Lumber Industry," 125–26; Report No. 61, Senate Committee on Territories, Admission of Washington Territory as a State, 49th cong., 1st sess., in Washington Territorial Papers (microfilm), RG 48, Federal Records Center, Seattle; Walker to W. Talbot, 2 September 1889, Walker Letterbooks.

65. Walker to W. Talbot, 12, 15, 21, 24, and 31 July 1889, Walker Letterbooks; Alan Hynding, *The Public Life of Eugene Semple: Promoter and Politician of the Pacific Northwest* (Seattle, 1973), 121–22; Robert C. Nesbit, *"He Built Seattle": A Biography of Judge Thomas Burke* (Seattle, 1961), 308–15.

66. Walker to W. Talbot, 12, 15, 21, 24, and 31 July 1889, Walker Letterbooks; H. G. Struve to Walker, 23 July 1889; Pope & Talbot to Puget Mill Co., 15 and 29 July and 3 and 8 August 1889; John L. Howard to E. M. Herrick, 20 July 1889, Pope & Talbot Records, University of Washington Library.

67. Hynding, *Public Life,* 122–23; Nesbit, *"He Built Seattle,"* 315–17; Roy O. Hoover, "The Public Land Policy of Washington State: The Initial Period, 1889–1912" (Ph.D. diss., Washington State University, 1967), 129–30; Pope & Talbot to Puget Mill Co., 3 and 8 August 1889, Pope & Talbot Records, University of Washington Library; Walker to W. Talbot, 3 and 15 August 1889, Walker Letterbooks.

68. Emerson to Heermans, 2 October 1891; to Simpson, 21 December 1891; to Holt, 9 February 1892; to R. B. Dyer, 6 April 1894, Emerson Letterbooks; Hynding, *Public Life,* 123–41; Nesbit, *"He Built Seattle,"* 317–42; Hoover, "Public Land Policy," 119–36.

69. Walker to W. Talbot, 21 July 1889; to C. F. A. Talbot, 20 November 1889, Walker Letterbooks; Ames to W. Talbot, 19 June 1916, Pope & Talbot Records, University of Washington Library; *West Coast Lumberman* 3(August 1892):1.

70. *West Coast Lumberman* 4(March 1893):6.

7: Memorial Times

1. C. D. Stimson to T. D. Stimson, 4 August 1893, Stimson Mill Co. Records, Manuscripts Collection, University of Washington Library.

2. *West Coast Lumberman* 4(August 1893):8.

3. George S. Long to F. S. Bell, 18 November 1920, Weyerhaeuser Timber Co. Records, Weyerhaeuser Co. Archives; Chauncey W. Griggs to Mrs. F. C. Williams, 28 May 1896, Chauncey W. Griggs Letterbooks, St. Paul & Tacoma Lumber Co. Records, Manuscripts Collection, University of Washington Library.

4. Alex Polson to Timothy Jerome, 24 May 1926, Merrill & Ring Lumber Co. Records, Manuscripts Collection, University of Washington Library.

5. Frederick C. Talbot to Cyrus Walker, 19 May and 21 June 1893, Pope & Talbot Records, Ames Coll., Manuscripts Collection, University of Washington Library; C. Stimson to brother, 10 and 12 June 1893, Stimson Mill Co. Records; Murray Morgan, *Puget's Sound: A Narrative of Early Tacoma and the Southern Sound* (Seattle, 1979), 275–78; Norman H. Clark, *Mill Town* (Seattle, 1970), 29–38.

6. Walker to W. H. Talbot, 5 and 17 July, 3 August and 23 October 1893, Cyrus Walker Letterbooks, Ames Coll., Manuscripts Collection, University of Washington Library; C. Stimson to brother, 10 June 1893; to T. Stimson, 19 September 1893, Stimson Mill Co. Records; *West Coast Lumberman* 4(September 1893):7.

7. *Thirteenth Census of the United States, 1910, Abstract of the Census with Supplement for Washington* (Washington, D.C., 1913), 568, 576–77.

8. Walker to W. Talbot, 17 July 1893; to A. W. Jackson, 1 August 1893, Walker Letterbooks.

9. W. Talbot to Walker, 17 August 1893; Pope & Talbot to Walker, 11 August 1893, Pope & Talbot Records; *West Coast Lumberman* 4(August 1893):10; George H. Emerson to A. M. Simpson, 1 September 1893; 18 and 25 January 1894; to William Thompson, 19 January 1894, George H. Emerson Letterbooks, Manuscripts Collection, University of Washington Library.

10. Emerson to Simpson, 1 September 1893 and 18 January 1894, Emerson Letterbooks.

11. Emerson to Simpson, 6 and 13 July and 23 October 1893, Emerson Letterbooks; Alexander N. MacDonald, "Seattle's Economic Development, 1880–1910" (Ph.D. diss., University of Washington, 1959), 99; *Pacific Lumber Trade Journal* 5(October 1899):10; Walker to W. Talbot, 21 July 1893, Walker Letterbooks; *West Coast Lumberman* 6(February 1895):18; John H. Cox, "Organizations of the Lumber Industry in the Pacific Northwest, 1889–1914" (Ph.D. diss., University of California, Berkeley, 1937), 8, 35–37, 47.

12. Robert E. Ficken, "The Port Blakely Mill Company, 1888–1903," *Journal of Forest History* 21(October 1977):209–10; H. D. Langille, "Canadian Lumber Competition," *American Forestry* 21(February 1915):137–39; *West Coast Lumberman* 6(August 1895):13; Emerson to Simpson, 27 July 1894, Emerson Letterbooks.

13. Peter Musser to M. G. Norton, 6 May 1893; to Laird, Norton Co., 18 May 1893, Laird, Norton Co. Records, Minnesota Historical Society; Frederick Weyerhaeuser to J. P. Weyerhaeuser, 11 April and 22 July 1893; W. T. McDonald to F. Weyerhaeuser, 6 May 1895; Weyerhaeuser Co. to McDonald, 15 May 1895, Weyerhaeuser Family Collection, Minnesota Historical Society.

14. *West Coast Lumberman* 4(August 1893):10; (September 1893):9; Francis Rotch to A. C. Govey, 14 June 1899, Simpson Logging/Timber Co. Records, Simpson Timber Co. Archives; Thomas D. Stimson, "A Record of the Family, Life and Activities of Charles D. Stimson," Manuscripts Collection, University of Washington Library, 13–4.

15. Griggs to C. M. Griggs, 3 July 1894; to A. G. Foster, 14 and 16 July 1894, Griggs Letterbooks; Walker to W. Talbot, 23 and 28 April 1894, Walker Letterbooks; Emerson to Samuel Perkins, 31 May 1894, Emerson Letterbooks.

16. Emerson to Richard Dabney and James Hood, 4 June 1894, Emerson Letterbooks.

17. Ficken, "Port Blakely Mill Company," 210–11; Walker to W. Talbot, 21 November 1893; 26 March 1895 and 14 July 1896, Walker Letterbooks.

18. Sol Simpson was not related to Asa Simpson. Robert E. Ficken, *Lumber and Politics: The Career of Mark E. Reed* (Seattle, 1979), 12–16; Sol G. Simpson to Mark E. Reed, 29 March 1905, Mark E. Reed Personal Papers, Simpson Timber Co. Archives.

19. In 1897, the Pacific Pine Lumber Company was succeeded by the Pacific Pine Company, a joint marketing agency operated by Pope & Talbot and Renton, Holmes & Company. Emerson to Perkins, 23 and 28 September 1895; to Simpson, 15 October 1895, Emerson Letterbooks; Griggs to C. H. Jones, 16 May 1895; to Foster, 15 May 1895, Griggs Letterbooks; C. Stimson to T. Stimson, 6 March 1896, Stimson Mill Co. Records; Pope & Talbot to Walker, 13 August 1895; Jackson to Walker, 8 and 21

August 1895, Pope & Talbot Records; Thomas R. Cox, *Mills and Markets: A History of the Pacific Coast Lumber Industry to 1900* (Seattle, 1974), 266–69.

20. Pope & Talbot to Walker, 16 October 1895, Pope & Talbot Records; Edwin T. Coman, Jr. and Helen M. Gibbs, *Time, Tide and Timber: A Century of Pope & Talbot* (Stanford, Cal., 1949), 411–12; Cox, *Mills and Markets,* 268–69.

21. N. B. Noble to W. H. Day, 17 January and 10 March 1896; D. P. Simons to Noble, 22 April 1896, O. H. Ingram Papers, State Historical Society of Wisconsin; Walker to Jackson, 29 June 1896, Walker Letterbooks; Cox, *Mills and Markets,* 270–71; Coman and Gibbs, *Time, Tide and Timber,* 412–13.

22. Griggs to S. L. Levy, 3 August 1897, Griggs Letterbooks; *West Coast Lumberman* 8(March 1897):205.

23. Polson to Jerome, 14 March 1932; to R. D. Merrill, 6 August 1932, Merrill & Ring Lumber Co. Records; Griggs to C. M. Griggs, 6 July 1896, Griggs Letterbooks.

24. Emerson to Simpson, 28 July 1896, Emerson Letterbooks; Walker to W. Talbot, 23 August 1896 and 3 April 1900, Walker Letterbooks; C. Stimson to T. Stimson, 20 August 1896, Stimson Mill Co. Records.

25. Emerson to J. W. Sanborn, 5 October 1896; to Simpson 17, 23, and 27 October 1896; to Perkins, 19 October 1896, Emerson Letterbooks; Walker to W. Talbot, 13 September 1896, Walker Letterbooks; Griggs to J. L. Farwell, 25 September 1896, Griggs Letterbooks; C. Stimson to T. Stimson, 20 August 1896, Stimson Mill Co. Records.

26. Emerson to Harry C. Heermans, 9 November 1896, Emerson Letterbooks; Walker to Jackson, 14 November 1896, Walker Letterbooks; Griggs to L. B. Royce, 11 November 1896, Griggs Letterbooks.

27. Walker to Jackson, 14 November 1896, Walker Letterbooks.

28. Forty percent of the lumber shipped by rail from western Washington in 1900 was sold in the eastern part of the state and in Oregon. Another thirty-eight percent was disposed of in Montana, North Dakota, Minnesota and Iowa. Because of the freight rate structure, virtually all of the lumber was sold west of the Mississippi. Seattle Chamber of Commerce, *Seattle, the Puget Sound Country and Western Washington, 1901: Their Resources and Opportunities* (Seattle, 1901), 21; Edmond S. Meany, Jr., "The History of the Lumber Industry in the Pacific Northwest to 1917" (Ph.D. diss., Harvard University, 1935), 145, 162; Griggs to F. W. Miller, 27 January 1898, Griggs Letterbooks; Emerson to Bess Hammond, 23 November 1899, Emerson Letterbooks; *Pacific Lumber Trade Journal* 5(August 1899):21; (October 1899), 9–10; (November 1899), 10; (December 1899), 10.

29. Pope & Talbot to Walker, 2 February and 30 December 1899; F. Talbot to Walker, 20 December 1899; Jackson to Walker, 28 April 1899, Pope & Talbot Records; Emerson to Simpson, 20 February 1899, Emerson Letterbooks.

30. Kramer A. Adams, "Blue Water Rafting: The Evolution of Ocean Going Log Rafts," *Forest History* 15(July 1971):17–18, 20–21; Pope & Talbot to Walker, 16 and 28 November 1893; 5 April, 13 September, 7 October, and 17 December 1897; Jackson to Walker, 4 November 1897, Pope & Talbot Records.

31. Renton, Holmes & Co. to Port Blakely Mill Co., 18, 22, 25, 26, and 28 September 1899, Port Blakely Mill Co. Records, Manuscripts Collection, University of Washington Library; F. Talbot to Walker, 19 September 1899; W. Talbot to Walker, 12 and 26 September 1899; Pope & Talbot to Walker, 31 December 1898 and 20 March 1899, Pope & Talbot Records; *Pacific Lumber Trade Journal* 5(October 1899):10; Adams, "Blue Water Rafting," 22–27.

32. Eugene Semple to daughter, 17 July and 17 August 1897, Eugene Semple Papers, Manuscripts Collection, University of Washington Library; Emerson to Simpson, 15 February 1898, Emerson Letterbooks; Walker to W. Talbot, 15 August and 14 and 24 December 1897, Walker Letterbooks.

33. G. F. Evans to Puget Lumber Co., 25 January 1898, Pope & Talbot Records; Walker to W. Talbot, 14 and 17 January 1898, Walker Letterbooks; L. Jensen to Renton, Holmes & Co., 29 March 1898, Port Blakely Mill Co. Records; Ficken, "Port Blakely Mill Company," 215; MacDonald, "Seattle's Economic Development," 137–45, 183–84.

34. Seattle Chamber of Commerce, *Seattle, the Puget Sound Country and Western Washington*, 20–21; Meany, "History of the Lumber Industry," 142; *Pacific Lumber Trade Journal* 5(November 1899):10.

35. Pope & Talbot to Walker, 6 August and 13 September 1898; F. Talbot to Walker, 25 February 1899, Pope & Talbot Records; Meany, "History of the Lumber Industry," 144–45.

36. Cox, *Mills and Markets*, 222–23; Pope & Talbot to Walker, 7 February 1899; J. H. Claiborne, Jr. to Pope & Talbot, 10 and 14 May and 5 and 16 June 1900, Pope & Talbot Records.

37. Claiborne to Pope & Talbot, 24 June, 10 and 22 July, and 9 August 1900; W. Talbot to Claiborne, 17 July 1900; F. Talbot to Walker, 9 July 1900, Pope & Talbot Records; *West Coast Lumberman* 11(July 1900):419.

38. R. Merrill to T. D. Merrill, 2 August 1898, Merrill & Ring Lumber Co. Records; *Pacific Lumber Trade Journal* 5(August 1899):21; (October 1899):21; (December 1899):21; (January 1900):24.

39. Emerson to E. J. Holt, 7 September 1895; 18 March 1897 and 18 May 1898; to Simpson, 15 October and 9 and 20 August 1898, Emerson Letterbooks.

40. Walker to F. Talbot, 17 December 1900; to W. Talbot, 9 September 1899, Walker Letterbooks; W. Talbot to Walker, 12 September 1899, Pope & Talbot Records; Emerson to Simpson, 13 July 1897, Emerson Letterbooks.

41. Griggs to H. F. Dimock, 22 October 1898, Griggs Letterbooks; Bureau of the Census, *Manufactures, 1905*, Part 3, *Special Reports on Selected Industries* (Washington, D.C., 1908), 617; Walker to W. Talbot, 16 March 1899, Walker Letterbooks; *Pacific Lumber Trade Journal* 5(October 1899):9.

8: Imperial Forests

1. Quoted in Ellis Lucia, *The Big Woods* (Garden City, N.Y., 1975), 19.

2. Frederick Jackson Turner, *The Significance of Sections in American History* (Gloucester, Mass., 1959 ed.), 243.

3. The holdings of Weyerhaeuser and the Southern Pacific Railroad, along with the remaining Northern Pacific lands, represented eleven percent of the privately owned timber in the United States. Bureau of Corporations, *The Lumber Industry*, Part 1, *Standing Timber* (Washington, D.C., 1913), 15–17, 208. Frederick Weyerhaeuser quoted in "Bunyan in Broadcloth—The House of Weyerhaeuser," *Fortune* 9(April 1934):174.

4. William Bancroft Hill, *Frederick Weyerhaeuser, Pioneer Lumberman* (Minneapolis, 1940), 17, 23, 25–27; Ralph W. Hidy, Frank Ernest Hill and Allan Nevins, *Timber and Men: The Weyerhaeuser Story* (New York, 1963), 3–7; "Bunyan in Broadcloth," 170.

5. R. G. Dun & Co. Credit Reports, 189:144, 263, Harvard Business School Library; Hidy, Hill, and Nevins, *Timber and Men*, Chaps. 1–11; Bureau of Corporations, *Standing Timber*, 101–4. The complainant, W. H. Day, is quoted in Charles E. Twining, *Downriver: Orrin H. Ingram and the Empire Lumber Company* (Madison, Wis., 1975), 216.

6. *Fifteenth Census of the United States, Manufactures: 1929*, vol. 2, *Reports by Industries* (Washington, D.C., 1933), 447; Wilson Compton, *The Organization of the Lumber Industry* (Chicago, 1916), 71, 117; Hidy, Hill, and Nevins, *Timber and Men*, 147, 208–9; *Pacific Lumber Trade Journal* 2(January 1897):10.

7. F. K. Weyerhaeuser oral history interview, Weyerhaeuser Family Collection, Minnesota Historical Society, 32–33; Annual Statement, 10 December 1900, Sound Timber Co. Records, Minnesota Historical Society; Minutes of Stockholders Meeting, Victoria Lumber & Mfg. Co. Ltd., 1 August 1905, Humbird Family Papers, University of British Columbia Library.

8. Minutes of Stockholders Meetings, 29 September 1900, 19 March 1901, and 18 March 1902, Coast Lumber Co. Records, Minnesota Historical Society; George S. Long to Charles H. Ingram, 11 January 1927, Weyerhaeuser Timber Co. Records, Weyerhaeuser Co. Archives; W. I. Ewart to O. H. Ingram, 21 July and 2 August 1899; 26 September and 6 December 1900; to stockholders, 18 and 19 July 1899, O. H. Ingram Papers, State Historical Society of Wisconsin.

9. *Timberman* 1(November 1899):11; (December 1899):12; (January 1900):4; R. D. Merrill to T. D. Merrill, 24 October 1899, Merrill & Ring Lumber Co. Records, Manuscripts Collection, University of Washington Library; W. H. Talbot to Cyrus Walker, 6 November 1899, Pope & Talbot Records, Manuscripts Collection, University of Washington Library.

10. The correct spelling of Weyerhaeuser was a problem for all concerned. The *Pacific Lumber Trade Journal* commented that "if Frederick Weyerhaeuser . . . should patronize a press clipping bureau and read all the items printed concerning his business he would probably be more amused over the variegated spelling given to his name than impressed by the recognized magnitude of his operations." Walker to W. Talbot, 16 March and 26 October 1899; to Pope & Talbot, 25 March 1899, Cyrus Walker Letterbooks, Ames Coll., Manuscripts Collection, University of Washington Library; *Pacific Lumber Trade Journal* 6(September 1900):9.

11. Walker to Pope & Talbot, 25 March 1899; to W. Talbot, 26 October and 21 November 1899, Walker Letterbooks; Pope & Talbot to Walker, 28 March 1899; Frederick C. Talbot to Walker, 30 October 1899, Pope & Talbot Records.

12. George H. Emerson to Chester A. Congdon, 26 October 1899; to Thomas Richardson, 9 November 1899 and 1 January 1900, George H. Emerson Letterbooks, Manuscripts Collection, University of Washington Library.

13. Emerson to Ewart, 21 November 1899; to Richardson, 9, 21, and 23 November and 11 December 1899; 1 January 1900; to E. J. Holt, 9 November 1899 and 17 February 1900; to Frederick Weyerhaeuser, 12 December 1899, Emerson Letterbooks.

14. C. S. Mellen to W. H. Phipps, 27 September 1899, Northern Pacific Railroad Records, Minnesota Historical Society.

15. A. E. Macartney to Ingram, 3 January 1900; F. Weyerhaeuser to Ingram, 27 December 1899, Ingram Papers; Mellen to C. H. Coster, 26 and 27 December 1899; to Phipps, 26 December 1899, Northern Pacific Railroad Records; Ross. R. Cotroneo, "The History of the Northern Pacific Land Grant, 1900–1952" (Ph.D. diss., University of Idaho, 1966), 247–50.

16. Mellen to Coster, 2 and 5 January 1900; to Edward D. Adams, 3 and 4 January 1900, Northern Pacific Railroad Records; Macartney to Ingram, 3 January 1900; W. H. Day to Ingram, 7 November 1899, Ingram Papers; Hidy, Hill, and Nevins, *Timber and Men*, 212–13.

17. Prior to 1900, the Northern Pacific had never sold more than 600,000 acres of land in a single year in Washington, Oregon and Idaho combined. Edmond S. Meany, Jr., "The History of the Lumber Industry in the Pacific Northwest to 1917" (Ph.D. diss., Harvard University, 1935), 223, 229; W. H. Boner to Ingram, 18 February 1900; J. G. Heim to N. B. Noble, 26 February 1900, Ingram Papers; Renton, Holmes & Co.

to Port Blakely Mill Co., 7 November 1899, Port Blakely Mill Co. Records, Manuscripts Collection, University of Washington Library; *Pacific Lumber Trade Journal* 7(July 1901):26.

18. Herbert Hunt, *Tacoma, Its History and Its Builders*, 3 vols. (Chicago, 1916):3:382–85; F. K. Weyerhaeuser interview, 34–36; Arch Whisnant oral history interview, 7 July 1960, Forest History Society, 44–45.

19. *West Coast Lumberman* 11(August 1900):455; Long to F. S. Bell, 21 February 1901; to James E. Long, 1 March 1906, Weyerhaeuser Timber Co. Records.

20. Long to J. Long, 1 March 1906; to John Campbell, 21 April 1900; to Jackson F. Kimball, 23 February 1909, Weyerhaeuser Timber Co. Records; F. Talbot to Fred Drew, 30 April 1900, Pope & Talbot Records.

21. *Timberman* 1(June 1900):17; Long to Bell, 27 March 1900; to F. Weyerhaeuser, 14 March and 7 June 1900; to S. T. McKnight, 3 April 1900; Memorandum Regarding Timber Lands, 15 August 1903, Weyerhaeuser Timber Co. Records.

22. F. Weyerhaeuser to Ingram, 11 December 1899 and 13 July 1900; Phipps to F. Weyerhaeuser, 7 July 1900, Ingram Papers; Long to F. Weyerhaeuser, 14 and 16 March, 7 and 21 June, and 3 and 18 July 1900; 5 October 1901; to Bell, 22 May 1900; to R. L. McCormick, 9 July 1900, Weyerhaeuser Timber Co. Records; Hidy, Hill, and Nevins, *Timber and Men*, 213–14.

23. Long to F. Weyerhaeuser, 23 October 1900 and 5 January 1901; to Ingram, 31 October 1900 and 27 April 1901; Memorandum Regarding Timber Lands, 15 August 1903, Weyerhaeuser Timber Co. Records; F. Weyerhaeuser to Ingram, 6 and 13 November 1900; Long to Ingram, 17 November 1900, Ingram Papers; *Pacific Lumber Trade Journal* 6(May 1900):18, Hidy, Hill, and Nevins, *Timber and Men*, 222–25; Bureau of Corporations, *The Lumber Industry*, Part 2, *Concentration of Timber Ownership in Selected Regions* (Washington, D.C., 1913), 30.

24. Long to F. Weyerhaeuser, 9 January 1901; to Hugh Bellas, 14 August 1901, Weyerhaeuser Timber Co. Records; *Pacific Lumber Trade Journal* 6(June 1900):14; (March 1901):10; 7(October 1901):9–10; 8(July 1902):30; *Timberman* 1(June 1900):17; *West Coast Lumberman* 11(August 1900):452.

25. Pope & Talbot also tried to sell the Grays Harbor Commercial Company to Weyerhaeuser. A. W. Jackson to Walker, 15 and 22 March 1902; W. Talbot to Walker, 19 October 1899, Pope & Talbot Records; Walker to Jackson, 18 March 1902, Walker Letterbooks; Robert E. Ficken, "Weyerhaeuser and the Pacific Northwest Timber Industry, 1899–1903," *Pacific Northwest Quarterly* 70(October 1979):152.

26. Long to Ewart, 23 October 1901; to Macartney, 30 December 1901 and 17 January 1902; to F. Weyerhaeuser, 13 January 1902 and 15 June 1903; to Weyerhaeuser Timber Co., St. Paul, 17 January 1902, Weyerhaeuser Timber Co. Records; *Pacific Lumber Trade Journal* 7(January 1902):34; Hidy, Hill, and Nevins, *Timber and Men*, 221–22.

27. Walker to Jackson, 23 March 1901 and 26 August 1902, Walker Letterbooks; Long to F. Weyerhaeuser, 28 May 1901; to Macartney, 5 July 1901; to McKnight, 30 November 1906, Weyerhaeuser Timber Co. Records; Hidy, Hill and Nevins, *Timber and Men*, 222–25.

28. Walker to Jackson, 23 March 1901 and 26 August 1902; to W. Talbot, 21 May 1902, Walker Letterbooks; F. Talbot to Drew, 30 April 1900; Jackson memorandum, 19 August 1902, Pope & Talbot Records; Long to F. Weyerhaeuser, 18 September 1901, Weyerhaeuser Timber Co. Records.

29. R. Merrill to T. Merrill, 1 May 1900, Merrill & Ring Lumber Co. Records; *Pacific Lumber Trade Journal* 6(January 1901):9; *West Coast Lumberman* 11(August 1900):451.

30. Simpson retained ownership of the South Bend and Knappton mills. Emerson to Richardson, 9 April 1900; to P. D. Norton, 10 April 1900; to Harry C. Heermans,

11 April, 9 May, and 16 June 1900; to C. H. Jones, 18 May, 20 June, 17 September, and 29 December 1900; to Charles B. Emerson, 8 June 1900; to A. M. Simpson, 11 February 1901, Emerson Letterbooks; Ficken, "Weyerhaeuser," 152.

31. R. Merrill to T. Merrill, 25 June 1899; to Clayton Hay, 14 February 1941, Merrill & Ring Lumber Co. Records; Charles P. LeWarne, *Utopias on Puget Sound, 1885–1915* (Seattle, 1975), 40–41, 51–54.

32. Statement of Polson Logging Co., 24 August 1903; T. Merrill to R. Merrill, 5 June, 1, 5, 15, and 23 August and 1, 22, and 25 September 1911, Merrill & Ring Lumber Co. Records; *Pacific Lumber Trade Journal* 9(March 1904):28. The Polson employee is quoted in Sam Churchill, *Don't Call Me Ma* (Garden City, N.Y., 1977), 49.

33. The Holmes-Campbell faction retained their interest in the Simpson Logging Company, but soon began negotiating a sale to Sol Simpson and A. H. Anderson. *Pacific Lumber Trade Journal* 8(February 1903):9; Robert E. Ficken, "The Port Blakely Mill Company, 1888–1903," *Journal of Forest History* 21(October 1977):217; W. Talbot to Edwin G. Ames, 16 July 1918; Ames to W. Talbot, 1 July 1914 and 11 July 1918; F. Talbot to Ames, 2 October 1913, Pope & Talbot Records.

34. "Marvelous Growth and Development of City of Smokestacks," *American Lumberman* (5 March 1910):56–57; *Pacific Lumber Trade Journal* 5(April 1900):9; 6(November 1900):12–13; 11(May 1905):16.

35. "Bloedel Donovan Lumber Company Mills," *American Lumberman* (19 July 1913):43–57; Long to John Blodgett, 25 August 1928, Weyerhaeuser Timber Co. Records; W. Talbot to Ames, 6 August 1918, Pope & Talbot Records.

36. Bureau of Corporations, *Standing Timber*, 205–9; John Ise, *The United States Forest Policy* (New Haven, 1920), 323–24; Compton, *Organization of the Lumber Industry*, 62–64; R. C. Bryant, "Lumbermen and Our National Development," *American Forests* 19(December 1913):946–51; Meany, "History of the Lumber Industry," 230–34.

37. *Pacific Lumber Trade Journal* 7(June 1901):9; George H. Emerson, "Logging on Grays Harbor," *Timberman* 8(September 1907):21.

9: Fits and Starts

1. *West Coast Lumberman* 15(June 1904):601–2.

2. George S. Long to D. N. King, 31 January 1908, Weyerhaeuser Timber Co. Records, Weyerhaeuser Co. Archives.

3. Edwin G. Ames to George P. Furber, 10 May 1906, Edwin G. Ames Papers, Ames Coll., Manuscripts Collection, University of Washington Library; Long to L. Heath, 23 April 1906, Weyerhaeuser Timber Co. Records.

4. Cyrus Walker to W. H. Talbot, 24 September 1902, Cyrus Walker Letterbooks, Ames Coll., Manuscripts Collection, University of Washington Library; Long to William Carson, 22 September 1908, Weyerhaeuser Timber Co. Records; Harry C. Heermans to Q. W. Wellington, 27 September 1904, Harry C. Heermans Papers, Manuscripts Collection, University of Washington Library.

5. Bureau of the Census, *Manufactures, 1905*, Part 3, *Special Reports on Selected Industries* (Washington, D.C., 1908), 588; Henry B. Steer, comp., *Lumber Production in the United States, 1799–1946*, USDA Misc. Pub. No. 669(Washington, D.C., 1948), 11, 196; *Pacific Lumber Trade Journal* 11(January 1906):20.

6. Long to W. G. Ingram, 17 July 1900, Weyerhaeuser Timber Co. Records; Long

to O. H. Ingram, 17 November 1900, O. H. Ingram Papers, State Historical Society of Wisconsin; Steer, *Lumber Production,* 11–12.

7. Frederick C. Talbot to Fred Drew, 19 October 1900, Pope & Talbot Records, Ames Coll., Manuscripts Collection, University of Washington Library; *Pacific Lumber Trade Journal* 6(October 1900):9–10; 9(January 1904):51–52; Edmond S. Meany, Jr., "The History of the Lumber Industry in the Pacific Northwest to 1917" (Ph.D. diss., Harvard University, 1935), 162; George H. Emerson to C. H. Jones, 2 June 1903, George H. Emerson Letterbooks, Manuscripts Collection, University of Washington Library.

8. Long to W. I. Ewart, 23 March 1906; to F. S. Bell, 28 November 1906, Weyerhaeuser Timber Co. Records; *Pacific Lumber Trade Journal* 12(May 1906):9; (June 1906):9; (August 1906):9; (December 1906):9; Fred E. Dickinson, "Development of Southern and Western Freight Rates on Lumber," *Southern Lumberman* 185(15 December 1952):244; Bureau of Corporations, *The Lumber Industry,* Part 1, *Standing Timber* (Washington, D.C., 1913), 84–85.

9. *Timberman* 2(September 1901):29; 7(November 1905):24B; *Pacific Lumber Trade Journal* 7(October 1901):11; 11(January 1906):17; 12(October 1906):9; Emerson to G. W. Pearce, 20 March 1900, Emerson Letterbooks.

10. When Frederick Weyerhaeuser proposed the erection of a statue on Puget Sound to honor his St. Paul neighbor, George Long responded that a meeting of mill owners had approved "a resolution, suggesting that the inscription on the monument be certain expressions which were far from complementary to Mr. Hill." Long to F. E. Weyerhaeuser, 27 April 1909, Weyerhaeuser Timber Co. Records; *Timberman* 8(September 1907):19; *Pacific Lumber Trade Journal* 10(October 1904):9; 11(October 1905):9; (February 1906):9; 12(September 1906):9.

11. Emerson to Jones, 25 November 1902 and 23 July 1903, Emerson Letterbooks; Dant and Russell, Inc. typescript history, Oregon Historical Society; Edwin T. Coman, Jr. and Helen M. Gibbs, *Time, Tide and Timber: A Century of Pope & Talbot* (Stanford, Cal., 1949), 265–68.

12. Meany, "History of the Lumber Industry," 151–52; Emerson to Jones, 27 November 1906, Emerson Letterbooks; F. Talbot to Walker, 27 November 1906, Pope & Talbot Records; Long to C. H. Davis, 17 October 1906, Weyerhaeuser Timber Co. Records.

13. Long to C. R. McCormick, 31 July 1906; to John W. Eddy, 22 December 1906, Weyerhaeuser Timber Co. Records; *Pacific Lumber Trade Journal* 12(August 1906):10; (October 1906):10; W. Talbot to Walker, 21 January and 9 February 1907, Pope & Talbot Records.

14. John H. Cox, "Organizations of the Lumber Industry in the Pacific Northwest, 1889–1914" (Ph.D. diss., University of California, Berkeley, 1937), 17–27, 31–34, 38–47; Emerson to E. J. Holt, 1 March 1895, Emerson Letterbooks.

15. Trade associations were also established among manufacturers on the Washington side of the Columbia River and in Oregon. *Pacific Lumber Trade Journal* 12(June 1906):9; John H. Cox, "Trade Associations in the Lumber Industry of the Pacific Northwest, 1899–1914," *Pacific Northwest Quarterly* 41(October 1950):290, 296–97, 301–2; Emerson to Jones, 27 November 1906, Emerson Letterbooks.

16. Renton, Holmes & Co. to Port Blakely Mill Co., 19 March, 1, 8, and 16 April, 29 May and 12 July 1901, Port Blakely Mill Co. Records, Manuscripts Collection, University of Washington Library; Walker to W. Talbot, 11 and 13 December 1902; to F. Talbot, 6 November 1902, Walker Letterbooks; Cox, "Trade Associations," 291–94.

17. *Timberman* 7(March 1906):20; Cox, "Trade Associations," 294–96; Long to Clair, 12 March and 26 April 1907, Weyerhaeuser Timber Co. Records; Bureau of Corporations, *The Lumber Industry,* Part 4, *Conditions in Production and Wholesale*

Distribution Including Wholesale Prices (Washington, D.C., 1914), 386–444; Harold
K. Steen, "Forestry in Washington to 1925" (Ph.D. diss., University of Washington,
1969), 189–98; R. C. Bryant, "The Lumber Industry," *American Forestry* 19(October
1913):674.

18. Ames to W. Talbot, 3 April 1913; W. Talbot to Ames, 8 April 1913; F. Talbot
to Ames, 8 October 1909, Pope & Talbot Records; Long to W. H. Boner, 29 August
1911, Weyerhaeuser Timber Co. Records.

19. *Timberman* 2(October 1901):5; *Pacific Lumber Trade Journal* 6(April 1901):9;
8(November 1902):11; Annual Report, 26 January 1912, West Coast Lumbermen's
Association Records, Oregon Historical Society; Walker to A. W. Jackson, 26 August
1902, Walker Letterbooks.

20. Ames to E. P. Blake, 1 March 1907; to R. H. Alexander, 2 June 1906, Ames
Papers; L. H. Pierson to Ames, 19 September 1908, Grays Harbor Commercial Co.
Records, Ames Coll., Manuscripts Collection, University of Washington Library; F.
Talbot to Ames, 29 July and 1 August 1910, Pope & Talbot Records; *Pacific Lumber
Trade Journal* 11(March 1906):9; (April 1906):9.

21. *Pacific Lumber Trade Journal* 11(June 1905):15; Long to Weyerhaeuser Timber
Co., Everett, 29 April 1905; to A. A. Baxter, 12 September 1905; to O. Ingram, 6
March 1906, Weyerhaeuser Timber Co. Records; Cox, "Trade Associations," 298–
300.

22. *Timberman* 9(January 1908):19.

23. Noting that the railroads, angered over Theodore Roosevelt's regulatory legislation, referred to empty cars as "Teddy Bears," a trade journal responded that "we
might return the comliment by naming these idle mills of the Pacific Northwest Jimmy
Hills." *Pacific Lumber Trade Journal* 13(September 1907):9–10; (February 1908):9;
14(January 1909):22; Emerson to Holt, 30 December 1907, Emerson Letterbooks.

24. *Pacific Lumber Trade Journal* 13(November 1907):9, 11–14; (December 1907):11–
16; (January 1908):9; (April 1908):9; 14(May 1908):9; (June 1908):17–18; (July 1908):24–
40; Emerson to Holt, 30 December 1907; to Austin E. Griffiths, 2 July 1908, Emerson
Letterbooks.

25. Long to F. E. Weyerhaeuser, 2 and 15 November and 23 December 1907; to
E. T. Abbott, 4 November 1907; to Boner, 7 November 1907; to Davis, 7 November
1907; to S. T. McKnight, 22 November 1907; to O. Ingram, 16 January 1908, Weyerhaeuser Timber Co. Records; Emerson to Jones, 31 October 1907, Emerson Letterbooks; W. Talbot to Walker, 19 November 1907, Pope & Talbot Records.

26. Steer, comp., *Lumber Production,* 12–13, 196; Montgomery Collins, "Pitfalls
of Timber Bond Issues," *Forestry Quarterly* 12(December 1914):548–58; Wilson
Compton, "The Present Conditions in the Lumber Industry," *Journal of Forestry* 15(April
1917):388, 390; Paul F. Sharp, "The War of the Substitutes: The Reaction of the Forest
Industries to the Competition of Wood Substitutes," *Agricultural History,* 23(October
1949):274–79; *American Forestry* 21(August 1915):876.

27. Emerson to Justin E. Emerson, 12 October 1910, Emerson Letterbooks; Long
to J. W. Blodgett, 11 April 1911, Weyerhaeuser Timber Co. Records; Ames to William
Walker, 6 and 20 April 1914, Ames Papers.

28. F. Talbot to Ames, 22 July 1910 and 3 May 1913; Ames to W. Talbot, 23 June
1913 and 19 February 1914; to F. Talbot, 13 September 1911 and 19 April 1912; W.
Talbot to Ames, 31 August 1914, Pope & Talbot Records; Everett Griggs to Ames,
24 August 1914, Ames Papers.

29. Emerson to A. M. Simpson, 21 April 1892, Emerson Letterbooks; *Timberman*
7(January 1906):17; 14(April 1913):30, 36; *Pacific Lumber Trade Journal* 18(January
1913):33; 19(July 1913):19.

30. Robert Dollar, "Lumber Trade and the Canal," *American Forestry* 20(July

1914):499–500; *Timberman* 16(February 1915):23; Ames to W. Walker, 23 and 28 March 1914; D. E. Skinner to Ames, 18 March 1914; W. E. Humphrey to Ames, 21 March 1914, Ames Papers; Ames to W. Talbot, 7 and 19 March and 8 May 1914; W. Talbot to Ames, 23 March 1914, Pope & Talbot Records.

31. *Timberman* 16(February 1915):51; 17(February 1916):32A.

32. Ames to J. J. Herlihy, 10 June 1913, Ames Papers; Long to D. B. Hanson, 27 October 1911, Weyerhaeuser Timber Co. Records; Ames to W. Talbot, 26 June 1913; to F. Talbot, 14 July 1913, Pope & Talbot Records.

33. Ames to F. Talbot, 3 October 1911, Pope & Talbot Records; Long to F. H. Thatcher, 21 November 1911 and 11 January 1912; to F. E. Weyerhaeuser and Bell, 14 December 1911; to F. E. Weyerhaeuser, 22 January 1912, Weyerhaeuser Timber Co. Records; H. J. Pierce to Clark L. Ring, 29 March 1912; to Alex Polson, 24 April 1912; Polson to Ring, 12 April 1912, Merrill & Ring Lumber Co. Records, Manuscripts Collection, University of Washington Library; *Timberman* 12(September 1911):47; 13(December 1911):64C; *Pacific Lumber Trade Journal* 17(September 1911):17; (January 1912):34.

34. Long to Bell, 8 November and 21 and 26 December 1911; to F. E. Weyerhaeuser and Bell, 14 December 1911; to F. E. Weyerhaeuser and Thatcher, 14 December 1911 and 22 January 1912; to F. E. Weyerhaeuser, 24 January 1912; to Frederick Weyerhaeuser, 26 January 1912; to Pierce, 7 March 1912, Weyerhaeuser Timber Co. Records; Ames to F. Talbot, 2 March 1912, Pope & Talbot Records; Pierce to Ring, 29 March 1912, Merrill & Ring Lumber Co. Records.

35. Long to F. E. Weyerhaeuser, 23 February, 14 March, 8 April, 3 June, and 24 July 1912; to F. E. Weyerhaeuser and Thatcher, 5 March 1912; to Thatcher, 12 and 25 March, 2, 15, and 29 April and 24 July 1912; to C. A. Weyerhaeuser, 26 March 1912; to Bell, 5 April 1912; to Pierce, 17 July 1912, Weyerhaeuser Timber Co. Records; Ames to F. Talbot, 4 March 1912, Pope & Talbot Records; Polson to R. D. Merrill, 6 May 1912, Merrill & Ring Lumber Co. Records; *Pacific Lumber Trade Journal* 18(August 1912):37.

36. Ames to W. Talbot, 15 and 20 July 1912; 20 February and 14 and 30 March 1914; W. Talbot to Ames, 17 February 1914, Pope & Talbot Records; Ames to Long, 4 December 1914; to E. J. Palmer, 24 November 1913; to Griggs, 3 March 1914; to Humphrey, 25 May 1914; to W. L. Jones, 19 June 1914, Ames Papers.

37. Ames to W. Talbot, 20 February 1914, Pope & Talbot Records; Ames to W. Walker, 14 October 1916, Ames Papers; *Timberman* 17(May 1916):28; (October 1916):27; 18(December 1916):48E; *West Coast Lumberman* 32(1 October 1917):19; Edwin E. Pratt, *The Export Lumber Trade of the United States,* USDC Misc. Series No. 67(Washington, D.C., 1918), 54–56, 104–6.

38. Burt P. Kirkland, "Continuous Forest Production of Privately Owned Timberlands as a Solution of the Economic Difficulties of the Lumber Industry," *Journal of Forestry* 15(January 1917):43–64; *American Forestry* 21(August 1915):876–77; (December 1915):1124–25; 23(February 1917):110–11; *Forestry Quarterly* 13(December 1915):590–91.

39. Long to Boner, 10 October 1912; to F. E. Weyerhaeuser, 24 July and 23 September 1912, Weyerhaeuser Timber Co. Records; W. Talbot to Ames, 20 March 1914, Pope & Talbot Records; Ralph W. Hidy, Frank Ernest Hill and Allan Nevins, *Timber and Men: The Weyerhaeuser Story* (New York, 1963), 274–78.

40. Ames to W. Walker, 21 October and 4 November 1912; to Emily T. Walker, 2 October 1913; to Alice B. Talbot, 17 November 1913, Ames Papers; *Timberman* 15(August 1914):42; 16(January 1915), 46; Stephen Dow Beckham, "Asa Mead Simpson, Lumberman and Shipbuilder," *Oregon Historical Quarterly* 68(September 1967):272.

41. *Timberman* 15(August 1914):44, 56, 68, 61; (September 1914):87; (October

254 *Notes*

1914):60–61, 64; Ames to Daniel Kelleher, 26 September 1914; to Alfred Rose, 21 December 1914; to W. R. C. Estes, 29 December 1914; W. Talbot to Ames, 25 September 1914, Ames Papers.

10: Gifford Pinchot Men

1. E. T. Allen, "What Protective Co-operation Did," *American Forestry* 16(November 1910):643.
2. George S. Long to R. W. Douglas, 29 August 1910, Weyerhaeuser Timber Co. Records, Weyerhaeuser Co. Archives.
3. Cyrus Walker to Frederick C. Talbot, 24 June and 13 and 17 September 1902; to W. H. Talbot, 15 and 19 September 1902, Cyrus Walker Letterbooks, Ames Coll., Manuscripts Collection, University of Washington Library; Stewart H. Holbrook, *Burning an Empire* (New York, 1944), 112, 115–16, 119; William T. Cox, "Recent Forest Fires in Oregon and Washington," *Forestry & Irrigation* 8(November 1902):469.
4. Long to Frederick Weyerhaeuser, 12 September 1902; to O. H. Ingram, 20 September 1902; to Thomas Irvine Lumber Co., 16 September 1902, Weyerhaeuser Timber Co. Records; W. Talbot to Walker, 15 September 1902, Pope & Talbot Records, Ames Coll., Manuscripts Collection, University of Washington Library; Ralph W. Andrews, *Timber!* (Seattle, 1968), 131; Harold K. Steen, "Forestry in Washington to 1925" (Ph.D. diss., University of Washington, 1969), 86–87.
5. Holbrook, *Burning an Empire,* 109–14; *Forestry & Irrigation* 8(December 1902):487.
6. W. H. Day to Ingram, 8 November 1900, O. H. Ingram Papers, State Historical Society of Wisconsin; George H. Emerson to Charles B. Emerson, 1 November 1900, George H. Emerson Letterbooks, Manuscripts Collection, University of Washington Library.
7. Stewart Holbrook, "Sawdust on the Wind," *American Heritage* 4(Summer 1953): 51; Charles P. Norcross, "Weyerhaeuser—Richer than John D. Rockefeller," *Cosmopolitan* 42(January 1907):255; Emerson Hough, "The Slaughter of the Trees," *Everybody's Magazine* 17(May 1908):588; Long to F. S. Bell, 5 September 1907, Weyerhaeuser Timber Co. Records.
8. *World Today* 21(February 1912):1734.
9. Edwin G. Ames to W. Talbot, 4 May 1914, Pope & Talbot Records; R. S. Kellogg to R. H. Downman, 6 September 1916; Downman to Kellogg, 12 August and 4 September 1916, National Forest Products Association Records, Forest History Society; *Pacific Lumber Trade Journal* 19(July 1913):19.
10. Roy E. Appleman, "Timber Empire from the Public Domain," *Mississippi Valley Historical Review* 26(September 1939):201–3; John Ise, *The United States Forest Policy* (New Haven, 1920), 184–85; Bureau of Corporations, *The Lumber Industry,* Part 1, *Standing Timber* (Washington, D.C., 1913), 239; Long to George M. Cornwall, 7 June 1910, Weyerhaeuser Timber Co. Records.
11. Roy O. Hoover, "The Public Land Policy of Washington State: The Initial Period, 1889–1912" (Ph.D. diss., Washington State University, 1967), 163–70, 193–205, 211–15, 217–34, 238–39; Steen, "Forestry in Washington," 27–28, 113–14, 169–70.
12. Long to Bell, 14 April 1910, Weyerhaeuser Timber Co. Records.
13. Long to Bell, 31 July and 27 August 1906; to F. E. Weyerhaeuser, 29 August 1907; to Weyerhaeuser Timber Co., St. Paul, 11 June 1909, Weyerhaeuser Timber

Co. Records; *Pacific Lumber Trade Journal* 10 (September 1904):9, *Report of the National Conservation Commission, February 1909* 3 vols. (Washington, D.C., 1909), 2:600.

14. Long to S. T. McKnight, 29 August 1907; to Clyde S. Martin, 30 August 1910, Weyerhaeuser Timber Co. Records; Ames to W. Talbot, 28 February 1914, Pope & Talbot Records; B. E. Fernow, "Taxation of Woodlands," in *Forestry Quarterly* 5(December 1907):373–75.

15. Ames to Howard Taylor, et al., 10 November 1913, Edwin G. Ames Papers, Ames Coll., Manuscripts Collection, University of Washington Library; Long to Bell, 27 March 1900 and 11 and 23 August 1910; to R. L. McCormick, 20 April 1908, Weyerhaeuser Timber Co. Records.

16. Ames to W. Talbot, 21 April and 6 November 1914, Pope & Talbot Records; Ames to B. R. Lewis, 6 November 1914, Ames Papers; Long to A. L. Miller, 14 January 1909; to F. E. Weyerhaeuser, 20 February 1911; to E. T. Allen, 1 February 1912; to Cornelius Gerber, 6 May 1912, Weyerhaeuser Timber Co. Records; *Timberman* 13(February 1912):18.

17. T. D. Merrill to R. D. Merrill, 17 June 1912, Merrill & Ring Lumber Co. Records, Manuscripts Collection, University of Washington Library; Emerson to Isabella H. Fitz, 31 October 1912, Emerson Letterbooks; *Timberman* 16(November 1914):24; 17(April 1916):28.

18. William B. Greeley, *Forests and Men* (Garden City, N.Y., 1951), 51; Arthur A. Ekirch, Jr., *Man and Nature in America* (New York, 1963), Chap. 7. Muir is quoted in Edwin Way Teale, ed., *The Wilderness World of John Muir* (Boston, 1954), 315.

19. Frederick Bancroft, ed., *Speeches, Correspondence and Political Papers of Carl Schurz*, 6 vols. (New York, 1913), 5:23; James E. Defebaugh, *History of the Lumber Industry of America*, 2 vols. (Chicago, 1906–7), 1:421–22.

20. Emerson to Harry C. Heermans, 4 March 1898; to A. M. Simpson, 17 March 1897; to John L. Wilson, 18 March 1897, Emerson Letterbooks; J. J. Donovan to Erastus Brainerd, 15 November 1909, Erastus Brainerd Papers, Manuscripts Collection, University of Washington Library. Seattle Chamber of Commerce quoted in Steen, "Forestry in Washington," 68.

21. Walker to W. Talbot, 13 July 1885, Pope & Talbot Records, Pope & Talbot Archives; H. K. Benson, "Industrial Resources of Washington," *Journal of Geography* 14(May 1916):354; Emerson to Simpson, 10 February 1896; to John L. Harris, 18 April 1896; to American Lumberman, 29 December 1899, Emerson Letterbooks.

22. Charles Lord Russell of Killowen, *Diary of a Visit to the United States of America in the Year 1883* (New York, 1910), 99–100; Lemuel Ely Quigg, *New Empires in the Northwest* (New York, 1889), 72; Jean Hazeltine, "Timber Conservation in Western Washington" (M.A. thesis, Ohio State University, 1952), 47.

23. James G. Swan, *The Northwest Coast, or, Three Years' Residence in Washington Territory* (1857; reprint, Seattle, 1972), 134; Charles M. Gates, ed., *Messages of the Governors of the Territory of Washington to the Legislative Assembly, 1854–1889* (Seattle, 1940), 66, 153, 169, 187; Steen, "Forestry in Washington," 22–27.

24. *Report of the National Conservation Commission*, 1:3; Gifford Pinchot, *The Fight for Conservation* (1910; reprint, Seattle, 1967), 50, 91–92; Samuel P. Hays, *Conservation and the Gospel of Efficiency: The Progressive Conservation Movement, 1890–1920* (Cambridge, Mass., 1959), 265–66.

25. Charles R. Van Hise, *The Conservation of Natural Resources in the United States* (New York, 1913), 210; Pinchot, *Fight for Conservation*, 14–17; *Proceedings of a Conference of Governors, May 13–15, 1908* (Washington, D.C., 1909), 9; Long to M. L. Erickson, 20 January 1908, Weyerhaeuser Timber Co. Records.

26. R. S. Kellogg, "What Forest Conservation Means," *Conservation* 15 (May 1909):283–86; *Forestry & Irrigation* 13(June 1907):276; Theodore Roosevelt to Gif-

ford Pinchot, 24 August 1906, in Elting E. Morison, ed., *The Letters of Theodore Roosevelt*, 8 vols. (Cambridge, Mass., 1951–54), 5:384.

27. Long to McKnight, 16 May 1903; to Allen, 22 December 1909; to M. E. Hay, 3 October 1910, Weyerhaeuser Timber Co. Records; J. E. Rhodes, "Lumbermen and Forestry," *American Forestry* 19(May 1913):318–20.

28. *Pacific Lumber Trade Journal* 11(September 1905):10–11; Long to D. P. Simons, Jr., 16 November 1908; to Victor Beckman, 20 March 1909; to E. G. Griggs, 23 March 1909, Weyerhaeuser Timber Co. Records; R. Merrill to E. D. Wetmore, 3 September 1937, Merrill & Ring Lumber Co. Records; Hays, *Conservation and the Gospel of Efficiency*, 29–35.

29. Long to J. E. Defebaugh, 19 September 1908; to F. Weyerhaeuser, 23 December 1902 and 20 January 1903; to F. E. Weyerhaeuser, 19 and 20 February and 9 and 12 March 1903; to McKnight, 16 May 1903; to McCormick, 2 and 13 May 1903, Weyerhaeuser Timber Co. Records.

30. Long to Cornwall, 20 September and 4 October 1902; to Beckman, 20 September 1902 and 8 June 1903; to McCormick, 23 September 1902; to A. E. Macartney, 26 January 1903 and 21 March 1905, Weyerhaeuser Timber Co. Records; Emerson to Long, 9 January 1903, Emerson Letterbooks; George T. Morgan, Jr., "Conflagration as Catalyst: Western Lumberman and American Forest Policy," *Pacific Historical Review* 47 (May 1978):171–76.

31. Long to D. P. Simons, 2 May 1906; to Fred G. Plummer, 15 January 1909, Weyerhaeuser Timber Co. Records; *Report of the State Forester to the State Board of Forest Commissioners, 1914* (Olympia, 1915), 15; H. J. Andrews and Robert W. Cowlin, *Forest Resources of the Douglas Fir Region*, PNFES Research Note No. 13(Portland, 1934), 5.

32. Long to Ingram, 6 May 1908; to Griggs, 11 December 1908, Weyerhaeuser Timber Co. Records; *Report of the National Conservation Commission*, 2:464–66; Hugo Winkenwerder, *Forestry in the Pacific Northwest* (Washington, D.C., 1928), 35.

33. Long to Bell, 24 April 1908; to Ingram, 6 May 1908; to Victor Thrane, 20 May 1908; to J. L. Bridge, 17 June 1908; to Macartney, 23 August 1910, Weyerhaeuser Timber Co. Records; Ames to W. Talbot, 7 March 1914, Pope & Talbot Records; Annual Report, 1909, Sound Timber Co. Records, Minnesota Historical Society.

34. Long to F. E. Weyerhaeuser, 17 February 1909; to J. P. McGoldrick, 11 and 19 February 1909; to Allen, 27 December 1909; 4 and 25 January, 3 June and 1 July 1910, and 17 June 1911; to Allison Laird, 14 June 1911, Weyerhaeuser Timber Co. Records; E. T. Allen, *Practical Forestry in the Pacific Northwest* (Portland, 1911).

35. Steen, "Forestry in Washington," 185–88, 223–32; *Report of the State Forester, 1914*, 19; Floyd Schmoe, *A Year in Paradise* (New York, 1959), 178; Long to Laird, 7 December 1910; to Griggs, 11 December 1908 and 17 December 1909; to Edward Hines Lumber Co., 23 December 1909, Weyerhaeuser Timber Co. Records.

36. Donovan to Richard A. Ballinger, 27 November 1909 and 25 January 1910, Richard A. Ballinger Papers, Manuscripts Collection, University of Washington Library; Hoover, "Public Land Policy," 174–79, 181–87; H. J. Bergman, "The Reluctant Dissenter: Governor Hay of Washington and the Conservation Problem," *Pacific Northwest Quarterly* 62(January 1971):27–33.

37. Long to Cornwall, 10 January 1910; to Allen, 14 January 1910; to S. L. Flewelling, 14 January 1910, Weyerhaeuser Timber Co. Records; *Timberman* 10(October 1909):19; 11(January 1910):19; Steen, "Forestry in Washington," 144–46.

38. B. E. Fernow, "Why Do Lumber Men Not Apply Forestry?" *American Forestry* 18(October 1912):613–15; M. Nelson McGeary, *Gifford Pinchot, Forester-Politician* (Princeton, N.J., 1960), 72–73, 84; Pinchot to Henry S. Graves, 31 December 1928, Gifford Pinchot Papers, Library of Congress.

39. Emerson to Griggs, 20 August 1910, Emerson Letterbooks; *American Forestry* 16(October 1910):569–88; *Pacific Lumber Trade Journal* 16(August 1910):39.

40. Hay to Donovan, 16 September 1910, M. E. Hay Papers, Eastern Washington State Historical Society; Ames to W. Talbot, 9 September 1910, Ames Papers; *Pacific Lumber Trade Journal* 16(September 1910): 26; Long to Allen, 24 August and 14 September 1910; to W. I. Ewart, 29 August 1910, Weyerhaeuser Timber Co. Records.

41. Long to Ewart, 29 August 1910, Weyerhaeuser Timber Co. Records; J. J. Donovan, "The Problem of our Logged Off Lands," *American Forestry* 18(July 1912):467–68.

42. Ames to W. E. Humphrey, 29 November 1911, Ames Papers; *Fifth Biennial Report of the Bureau of Labor Statistics and Factory Inspection, 1905–1906* (Olympia, 1907), 239–44; Long to M. T. Milroy, 5 April 1902; to Boner, 19 December 1911, Weyerhaeuser Timber Co. Records.

43. Cloice R. Howd, *Industrial Relations in the West Coast Lumber Industry*, BLS Bulletin No. 349(Washington, D.C., 1924), 41–42; Rexford Tugwell, "The Casual of the Woods," *Survey* 44(3 July 1920):473; T. Merrill to R. Merrill, 29 May 1923, Merrill & Ring Lumber Co. Records; Ames to W. Talbot, 12 March 1918, Pope & Talbot Records.

44. James Rowan, *The I.W.W. in the Lumber Industry* (Seattle, n.d.), 10; Howd, *Industrial Relations*, 42–43; Edmond S. Meany, Jr., "The History of the Lumber Industry in the Pacific Northwest to 1917" (Ph.D. diss, Harvard University, 1935), 282; Frank Hobi Reminiscences, Manuscripts Collection, University of Washington Library, 16.

45. *Timberman* 14(December 1912):23; *Fifth Biennial Report of the Bureau of Labor Statistics*, 168; Andrew W. Prouty, "Logging with Steam in the Pacific Northwest: The Men, the Camps, and the Accidents, 1885–1918" (M.A. thesis, University of Washington, 1973), 142; Robert Cantwell, *The Land of Plenty* (New York, 1934), 252; *Pacific Lumber Trade Journal* 18(November 1912):19.

46. Walker to F. Talbot, 2 September 1901, Walker Letterbooks; F. Talbot to Walker, 13 August 1901; Ames to W. Talbot, 20 August 1913, Pope & Talbot Records; Long to F. E. Weyerhaeuser, 13 May 1903, Weyerhaeuser Timber Co. Records; Francis Rotch to Ridgeley Force, 5 September 1905, Simpson Logging/Timber Co. Records, Simpson Timber Co. Archives; L. T. Murray, Sr. oral history interview, Forest History Society, 29.

47. Guy W. Williams, *Logger-Talk: Some Notes on the Jargon of the Pacific Northwest Woods* (Seattle, 1930), 17–18, 29; Austin Cary, "How Lumbermen in Following Their Own Interest Have Served the Public," *Journal of Forestry* 15(March 1917):272–73. Labor organizer quoted in Melvyn Dubofsky, *We Shall Be All: A History of the Industrial Workers of the World* (Chicago, 1969), 128.

48. Howd, *Industrial Relations*, 55–56; *Fifth Biennial Report of the Bureau of Labor Statistics*, 101–2, 191–96; *Sixth Biennial Report of the Bureau of Labor Statistics and Factory Inspection, 1907–1908* (Olympia, 1908), 50, 131–33; Vernon H. Jensen, *Lumber and Labor* (New York, 1945), 117–19; Walker to W. Talbot, 17 May 1901, Walker Letterbooks.

49. *The Founding Convention of the I.W.W.: Proceedings* (New York, 1969), 5–6. Constitution quoted in Ralph Chaplin, *Wobbly: The Rough-and-Tumble Story of an American Radical* (Chicago, 1948), 148.

50. Archie Binns, *The Timber Beast* (New York, 1944), 232. T-Bone Slim quoted in Robert L. Tyler, "The Rise and Fall of an American Radicalism: The I.W.W.," *Historian* 19(November 1956):55.

51. *Timberman* 8(March 1907):19; Ames to R. H. Alexander, 15, 16, and 18 March, 1907, Ames Papers; W. Talbot to Walker, 7, 12, 22, and 27 March 1907, Pope & Talbot Records; Howd, *Industrial Relations*, 64–65.

52. Emerson to C. H. Jones, 23 June 1906, Emerson Letterbooks; Ames to Mark E. Reed, 6 December 1911, Ames Papers; Stewart H. Holbrook, "Ghost Towns Walk," *American Forests* 43(May 1937):217; *Pacific Lumber Trade Journal* 10(May 1904):22; (August 1904):9; 13(September 1907):49.

53. Emerson to E. J. Holt, 23 February and 2 March 1891; to Simpson, 5 September 1898; to R. B. Dyer, 22 June 1891, Emerson Letterbooks; Walker to W. Talbot, 4 August 1887, Walker Letterbooks; Renton, Holmes & Co. to Port Blakely Mill Co., 20 June 1888, Port Blakely Mill Co. Records, Manuscripts Collection, University of Washington Library.

54. Joseph F. Tripp, "Progressive Labor Laws in Washington State (1900–1925)" (Ph.D. diss., University of Washington, 1973), 33–35; *Pacific Lumber Trade Journal* 6(April 1901):15; 10(May 1904):10; (August 1904:9–10; Emerson to Simpson, 5 September 1898; to Jones, 27 June 1904, Emerson Letterbooks; Ames to W. Talbot, 20 May 1911, Pope & Talbot Records.

55. Ames to Griggs, 11 January 1911; to Hay, 15 March 1911, Ames Papers; Ames to W. Talbot, 15 and 25 March, 18 April and 20, 27, and 29 May 1911, Pope & Talbot Records; Tripp, "Progressive Labor Laws," 31–32, 38–44, 46–51.

56. Ames to F. Talbot, 20 March and 1, 2, and 4 April 1912, Pope & Talbot Records; Alex Polson to R. Merrill, '29 March and 4, 15, and 19 April 1912, Merrill & Ring Lumber Co. Records; Arch Whisnant oral history interview, 1 November 1958, Forest History Society, 12–15.

57. Ames to F. Talbot, 15 February, 17 March, and 12 May 1913; to W. Talbot, 18 April and 2, 6, 9, and 12 June 1913; F. Talbot to Ames, 18 February and 20 March 1913; W. Talbot to Ames, 6 June 1913, Pope & Talbot Records; Ames to Alexander, 16 June 1913, Ames Papers; Howd, *Industrial Relations,* 57–61, 66–68.

58. Emerson to A. J. West, 22 June 1904, Emerson Letterbooks; Ames to W. Talbot, 7 April 1914, Pope & Talbot Records.

11:   The Spawn of Hate

1. Ray Stannard Baker and William E. Dodd, eds., *The Public Papers of Woodrow Wilson,* 6 vols. (New York, 1925–27), 5:122.

2. *Timberman* 18(July 1917):71.

3. John Dos Passos to Edmund Wilson, 14 January 1931, in Townsend Ludington, ed., *The Fourteenth Chronicle: Letters and Diaries of John Dos Passos* (Boston, 1973), 398.

4. W. H. Talbot to Edwin G. Ames, 26 September 1914; Ames to Frederick C. Talbot, 19 and 30 December 1914; to W. Talbot, 26 March 1915, Pope & Talbot Records, Manuscripts Collection, University of Washington Library; *Timberman* 16(February 1915):63; Bristow Adams, "The War and the Lumber Industry," in *American Forestry* 20(September 1914):617–25; R. C. Bryant, "The European War and the Lumber Trade," *American Forestry* 20(December 1914), 881–86.

5. Ames to Alfred K. Ames, 23 June 1915, Edwin G. Ames Papers, Manuscripts Collection, University of Washington Library; Ames to F. Talbot, 8 February and 31 July 1915, Pope & Talbot Records; *Timberman* 16(October 1915):56, 67; 17(December 1915):68–70, 72; (April 1916):68; Henry B. Steer, comp., *Lumber Production in the United States, 1799–1946,* USDA Misc. Pub. No. 669(Washington, D.C., 1948), 11–13.

6. *Pacific Lumber Trade Journal* 18(February 1913):19; *Timberman* 15(December

1914):53, 65; 16(March 1915):41; (October 1915):67; 17(March 1916):27; (July 1916):75; Timothy Jerome to R. D. Merrill, 22 January 1917, Merrill & Ring Lumber Co. Records, Manuscripts Collection, University of Washington Library.

7. Ames to F. Talbot, 29 January 1916; to W. Talbot, 18 and 21 November 1916, Pope & Talbot Records; Joseph F. Tripp, "Progressive Labor Laws in Washington State (1900–1925)" (Ph.D. diss.,. University of Washington, 1973), 127–43.

8. *Timberman* 18(November 1916):80P; Ames to W. Talbot, 10, 13, and 21 November 1916; W. Talbot to Ames, 9 November 1916, Pope & Talbot Records; Norman H. Clark, *Mill Town* (Seattle, 1970), Chaps. 9–10.

9. Ames to R. C. Fuller, 13 June 1917; to Richard Condon, 1 May 1917; to Everett Griggs, 22 September 1917, Ames Paper; Ames to W. Talbot, 23 May 1917, Pope & Talbot Records; R. C. Bryant, "The War and the Lumber Industry," *Journal of Forestry* 17(February 1919):126–27.

10. George S. Long to R. H. Downman, 5 (two letters of this date) and 28 June 1917; to Austin Cary, 7 July 1917, Weyerhaeuser Timber Co. Records, Weyerhaeuser Co. Archives; Jerome to Clark L. Ring, 15 June 1917, Merrill & Ring Lumber Co. Records; Ames to W. Talbot, 28 May and 5, 6, and 7 June 1917, Pope & Talbot Records; *Timberman* 18(April 1917):37–39.

11. Long to Howard F. Weiss, 14 November 1917, Weyerhaeuser Timber Co. Records.

12. T. J. Humbird to R. M. Weyerhaeuser, 21 March 1917; Huntington Taylor to William Carson, 5 July 1917, Edward Rutledge Timber Co. Records, University of Idaho Library; Ames to W. Talbot, 27 and 30 June 1917, Pope & Talbot Records.

13. Ames to W. Talbot, 21, 23, 24, 25, 27, and 28 July and 4 August 1917, Pope & Talbot Records; Ames to C. R. Cranmer, 23 July 1917, Ames Papers; Long to F. E. Weyerhaeuser, 30 July 1917, Weyerhaeuser Timber Co. Records; Robert L. Tyler, *Rebels of the Woods: The I.W.W. in the Pacific Northwest* (Eugene, 1967), 93–94.

14. Thorpe Babcock to C. H. Jones, 20 July 1917, Thorpe Babcock Papers, Forest History Society; Long to J. F. Kimball, 17 July 1917; to O. Bystrom, 30 July 1917, Weyerhaeuser Timber Co. Records; Carleton Parker to M. P. Goodner, 13 October 1917, Brice P. Disque Papers, Manuscripts Collection, University of Washington Library; *Bureau of Labor Eleventh Biennial Report, 1917–1918* (Olympia, 1918), 66–67.

15. Ames to W. Talbot, 7, 26, and 30 July 1917; W. Talbot to Ames, 9 July and 10 August 1917, Pope & Talbot Records; Jerome to Ring, 10 and 30 July 1917; to Alex Polson, 7 July 1917; to T. D. Merill, 3 August 1917; Polson to Jerome, 30 July 1917, Merrill & Ring Lumber Co. Records; Mark E. Reed to T. P. Fisk, 25 March 1922, Simpson Logging Co. Records, Simpson Timber Co., Shelton.

16. *West Coast Lumberman* 32(15 August 1917):19; 33(15 December 1917), 19; J. P. Weyerhaeuser to L. S. Case, 30 July 1917, Weyerhaeuser Family Collection, Minnesota Historical Society; Ames to W. Talbot, 11 August 1917, Pope & Talbot Records; Jerome to Ring, 24 September 1917, Merrill & Ring Lumber Co. Records; J. H. Bloedel to Henry Suzzallo, 16 January 1918, Henry Suzzallo Papers, University of Washington Archives.

17. Long to F. E. Weyerhaeuser, 30 July 1917; to Mrs. E. T. Allen, 18 July 1917, Weyerhaeuser Timber Co. Records; Jerome to Ring, 10 July 1917, Merrill & Ring Lumber Co. Records; Ames to W. Talbot, 23, 26, and 30 July 1917, Pope & Talbot Records.

18. T. Merrill to Harrison Musgrave, 29 September 1917; Polson to Jerome, 23 July and 9 August 1917, Merrill & Ring Lumber Co. Records; Ames to W. Talbot, 3 August 1917; W. Talbot to Ames, 10 August 1917, Pope & Talbot Records; Cloice R. Howd, *Industrial Relations in the West Coast Lumber Industry*, BLS Bull. No. 349 (Washington, D.C., 1924), 74.

19. Ames to W. Talbot, 6 and 20 August 1917; W. Talbot to Ames, 10 and 17 August 1917, Pope & Talbot Records; Long to F. E. Weyerhaeuser, 30 July and 16 August 1917, Weyerhaeuser Timber Co. Records; J. Weyerhaeuser to Case, 30 July 1917, Weyerhaeuser Family Collection.

20. Long to F. E. Weyerhaeuser, 30 July and 16 August 1917; to W. W. Warren, 24 July 1917; to W. H. Boner, 31 July 1917; to Mrs. E. T. Allen, 31 July 1917, Weyerhaeuser Timber Co. Records; J. Weyerhaeuser to Case, 30 July 1917, Weyerhaeuser Family Collection; Parker to Walter Lippmann, 27 July 1917; Newton D. Baker to Parker, 28 July 1917; Parker to Baker, 29 July 1917, Disque Papers, University of Washington Library.

21. J. Weyerhaeuser to Humbird, 1 August 1917, Weyerhaeuser Family Collection; George A. Stephenson to R. Merrill, 1 September 1917, Merrill & Ring Lumber Co. Records; *West Coast Lumberman* 32(1 August 1917):19; (15 August 1917):32.

22. Ames to W. Talbot, 24 July and 9, 10, 11, and 22 August 1917, Pope & Talbot Records; Harold M. Hyman, *Soldiers and Spruce: Origins of the Loyal Legion of Loggers and Lumberman* (Los Angeles, 1963), 66–76; Melvyn Dubofsky, *We Shall Be All: A History of the Industrial Workers of the World* (Chicago, 1969), 411–14.

23. Polson to Jerome, 23 July 1917; Jerome to Ring, 17 July 1917, Merrill & Ring Lumber Co. Records; *Timberman* 18(July 1917):29; (August 1917):29; *West Coast Lumberman* 33(15 October 1917):19.

24. Long to Irvine, 6 August 1917; to F. E. Weyerhaeuser, 16 August 1917, Weyerhaeuser Timber Co. Records; Ring to Jerome, 16 August 1917; to R. Merrill, 28 July 1917, Merrill & Ring Lumber Co. Records; *West Coast Lumberman* 32(15 August 1917):20; *Timberman* 18(July 1917):31.

25. R. Merrill to Ring, 17 and 21 September 1917; to E. D. Wetmore, 28 September 1917, Merrill & Ring Lumber Co. Records; Ames to F. Talbot, 19 and 25 September 1917, Pope & Talbot Records; Tyler, *Rebels of the Woods*, 97–98.

26. Jerome to Ring, 24 September 1917; Polson to R. Merrill, 22 September 1917, Merrill & Ring Lumber Co. Records; Suzzallo to Taylor, 13 November 1917, Suzzallo Papers; *Timberman* 19(November 1917):29; (December 1917):29.

27. Long to J. J. Donovan, 23 October 1917; to E. T. Allen, 1 November 1917; to W. W. Seymour, 1 April 1918, Weyerhaeuser Timber Co. Records; Ames to F. Talbot, 27 October 1917, Pope & Talbot Records; Brice P. Disque to Chief Signal Officer, 27 November 1917, Disque Papers, University of Washington Library; Hyman, *Soldiers and Spruce*, Chap. 6.

28. Long to E. A. Selfridge, 18 October 1917; to Boner, 24 October 1917, Weyerhaeuser Timber Co. Records; Ames to W. Talbot, 22 December 1917, Pope & Talbot Records; Logbook, 29 October and 5 and 15 November 1917, Disque Papers, University of Washington Library; Hyman, *Soldiers and Spruce*, Chap. 6.

29. The colonel was, according to a trade journal, the "lumber Napoleon." *West Coast Lumberman* 33(15 November 1917):19; Jerome to R. Merrill, 11 and 19 December 1917; R. Merrill to T. Merrill, 6 December 1917, Merrill & Ring Lumber Co. Records; Ames to F. Talbot, 27 November 1917; to W. Talbot, 22 December 1917, Pope & Talbot Records; *Timberman* 19(June 1918):27.

30. Disque to Neil M. Clark, 30 April 1919, Disque Papers, University of Washington Library; Disque to Suzzallo, 8 January 1918, Suzzallo Papers; Edward B. Mittelman, "The Loyal Legion of Loggers and Lumberman: An Experiment in Industrial Relations," *Journal of Political Economy* 31(June 1923):338.

31. Ironically, operators east of the Cascades, where the strike had started, decided to go to eight hours at the beginning of 1918, hoping to take advantage of rising prices through resumption of full-scale production. Ames to W. Talbot, 12 December 1917,

Pope & Talbot Records; Diary, 25 January 1918, Brice P. Disque Papers, University of Oregon Library; Long to Disque, 21 November 1917, Weyerhaeuser Timber Co. Records.

32. William C. Butler to Nicholas Murray Butler, 27 February 1918, Nicholas Murray Butler Papers, Columbia University Library; Diary, 31 December and 7, 10, 11, 15, 22, 29, and 30 January 1918, Disque Papers, University of Oregon Library.

33. Disque to Donovan, 10 January 1918; Donovan to Disque, 19 January 1918; Diary, 1 February 1918, Disque Papers, University of Oregon Library.

34. Ames to W. Talbot, 2, 11, and 22 February 1918; W. Talbot to Ames 15, 20, and 26 February 1918, Pope & Talbot Records; Suzzallo and Parker to Baker, 26 January 1918, Suzzallo Papers; Diary, 2, 5, and 27 February and 5 April 1918, Disque Papers, University of Oregon Library.

35. Long to F. E. Weyerhaeuser, 8 March 1918, Weyerhaeuser Timber Co. Records; Jerome to W. F. MacPherson, 2 March 1918; T. Merrill to E. H. Eddy, 2 March 1918, Merrill & Ring Lumber Co. Records; Ames to W. Talbot, 2 and 5 March 1918; W. Talbot to Ames, 8 and 11 March 1918, Pope & Talbot Records; *Timberman* 19(March 1918):33–34, 59.

36. *West Coast Lumberman* 33(1 March 1918):19; Disque to Ralph Hayes, 23 August 1932, Disque Papers, University of Oregon Library.

37. *West Coast Lumberman* 33(5 March 1918):33; *Timberman* 19(May 1918):74; (September 1918), 27; (October 1918), 31; Wesley C. Mitchell, *History of Prices During the War* (Washington, D.C., 1919), 64–65; Bryant, "War and the Lumber Industry," 130; Ames to W. Talbot, 28 September and 4 October 1918, Pope & Talbot Records; Donovan to Disque, 19 September and 21 October 1918, Disque Papers, University of Oregon Library.

38. W. Butler to N. Butler, 27 February 1918, Butler Papers; T. Merrill to A. M. Marshall, 25 January and 12 October 1918, Merrill & Ring Lumber Co. Records; Ames to W. Talbot, 18 and 23 July and 10 and 26 August 1918, Pope & Talbot Records.

39. Howd, *Industrial Relations,* 97–101; Ames to W. Talbot, 11 and 15 July 1918; W. Talbot to Ames, 6 August 1918, Pope & Talbot Records; Donovan to Disque, 16 August 1918, Disque Papers, University of Oregon Library; Hyman, *Soldiers and Spruce,* Chaps. 13–14.

40. Disque to Donovan, 25 and 31 July 1918; Diary, 13 and 27 March, 8 June, 10 July, and 15 and 22 August 1918, Disque Papers, University of Oregon Library; Spruce Production Division Bull. No. 68, 8 August 1918, Puget Mill Co. Records, Ames Coll., Manuscripts Collection, University of Washington Library.

41. Diary, 6 April, 28 June, and 15 August 1918, Disque Papers, University of Oregon Library; *Timberman* 19(February 1918):35; (June 1918):34; (July 1918):42; 20(November 1918):30; Steer, *Lumber Production,* 57, 59.

42. Polson to R. Merrill, 19 April 1918; to Jerome, 18 October 1918; Jerome to Ring, 21 January and 7, 18, and 21 November 1918, Merrill & Ring Lumber Co. Records; Diary, 11 and 19 April, 10 July and 21 October 1918; Disque to J. D. Reardon, 4 August 1919; John D. Ryan to Disque, 10 December 1918, Disque Papers, University of Oregon Library; Ames to W. Talbot, 30 September and 5 October 1918, Pope & Talbot Records; *Timberman* 20(August 1919):30, 44; (September 1919):40–41, 48B–48F.

43. Long to Allen, 26 November 1918, Weyerhaeuser Timber Co. Records; Donovan to Disque, 6 November 1918; Diary, 22 November 1918, Disque Papers, University of Oregon Library; R. Merrill to Robert Durney, 27 November 1918; Jerome to Ring, 18 November 1918; to T. Merrill, 9 December 1918, Merrill & Ring Lumber Co. Records; *Timberman* 20(November 1918); 31; (December 1918):33.

44. *West Coast Lumberman* 35(15 December 1918):19.

45. W. Talbot to Ames, 11 October 1918, Pope & Talbot Records; *West Coast Lumberman* 35(15 November 1918):19; (1 December 1918), 18; Long to Allen, 26 November 1918, Weyerhaeuser Timber Co. Records.

12:   An Epoch of Unrest

1. George S. Long to George S. Gardiner, 8 December 1919, Weyerhaeuser Timber Co. Records, Weyerhaeuser Archives.
2. "I hope Mr. Polson did not offer 'The Loggers' Prayer' to the Lord," quipped Clark Ring, "for it would be apt to queer the whole industry." Alex Polson to Timothy Jerome, 18 November 1918; Clark L. Ring to Jerome, 5 December 1918, Merrill & Ring Lumber Co. Records, Manuscripts Collection, University of Washington Library.
3. Edwin G. Ames to W. H. Talbot, 12 November 1918, Pope & Talbot Records, Ames Coll., Manuscripts Collection, University of Washington Library; Long to E. T. Allen, 11 January 1921, Weyerhaeuser Timber Co. Records.
4. J. P. Weyerhaeuser to R. M. Weyerhaeuser, 19 November 1918, Weyerhaeuser Family Collection, Minnesota Historial Society; Long to W. H. Boner, 15 February 1919; to Allen, 26 November 1918, Weyerhaeuser Timber Co. Records; *West Coast Lumberman* 35(1 December 1918):18.
5. *Timberman* 19(23 August 1918):28–29; 20(December 1918) 29–30, 45; Convention of the Loyal Legion of Loggers and Lumberman, Proceedings, 6–7 January 1919, Oregon Historical Society; Claude W. Nichols, Jr., "Brotherhood in the Woods: The Loyal Legion of Loggers and Lumbermen, A Twenty Year Attempt at 'Industrial Cooperation'" (Ph.D. diss., University of Oregon, 1959), 96–112.
6. Ames to R. C, Fuller, 31 January 1919 and 6 January 1920, Edwin G. Ames Papers, Ames Coll., Manuscripts Collection, University of Washington Library; Ames to Talbot, 6, 7, 10, and 14 February 1919, Pope & Talbot Records; Robert L. Friedheim, *The Seattle General Strike* (Seattle, 1964), Chaps. 3–6.
7. Ames to Talbot, 14 and 19 February 1919, Pope & Talbot Records; Ames to Fuller, 6 January 1920 and 2 October 1923; to Long, 18 October 1919, Ames Papers.
8. Ames to Talbot, 19 February 1919, Pope & Talbot Records; Agent No. 17 Reports, 7 June, 15 October, 11 and 30 November, and 22 December 1919, Broussais C. Beck Papers, Manuscripts Collection, University of Washington Library; Jerome to Ring, 6 August 1919; to Polson, 8 June 1921; T. D. Merrill to Jerome, 21 June 1921, Merrill & Ring Lumber Co. Records; *West Coast Lumberman* 36(15 July 1919):19.
9. Ames to Talbot, 19 February, 20 March, and 15 August 1919, Pope & Talbot Records; Polson to Jerome, 21 August 1919, Merrill & Ring Lumber Co. Records; Stewart H. Holbrook to M. T. Dunten, 15 December 1925, Stewart H. Holbrook Papers, Manuscripts Collection, University of Washington Library.
10. It became an article of Wobbly faith that the massacre and the trial were part of a conspiracy by "the Lumber Trust." Polson to Jerome, 23 January and 19 February 1920, Merrill & Ring Lumber Co. Records; Agent No. 17 Reports, 9 and 16 February and 7 June 1920; Agent No. 106 Report, 17 April 1920, Beck Papers.
11. *Timberman* 20(May 1919):29; Henry B. Steer, comp., *Lumber Production in the United States, 1799–1946*, USDA Misc. Pub. No. 669(Washington, D.C., 1948), 13–14, 197; Dwight Hair, *Historical Forestry Statistics of the United States*, USFS Stat. Bull. No. 228(Washington, D.C., 1958), 22; Long to Boner, 24 June and 7 August 1919, Weyerhaeuser Timber Co. Records.
12. Ames to Talbot, 2 January, 6 September and 23 October 1919; 24 January 1920;

Talbot to Ames, 10 September 1919, Pope & Talbot Records; Ames to W. H. Furber, 5 May 1919, Ames Papers.

13. *Timberman* 21(February 1920):101; Nelson C. Brown, "Recent Developments in Lumber Distribution," *Journal of Forestry* 22(January 1924):62–64.

14. *West Coast Lumberman* 35(15 January 1919):19; 36(1 June 1919):18; (15 August 1919):19; 37(1 October 1919):19; 40(15 June 1921):19; Polson to Jerome, 21 May 1920, Merrill & Ring Lumber Co. Records.

15. R. D. Merrill to Polson, 12 August 1921; Polson to Jerome, 7 and 8 November 1921; T. Merrill to Jerome, 5 September 1921, Merrill & Ring Lumber Co. Records; Ames to E. B. Curtis, 25 July 1921; to A. K. Ames, 12 November 1921, Ames Papers.

16. R. C. Hall, "The Federal Income Tax and Forestry," *Journal of Forestry* 21(October 1923):554–55; Long to Allen, 20 February 1919, Weyerhaeuser Timber Co. Records; Ames to Talbot, 26 July 1919, Pope & Talbot Records; Elwood R. Maunder, ed., "Memoirs of a Forester: An Excerpt from Oral History Inteviews with David T. Mason," *Forest History* 10(January 1967):34.

17. Carl M. Stevens, "The Forest Industries and the Income Tax," *Journal of Forestry* 18(April 1920):334; Maunder, ed., "Memoirs of a Forester," 34; Wilson Compton to L. C. Boyle, 23 May 1919; Allen to Compton, 19 August and 11 September 1919; Robert Ash to Compton, 20 June 1919, National Forest Products Association Records, Forest History Society; Ames to Talbot, 26 July 1919, Pope & Talbot Records.

18. Ring to Jerome, 9 March 1920; Jerome to Ring, 4 December 1919, Merrill & Ring Lumber Co. Records; Long to Allen, 28 November 1919, Weyerhaeuser Timber Co. Records; *West Coast Lumberman* 37(1 October 1919):19.

19. Ames to Talbot, 11 and 14 October 1919; F. L. Dettman to Talbot, 13 October 1919; Talbot to Ames, 21 October 1919, Pope & Talbot Records; Long to Stiles W. Burr, 30 September 1919; to William P. Hopkins, 16 October 1919, Weyerhaeuser Timber Co. Records; Stevens, "Forest Industries and the Income Tax," 334–47; Maunder, ed., "Memoirs of a Forester," 34–35.

20. T. Merrill to Jerome, 28 September 1921; Jerome to Polson, 1 and 29 June 1922; to T. Merrill, 5 May and 1 August 1922 and 1 May 1923, Merrill & Ring Lumber Co. Records; Dettman to Talbot, 13 October 1919, Pope & Talbot Records.

21. Ames to Talbot, 31 July 1919, Pope & Talbot Records; Long to Burr, 20 October 1919 and 14 December 1921, Weyerhaeuser Timber Co. Records; Mark E. Reed to Polson, 10 June 1921; Polson to Jerome, 4 November 1921; Jerome to Polson, 9 January 1922, Merrill & Ring Lumber Co. Records.

22. Long to F. E. Weyerhaeuser, 22 April 1920; to Allen, 9 December 1919, Weyerhaeuser Timber Co. Records; J. Weyerhaeuser to F. E. Weyerhaeuser, 7 April 1920, Weyerhaeuser Family Collection; Jerome to Polson, 20 June 1923; R. Merrill to T. Merrill, 3 August 1923, Merrill & Ring Lumber Co. Records.

23. Stevens, "Forest Industries and the Income Tax," 337; Hall, "Federal Income Tax and Forestry," 555–56; Long to Burr, 4 January 1921, Weyerhaeuser Timber Co. Records; Ring to William B. Mershon, 2 June 1924, Merrill & Ring Lumber Co. Records.

24. *Timberman* 21(August 1920):33; (September 1920),33; 22(January 1921):33; 23(January 1922):25; Ames to Talbot, 27 August 1920, Pope & Talbot Records; Steer, comp., *Lumber Production*, 13–14, 197; A. H. Landram to Leonard Howarth, 8 November 1920, St. Paul & Tacoma Lumber Co. Records, Manuscripts Collection, University of Washington Library.

25. Ames to Talbot, 17 November and 30 December 1920, Pope & Talbot Records; Ames to Millard T. Hartson, 3 November 1920; to Joseph W. Fordney, 4 November 1920; to Fuller, 4 November 1920, Ames Papers; Jerome to Polson, 3 November 1920, Merrill & Ring Lumber Co. Records.

26. Long to James E. Long, 29 November 1922; to L. S. Case, 27 September 1922, Weyerhaeuser Timber Co. Records; Case to J. Weyerhaeuser, 9 June 1922, Weyerhaeuser Family Collection; Ames to Talbot, 29 November 1920, Pope & Talbot Records; Steer, comp., *Lumber Production*, 14–15, 196–67; Hair, *Historical Forestry Statistics*, 22.

27. *Timberman* 23(March 1922):85; 25(February 1924):41; Ames to A. G. Harms, 5 December 1922, Pope & Talbot Records; Ames to Furber, 2 January 1923; to F. S. Ames, 4 April 1923, Ames Papers; Long to J. Long, 19 January 1924, Weyerhaeuser Timber Co. Records; Brown, "Recent Developments in Lumber Distribution," 62–63.

28. Long to F. E. Weyerhaeuser, 22 April 1920; to William Carson, 27 April and 12 August 1921; 11 May 1923; to W. L. McCormick, 8 January 1923; to J. Long, 19 January and 25 February 1924; 26 December 1925; to F. S. Bell, 16 September 1925, Weyerhaeuser Timber Co. Records; Ralph W. Hidy, Frank Ernest Hill and Allan Nevins, *Timber and Men: The Weyerhaeuser Story* (New York, 1963), 286–87, 398–400.

29. Ames to Talbot, 8 August 1921, Pope & Talbot Records; Long to George Cornwall, 22 June 1921; to M. B. Nelson, 25 November 1925, Weyerhaeuser Timber Co. Records; *West Coast Lumberman* 39(1 January 1921):19; 42(1 June 1922):19; (1 August 1922):19; *Timberman* 22(June 1921):38; 23(May 1922):29.

30. Dropped from official price lists after complaints from Japan, the designation "Jap Square" continued in common usage. *West Coast Lumberman* 41(15 November 1921):19; 45(1 March 1924):30–31; Long to Peter Connacher, 8 February 1922, Weyerhaeuser Timber Co. Records; Ames to Harms, 29 February 1924, Pope & Talbot Records; Brown, "Recent Developments in Lumber Distribution," 64.

31. Ames to Furber, 2 January 1923; to F. Ames, 4 April 1923, Ames Papers; Long to J. Long, 29 November 1922; to F. G. Wisner, 19 October 1923, Weyerhaeuser Timber Co. Records; *Timberman* 24(October 1923).

32. Ames to Talbot, 9 September, 29 and 30 November and 18 and 22 December 1920; 28 January, 26 February, 24 March and 11 and 18 May 1921, Pope & Talbot Records; Ames to A. Ames, 12 November 1921, Ames Papers; J. Weyerhaeuser to Samuel S. Davis, 9 June 1920, Weyerhaeuser Family Collection.

33. Ames to Talbot, 20 November 1920 and 26 February and 29 April 1921, Pope & Talbot Records; William C. Ruegnitz to Helen Wood, 21 February 1934, William C. Ruegnitz Papers, Manuscripts Collection, University of Washington Library.

34. Ames to Talbot, 2, 3, 17, 29, and 31 March: 1, 4, 5, 7, 11, 18, and 28 April; 5, 10, 12, and 26 May, and 11 and 23 June 1922; 2 and 6 April 1923, Pope & Talbot Records; Long to Ames, 21 April 1922, Weyerhaeuser Timber Co. Records.

35. Ames to Talbot, 2 and 3 April, 2 May and 15 and 29 June 1923, Pope & Talbot Records; Ames to Neil Cooney, 16 March 1923; to Zeb Mayhew, 3 May 1923; to Fuller, 3 July 1923; to A. J. Hendry, 3 July 1923, Ames Papers; *West Coast Lumberman* 44(1 May 1923):33; (1 June 1923):19; Robert L. Tyler, *Rebels of the Woods: The I.W.W. in the Pacific Northwest* (Eugene, 1967), 201–6.

36. J. E. Hellenius to Ruegnitz, 11 December 1926; 9 January and 28 February 1927; J. A. Ziemer to Ruegnitz, 12 November 1930, Ruegnitz Papers.

37. *Journal of Forestry* 16(January 1918):131–32; (March 1918):359–61; B. E. Fernow, "Forestry and the War," *Journal of Forestry* 16(February 1918):149–54; P. S. Lovejoy, "Review of Lumber Industry Affairs," in *Journal of Forestry* 17(March 1919):245–46.

38. "Forest Devastation: A National Danger and a Plan to Meet It," in *Journal of Forestry* 17(December 1919):911–45; Gifford Pinchot, "The Economic Significance of Forestry," *North American Review* 213(February 1921):162; Frederick E. Olmsted, "The Year's Accomplishments," *Journal of Forestry* 18(February 1920):96–97; Long to Fordney, 10 December 1920, Weyerhaeuser Timber Co. Records.

39. Gifford Pinchot, "Where We Stand," *Journal of Forestry* 18(May 1920):442,

and "Economic Significance of Forestry," 162; *Journal of Forestry* 18(May 1920):577–78; "An Answer to Dr. Compton's Fourteen Points," *Journal of Forestry* 17(December 1919), 946–54.

40. William B. Greeley to John W. Blodgett, 6 June 1924, William B. Greeley Papers, University of Oregon Library; George T. Morgan, Jr., *William B. Greeley: A Practical Forester, 1879–1955* (St. Paul, 1961), Chap. 4; Long to J. Long, 17 March 1920; to Allen, 10 May, 21 June, and 29 July 1920, Weyerhaeuser Timber Co. Records; Donald C. Swain, *Federal Conservation Policy, 1921–1933* (Berkeley, 1963), 12–13, 15–17.

41. Gifford Pinchot to George W. Butz, Jr., 1 August 1921, Gifford Pinchot Papers, Library of Congress; Long to Arthur Bolling Johnson, 19 July 1920; to George F. Lindsay, 16 September 1920; to C. L. Hamilton, 30 November 1920; to Fordney, 10 December 1920, Weyerhaeuser Timber Co. Records; Talbot to Ames, 1 August 1921, Pope & Talbot Records.

42. Long to J. E. Rhodes, 20 November 1920; to Hamilton, 30 November 1920; to Allen, 23 December 1920, Weyerhaeuser Timber Co. Records; Morgan, *Greeley*, 45–54; Ames to Talbot, 13 and 27 July 1921, Pope & Talbot Records.

43. *West Coast Lumberman* 46(15 June 1924):19; Swain, *Federal Conservation Policy*, 14–15; Harold K. Steen, *The U.S. Forest Service: A History* (Seattle, 1976) 185–93; Morgan, *Greeley*, 57–58.

44. H. H. Chapman, "The Profession of Forestry and Professional Ethics," *Journal of Forestry* 21(May 1923):452–57; Frederick E. Olmsted, "Professional Ethics," *Journal of Forestry* 20(February 1922), 106–12; George L. Drake, "The U.S. Forest Service, 1905–1955: An Industry Viewpoint," *Journal of Forestry* 53(February 1955):117.

45. According to Stewart Holbrook, the term "Greeleyism" was invented to describe the move of Forest Service personnel into the private sector. Pinchot to Raphael Zon, 26 December 1928, Pinchot Papers; Long to J. D. Tennant, 3 May 1929, Weyerhaeuser Timber Co. Records; Morgan, *Greeley*, 59–63; Stewart Holbrook, "Greeley Went West," *American Forests* 64(March 1958):59–60.

46. George P. Ahern, *Deforested America* (Washington, D.C., 1928), 39; David T. Mason, "Putting the Brakes on Lumber Production," typescript, Washington State University Library; Rodney C. Loehr, ed., *Forests for the Future: The Story of Sustained Yield as Told in The Diaries and Papers of David T. Mason, 1907–1950* (St. Paul, 1952), 46, 66–67, 69, 73; David T. Mason to Reed, 7 October 1922; Reed to Mason, 18 April 1927, Simpson Logging Co. Records, Simpson Timber Co. Shelton.

47. *Journal of Forestry* 24(April 1926):331; James Stevens to Robert Case, 13 July 1944, West Coast Lumbermen's Association Records, Oregon Historical Society; Henry Clepper, "The Ten Most Influential Men in American Forestry," *American Forests* 56(May 1950):10–11, 30, 37–39.

48. *West Coast Lumberman* 47(1 October 1924):19.

13: The Law of Supply and Demand

1. W. B. Greeley, "How the Pacific Northwest Lumber Industry Can Be Made Prosperous," *West Coast Lumberman* 54(15 June 1928):15.

2. National City Co. Report on the Douglas Fir Lumber Industry, 1927, National Forest Products Association Records, Forest History Society.

3. Edwin G. Ames to W. H. Talbot, 3 and 16 October 1922; 26 April and 1 August 1924, Pope & Talbot Records, Ames Coll., Manuscripts Collection, University of Washington Library; *West Coast Lumberman* 43(15 November 1922):26.

4. Talbot to Ames, 14 and 19 May and 11 September 1925; Ames to Talbot, 18 August 1925; to A. G. Harms, 3 November 1925 (two letters of this date) and 16 January 1926, Pope & Talbot Records; Edwin T. Coman, Jr. and Helen M. Gibbs, *Time, Tide and Timber: A Century of Pope & Talbot* (Stanford, Cal., 1949), Chaps. 22–24.

5. *Fifteenth Census of the United States, Manufacturers: 1929*, Vol. 2, *Reports by Industries* (Washington, D.C. 1933), 447; William B. Greeley to George H. Kelly, 9 February 1920, William B. Greeley Papers, University of Oregon Library; George S. Long to E. T. Allen, 11 August 1923, Weyerhaeuser Timber Co. Records, Weyerhaeuser Co. Archives; Ames to Talbot, 4 August 1922 and 19 September 1923, Pope & Talbot Records.

6. *West Coast Lumberman* 43(15 August 1922):24; Ames to Talbot, 28 November 1919 and 3 and 26 March 1920, Pope & Talbot Records; Long to J. P. Weyerhaeuser, 30 January 1919; to F. H. Thatcher, 30 September and 30 October 1919; to F. S. Bell, 22 November 1919; to F. E. Weyerhaeuser, 27 February 1920, Weyerhaeuser Timber Co. Records.

7. Ames to Talbot, 1 January 1921; 17 February, 26 May and 3 and 16 October 1922; 15 and 26 January, 4 May and 9 July 1923; 26 April, 15 July and 1 August 1924, Pope & Talbot Records; John M. McClelland, Jr., *Longview: The Remarkable Beginnings of a Modern Western City* (Portland, 1949), Chaps. 6–7.

8. *Timberman* 24(April 1923):25; Greeley to Kelly, 9 February 1920, Greeley Papers.

9. T. D. Merrill to R. D. Merrill, 23 August 1922 and 22 November 1928; Timothy Jerome to Alex Polson, 2, 3, and 9 February and 12 April 1927; 9 August 1929; Polson to Jerome, 3, 9, and 10 February 1927; to R. Merrill, 20 June 1928; R. Merrill to T. Merrill, 12 and 28 November 1928; to Corydon Wagner, 13 February and 3 April 1929, Merrill & Ring Lumber Co. Records, Manuscripts Collection, University of Washington Library.

10. W. B. Greeley, "Forest Management on Federal Lands," *Journal of Forestry* 23(March 1925):227–28; F. R. Archer to Commissioner of Indian Affairs, 11 December 1912, Wesley L. Jones Papers, Manuscripts Collection, University of Washington Library; J. P. Kinney to Greeley, 3 April 1929, Merrill & Ring Lumber Co. Records; George H. Emerson to C. F. White, 9 June 1903, George H. Emerson Letterbooks, Manuscripts Collection, University of Washington Library.

11. Greeley to R. Merrill, 8 April 1929; Kinney to Greeley, 3 April 1929, Merrill & Ring Lumber Co. Records; J. P. Kinney, "Forestry Administration on Indian Reservations," *Journal of Forestry* 19(December 1921):836–39; Charles H. Burke to Miller, Wilkinson & Miller, 10 June 1927, Taholah Indian Agency Records, RG 75, Federal Records Center, Seattle.

12. Kinney to Greeley, 3 April 1929, Merrill & Ring Lumber Co. Records; Henry Steer to Mrs. W. L. Montgomery, 20 February 1923, to Commissioner of Indian Affairs, 26 December 1922; 2 October 1928 and 18 June 1929; W. B. Sams to Commissioner of Indian Affairs, 18 June 1929, Taholah Indian Agency Records; L. H. Brewer to Wesley L. Jones, 1 December 1917; Burke to Jones, 29 September 1922, Jones Papers.

13. The available evidence also suggests that arrangements among the bidders also restricted the competition. Steer to Charles E. Coe, 7 March 1923, Taholah Indian Agency Records; Polson to R. Merrill, 7 July and 31 October 1922; to Jerome, 31 October 1922; 13 July 1923 and 3 March 1925; Jerome to Clark L. Ring, 21 July 1923, Merrill & Ring Lumber Co. Records; B. J. Wooster to Paul Smith, 2 November 1920

and 20 March and 31 October 1922, Aloha Lumber Co. Records, Manuscripts Collection, University of Washington Library.

14. George P. Ahern, *Forest Bankruptcy in America*, 2nd ed. (Washington, D.C., 1934), 271–72; Henry B. Steer, comp., *Lumber Production in the United States, 1799–1946*, USDA Misc. Pub. No. 669(Washington, D.C., 1948), 61, 73.

15. Richard H. Buskirk, "A Description and Critical Analysis of the Marketing of Douglas Fir Plywood" (D.B.A. diss., University of Washington, 1955), 9–18; Robert W. Cour, *The Plywood Age* (Portland, 1955), 34–39.

16. Lyman Horace Weeks, *A History of Paper-Manufacturing in the United States, 1690–1916* (New York, 1916), Chaps. 9 and 11; David C. Smith, *History of Paper-making in the United States, 1691–1969* (New York, 1970), 135–36; John A. Guthrie, *The Newsprint Paper Industry: An Economic Analysis* (Cambridge, Mass., 1941), 5.

17. John A. Guthrie, *The Economics of Pulp and Paper* (Pullman, Wa., 1950), 7–9; Weeks, *History of Paper-Manufacturing*, 342–43; Albert L. Seeman, "Economic Adjustments on the Olympic Peninsula," in *Economic Geography* 8(July 1932):307–8.

18. R. L. Fromme to Thomas T. Aldwell, 13 December 1919, Thomas T. Aldwell Papers, Manuscripts Collection, University of Washington Library; Thornton T. Munger, "The Pulpwood Forests of Oregon and Washington," *Pacific Pulp and Paper Industry* 8(September 1934):20–23; William B. Greeley, *Forests and Men* (Garden City, N.Y., 1951), 123.

19. Russell F. Erickson, *The Story of Rayonier Incorporated* (New York, 1963), 8–10; Reed O. Hunt, *Pulp, Paper and Pioneers: The Story of Crown Zellerbach Corporation* (New York, 1961), 14–18.

20. Although legally distinct, the Shelton, Hoquiam, and Port Angeles mills operated under a joint management, with Mills as president of each firm. Mark E. Reed to R. M. Calkins, 25 November 1929, Mark E. Reed Papers, Manuscripts Collection, University of Washington Library; Claude Adams, *History of Papermaking in the Pacific Northwest* (Portland, 1951), 31–33.

21. *Timberman* 27(July 1926):36; Charles E. Calhoun, "Financing the Pulp and Paper Industry of the Pacific Coast" (M.B.A. thesis, University of Washington, 1930), 2, 195–96; Robert E. Ficken, *Lumber and Politics: The Career of Mark E. Reed* (Seattle, 1979), 143–47.

22. Steer, comp., *Lumber Production*, 14–15, 196; Ames to Talbot, 25 November 1924, Pope & Talbot Records; Long to William Carson, 5 and 18 March 1924; to C. M. Kellogg, 25 June 1924; to H. H. Irvine, 31 December 1925; to T. J. Humbird, 18 July 1927, Weyerhaeuser Timber Co. Records; W. B. Greeley, "The West Coast Problem of Stabilizing Lumber Production," *Journal of Forestry* 28(February 1930):193.

23. Long to R. C. Lilly, 12 October 1923; to Bell, 21 April 1926, Weyerhaeuser Timber Co. Records; Ames to Talbot, 12 January and 1 May 1925; 7 April and 19 October 1926, Pope & Talbot Records; McClelland, *Longview*, 128.

24. Ames to Talbot, 1 July and 8 and 11 November 1926; to Harms, 11 and 18 October 1926 and 11 February 1927, Pope & Talbot Records; Polson to R. Merrill, 11 and 14 December 1926, Merrill & Ring Lumber Co. Records; Coman and Gibbs, *Time, Tide and Timber*, Chaps. 24–25.

25. *West Coast Lumberman* 45(1 March 1924):19; 46(15 September 1924):19; 47(15 October 1924):19; (1 November 1924):20; (15 November 1924):19; Jerome to Ring, 10 July and 4 November 1924, Merrill & Ring Lumber Co. Records.

26. Polson to R. Merrill, 26 January, 24 March, 7 June, and 11 December 1926, Merrill & Ring Lumber Co. Records; Long to James D. McCormick, 12 November 1927 and 31 January 1928; to L. P. Lewis, 21 October 1929; to C. E. Dant, 4 September 1929, Weyerhaeuser Timber Co. Records; Polson to Reed, 26 October 1928, Simpson Logging Co. Records, Simpson Timber Co., Shelton; Ficken, *Lumber and Politics*, 133–42.

27. Greeley, "West Coast Problem," 193; Long to William C. Butler, 6 July 1925; to Irvine, 29 March 1926; to F. H. Farwell, 23 March 1926; to James G. Eddy, 16 August 1926, Weyerhaeuser Timber Co. Records; Jerome to Ring, 17 April 1924, Merrill & Ring Lumber Co. Records; Ames to Talbot, 11 November 1924 and 4 May 1925, Pope & Talbot Records.

28. With an annual output of 330 million feet, Long-Bell was the largest producer in one location. Long to John S. Owen, 17 December 1927, Weyerhaeuser Timber Co. Records; *Timberman* 31(January 1930):44, 46; (February 1930):46; Ralph W. Hidy, Frank Ernest Hill, and Allan Nevins, *Timber and Men: The Weyerhaeuser Story* (New York, 1963), 395, 403–12.

29. Long to F. E. Weyerhaeuser, 24 September 1926, 15 and 21 February, and 17 May 1928, Weyerhaeuser Timber Co. Records; Ames to Talbot, 4 March 1927, Pope & Talbot Records; *West Coast Lumberman* 51(15 October 1926):19; (15 November 1926):19.

30. *West Coast Lumberman* 52(1 April 1927):19; Everett G. Griggs, "Self-Government in the Lumber Business," *West Coast Lumberman* 50(1 June 1926):35, 50.

31. Long to Humbird, 19 July 1929, Weyerhaeuser Timber Co. Records; Humbird to Long, 11 July 1929, Weyerhaeuser Family Collection, Minnesota Historical Society; Reed to W. A. Whitman, 7 November 1929, Simpson Logging Co. Records; *West Coast Lumberman* 52(15 May 1927):19.

32. Wagner to R. Merrill, 25 April, 7 and 17 May, and 25 June 1928; 18 January, and 28 September 1919; n.d.; R. Merrill to Wagner, 30 April 1928, Merrill & Ring Lumber Co. Records; *West Coast Lumberman* 55(15 December 1928):17; 56(May 1929):39; Coman and Gibbs, *Time, Tide and Timber*, 301.

33. Long to Humbird, 6 March 1926; to Huntington Taylor, 24 June 1927, Weyerhaeuser Timber Co. Records; F. K. Weyerhaeuser oral history interview, Weyerhaeuser Family Collection, 110–12; *Timberman* 31(August 1930):35; Hidy, Hill, and Nevins, *Timber and Men*, 366–68.

34. Long to Polson, 12 April 1926; to F. E. Weyerhaeuser, 2 February 1926; to Carson, 23 March and 10 December 1926; to Wilson Compton, 21 June 1926; to F. C. Knapp, 10 May 1927; to J. D. Tennant, 8 November 1927, Weyerhaeuser Timber Co. Records; *West Coast Lumberman* 52(1 June 1927):19; Wilson Compton, "Lumber—an Old Industry—and the New Competition," *Harvard Business Review* 10(January 1932):167–68.

35. Long to Neil Cooney, 4 March 1926; to Henry Schott, 5 March 1926; to Compton, 18 February 1927; to C. R. Musser, 10 May 1927; to John W. Blodgett, 25 August 1928, Weyerhaeuser Timber Co. Records; *West Coast Lumberman* 49(1 February 1926):19.

36. Long to Charles S. Keith, 8 April 1926; to Greeley, 18 February 1928; to Ingram, 2 July 1928; to Blodgett, 25 August 1928, Weyerhaeuser Timber Co. Records; *West Coast Lumberman* 53(1 March 1928):17; 54(1 July 1928):20.

37. Sams to Commissioner of Indian Affairs, 26 May 1924 and 18 June and 26 September 1925, Taholah Indian Agency Records.

38. Long to E. S. Collins, 8 January 1929; to Greeley, 16 January 1929; to F. E. Weyerhaeuser, 20 September and 25 November 1929; to R. W. Condon, 8 May 1929; to Reed, 29 October and 27 November 1929, Weyerhaeuser Timber Co. Records; Ficken, *Lumber and Politics*, Chap. 6.

39. *West Coast Lumberman* 56(April 1929):7–8, 34; (June 1929):7, 19, 27; *Timberman* 30(April 1929):48, 178; R. Merrill to Robert Wetmore, 6 December 1929, Merrill & Ring Lumber Co. Records.

40. Polson to R. Merrill, 5 and 12 April, 1 May, 21 June, and 9 July 1929; to Jerome, 6, 8, 9, 11, 16, and 22 May 1929, Merrill & Ring Lumber Co. Records; Steer

to Frank H. Lamb, 25 April 1931, J. P. Kinney Papers, Cornell University Library; *West Coast Lumberman* 56(July 1929):33; (September 1929):7.
41. Quoted in newspaper clipping, Washington State Biography File, Northwest Collection, University of Washington Library.

14: Old Man Depression

1. Alex Polson to Timothy Jerome, 5 June 1932, Merrill & Ring Lumber Co. Records, Manuscripts Collection, University of Washington Library.
2. Everett Griggs to R. G. Brownell, 9 December 1931, St. Paul & Tacoma Lumber Co. Records, Manuscripts Collection, University of Washington Library.
3. *West Coast Lumberman* 57(September 1930):24; George S. Long to F. S. Bell, 2 May 1921 and 16 November 1928; to T. J. Humbird, 24 January 1929; to S. T. McKnight, 30 July 1929, Weyerhaeuser Timber Co. Records, Weyerhaeuser Co. Archives.
4. *American Lumberman* (23 November 1929), 29; (14 December 1929), 32; *West Coast Lumberman* 56(November 1929):92; 57(January 1930):13–15, 21, 58; (February 1930):54a; (March 1930):34; (July 1930):11, 41; *Timberman* 31(December 1929):36.
5. T. D. Merrill to Walter Nordhoff, 7 December 1929; to R. D. Merrill, 12 and 23 June 1930; Polson to Jerome, 23 June 1930; to R. Merrill, 21 June 1930; Jerome to E. D. Wetmore, 6 August 1930, Merrill & Ring Lumber Co. Records; Long to William Carson, 6 December 1929; F. R. Titcomb to Long, 13 February 1930, Weyerhaeuser Timber Co. Records.
6. Clark I. Cross, "Factors Influencing the Abandonment of Lumber Mill Towns in the Puget Sound Region" (M.S. thesis, University of Washington, 1946), 38; Henry B. Steer, comp., *Lumber Production in the United States, 1799–1946*, USDA Misc. Pub. No. 669(Washington, D.C., 1948), 15–16, 196–97; T. Merrill to E. H. Eddy, 16 October 1931, Merrill & Ring Lumber Co. Records; *American Lumberman* (11 June 1932):67; William C. Butler to Nicholas Murray Butler, 11 October and 11 December 1931, Nicholas Murray Butler Papers, Columbia University Library.
7. *Measurement of the Social Performance of Business*, TNEC Monograph No. 7(Washington, D.C., 1940), 156; *Fourth Report of the Department of Labor and Industries, 1932* (Olympia, 1933), 6; *West Coast Lumberman* 58(20 May 1931):53; Jerome to Stiles W. Burr, 11 August 1931 and 16 April 1932, Merrill & Ring Lumber Co. Records.
8. *Division of Forestry Annual Reports, 1928 to 1932* (Olympia, 1933), 42; Edwin T. Coman, Jr. and Helen M. Gibbs, *Time, Tide and Timber: A Century of Pope & Talbot* (Stanford, Cal., 1949), 294–95; John M. McClelland, Jr., *Longview: The Remarkable Beginnings of a Modern Western City* (Portland, 1949), 118–19, 130–33; T. Merrill to Jerome, 11 January 1930; Jerome to T. Merrill, 14 January 1930 and 10 July 1931; to Clark L. Ring, 11 May 1931 and 20 January 1932; R. Merrill to Ring, 29 June 1931 and 21 January and 1 February 1932; to Robert Polson, 28 January 1932; to Theodore B. Bruener, 23 February 1932; to Francis P. Sears, 10 October 1932, Merrill & Ring Lumber Co. Records.
9. Annual Report, 1938, Rayonier Incorporated Records, ITT Rayonier, Hoquiam; E. M. Mills to A. N. Parrett, 12 November 1942, Rayonier Incorporated Records, ITT Rayonier, Shelton. The Mills associate is quoted in notes found in Rayonier Incorporated Records, Hoquiam.
10. John C. Higgins to Russell M. Pickens, 5 June 1930, Rayonier Incorporated

Records, Shelton; *Pacific Pulp and Paper Industry* 4(July 1930):40; 5(February 1931):39; Annual Report, 1938, Rayonier Incorporated Records, Hoquiam; Russell T. Erickson, *The Story of Rayonier Incorporated* (New York, 1963), 13–16.

11. *West Coast Lumberman* 56(March 1929):7; 58(February 1931):11–12; (May 1931):38–39; (July 1931):29; 59(January 1932):26; (May 1932): 9–10; 44–45; (June 1932), 41; R. Y. Stuart, "National Forest Timber and the West Coast Lumber Industry," *Journal of Forestry* 31(January 1933):45–50; Daniel W. Cairney, "The Effect of Some Economic Disturbances on Lumber Trade of Washington and British Columbia" (M.S. thesis, University of Washington, 1935) 8, 10–12, 17–19, 22.

12. *West Coast Lumberman* 57(December 1930):14–15, 26; 59(July 1932):7; (August 1932):5, 21; (September 1932):5; (October 1932):11; Griggs to A. C. Dutton, 10 April 1931, St. Paul & Tacoma Lumber Co. Records.

13. E. T. Allen to A. W. Laird, 23 May 1930; to Wilson Compton, 17 November 1930; Compton to Allen, 4 December 1930, Western Forestry & Conservation Association Records, Oregon Historical Society; Franklin W. Reed, "The United States Timber Conservation Board: Its Origin and Organization; Its Purpose and Progress," *Journal of Forestry* 29(December 1931):1202–5; Wilson Compton, "Lumber—An Old Industry—and the New Competition," *Harvard Business Review* 10(January 1932): 168–69. Compton quoted in American Tree Association, *Forestry Almanac, 1933* (Washington, D.C., 1933), 116.

14. Allen to Laird, 23 February and 9 April 1931; to Ripley Bowman, 20 April and 5 May 1931; Bowman to Allen, 9 May 1932, Western Forestry & Conversation Association Records; American Tree Association, *Forestry Almanac, 1933,* 82–97.

15. W. B. Greeley to Thorpe Babcock, 15 June 1930, Thorpe Babcock Papers, Forest History Society; *Timberman* 32(October 1931):17–18; *West Coast Lumberman* 59(April 1932):17; Titcomb to A. L. Raught, 2 August 1930, Weyerhaeuser Timber Co. Records; A. H. Landram to R. A. Gore, 3 November 1930; to Griggs, 17 November 1930, St. Paul & Tacoma Lumber Co. Records; R. Merrill to Polson, 2 March 1931, Merrill & Ring Lumber Co. Records.

16. *Timberman* 33(April 1932):15; (May 1932):14; *West Coast Lumberman* 59(June 1932):9; (December 1932):5–6; Griggs to Herman Dierks, 16 March 1931, St. Paul & Tacoma Lumber Co. Records.

17. One supporter of advertising, R. D. Merrill, urged the industry to purchase commercials on "Amos 'n Andy," the nation's most popular radio program. "They have done wonders for 'Pepsodent,'" he explained, "and the dentists all say that this is the poorest dentrifice on the market." *West Coast Lumberman* 57(August 1930):11; 58(June 1931):41; *Four L Lumber News* 14(1 March 1932):5; R. Merrill to R. W. Wetmore, 23 May 1930; Polton to R. Merrill, 9 June 1930, Merill & Ring Lumber Co. Records; Robert E. Ficken, *Lumber and Politics: The Career of Mark E. Reed* (Seattle, 1979), 193.

18. Titcomb to F. E. Weyerhaeuser, 25 September 1930, F. E. Weyerhaeuser to Titcomb, 20 September 1930; to J. D. Tennant, 20 September 1930, F. R. Titcomb Papers, Weyerhaeuser Co. Archives; Jerome to T. Merrill, 14 June 1930; R. Merrill to R. Wetmore, 23 May 1930; to Ring, 7 and 19 June 1930; to Polson, 7 June 1930; to Mark E. Reed, 29 May and 15 September 1930, Merrill & Ring Lumber Co. Records; C. H. Kreienbaum to Reed, 20 June 1930, Reed Mill Co. Records, Simpson Timber Co., Shelton.

19. Jerome to T. Merrill, 21 February, 28 May and 2 July 1931; R. Merrill to Polson, 2 February 1931; to T. Merrill, 20 February 1931; to Ring, 18 May 1931; Polson to R. Merrill, 24 July 1931, Merrill & Ring Lumber Co. Records; Minot Davis to F. E. Weyerhaeuser, 24 February 1931; Titcomb to F. Bell, 13 March 1931, Titcomb Papers.

20. F. S. Bell, a member of the original syndicate of investors, had succeeded J. P. Weyerhaeuser as president of the company in 1928. Reed to Carl M. Stevens, 3 August 1931, Simpson Logging Co. Records, Simpson Timber Co., Shelton; Titcomb to Laird Bell, 2 November 1931; F. Bell to Titcomb, 2 December 1931, Titcomb Papers; Stevens to R. Merrill, 11 September, 9 October and 9 November 1931; T. Merrill to R. Wetmore, 28 September 1931, Merrill & Ring Lumber Co. Records.

21. Stevens to R. Merrill, 18 November 1931; R. Merrill to Ring, 30 August, 19 September and 27 October 1932, Merrill & Ring Lumber Co. Records; J. L. Bridge to Sound Timber Co., 6 January 1933, Sound Timber Co. Records, Minnesota Historical Society; W. B. Greeley, "The West Coast Problem of Stabilizing Lumber Production," *Journal of Forestry* 28(February 1930):196.

22. *West Coast Lumberman* 58(December 1931):38; N. O. Nicholson to J. P. Kinney, 15 July 1932, Taholah Indian Agency Records, RG 75, Federal Records Center, Seattle.

23. Polson to R. Merrill, 30 December 1930; T. Merrill to Jerome, 13 January 1931; to R. Merrill, 26 November 1930, Merrill & Ring Lumber Co. Records; C. H. Ingram to R. R. Macartney, 28 June 1930, Weyerhaeuser Timber Co. Records; A. D. Chisholm to W. C. Ruegnitz, 2 August 1930; J. J. Donovan to Ruegnitz, 20 August 1930 and 11 March 1931, William C. Ruegnitz Papers, Manuscripts Collection, University of Washington Library; Stewart H. Holbrook to M. T. Dunten, 17 April 1931, Stewart H. Holbrook Papers, Manuscripts Collection, University of Washington Library.

24. W. Butler to N. Butler, 11 December 1931, Butler Papers; Polson to R. Merrill, 11 and 15 June and 3 July 1931, Merrill & Ring Lumber Co. Records; C. H. Kreienbaum and Elwood R. Maunder, "Forest Management and Community Stability: The Simpson Experience," *Forest History* 12(July 1968):11–12.

25. Ruegnitz to J. W. Lewis, 22 July 1932; to F. P. Foisie, 17 June 1932; to Ingram 23 June 1923; Chisholm to Ruegnitz, 5 September 1930 and 29 July and 5 August 1931, Ruegnitz Papers.

26. Ruegnitz to Helen Wood, 21 February 1934; to Donovan, 1 and 11 October 1930; to Tennant, 24 April 1931; to Ingram, 7 July 1931; to Raught, 28 October 1931, Ruegnitz Papers.

27. Ruegnitz to Ingram, 24 July 1931 and 20 June 1932; to Tennant, 15 September 1931; to Dunten, 2 July 1931, Ruegnitz Papers, Holbrook to Dunten, 19 November 1931, Holbrook Papers; *Four L Lumber News* 14(1 January 1932):5; (1 April 1932):3.

28. Ruegnitz to Ingram, 9 October 1931; to J. R. Spurgeon, 18 August 1932; Titcomb to Ruegnitz, 29 July 1931 and 25 June 1932; Griggs to Ruegnitz, 19 July 1932 (two letters of this date), Ruegnitz Papers; *Four L Lumber News* 14(1 August 1932):3; *West Coast Lumberman* 59(December 1932):28.

29. Polson to R. Merrill, 3 July 1931; 1 February, 27 July and 6 and 20 August 1932; to Jerome, 14 March 1932; R. Merrill to R. Polson, 11 February 1932; to John W. Blodgett, 18 April 1932; to C. S. Battle, 18 May 1932, Merrill & Ring Lumber Co. Records.

30. Polson to R. Merrill, 3 July 1931; 1 February and 27 September 1932; R. Merrill to Battle, 15 July 1932, Merrill & Ring Lumber Co. Records; Reed to Harold L. Ickes, 16 March 1932, Mark E. Reed Papers, Manuscripts Collection, University of Washington Library.

31. A Democrat, a onetime classmate of Franklin Roosevelt, and a rising force in the affairs of the Weyerhaeuser Timber Company, Laird Bell was a notable exception to the industry's Republican stance. W. Butler to N. Butler, 31 May 1932, Butler Papers; Franklin D. Roosevelt to Miller Freeman, 6 September 1932, in *West Coast Lumberman* 59(October 1932):5; R. Merrill to E. Wetmore, 11 and 27 October 1932; to R. Wetmore, 9 July 1932; to Polson, 28 September 1932; to Jarl M. Hanson, 27

September 1932; to Stevens, 14 October 1932; Merrill & Ring Lumber Co. Records; Reed to C. B. Blethen, 15 July 1932, Reed Papers.

32. Polson to R. Merrill, 28 July and 20 August 1932; R. Merrill to Polson, 28 September 1932, Merrill & Ring Lumber Co. Records; J. H. Bloedel to Griggs, 2 August 1932, St. Paul & Tacoma Lumber Co. Records; R. M. Weyerhaeuser to J. P. Weyerhaeuser, 4 November 1932, Weyerhaeuser Family Collection, Minnesota Historical Society; *Timberman* 33(June 1932):14; Ficken, *Lumber and Politics*, 199–203.

33. R. Merrill to Battle, 11 November 1932; Burr to Jerome, 18 November 1932, Merrill & Ring Lumber Co. Records.

34. R. Merrill to Albert Johnson, 12 December 1932; to Burr, 9 December 1932; Jerome to W. M. Leuthold, 2 March 1933, Merrill & Ring Lumber Co. Records; *American Lumberman* (21 January 1933):12.

15:   On the Road to Recovery

1. *West Coast Lumberman* 60(January 1933):21.

2. Timothy Jerome to A. C. Stubb, 4 January 1934, Merrill & Ring Lumber Co. Records, Manuscripts Collection, University of Washington Library.

3. Code of Fair Competition for the Lumber and Timber Products Industries, 19 August 1933, Pacific Northwest Loggers Association Records, Manuscripts Collection, University of Washington Library; *American Forests* 39(September 1933):405; *New York Lumber Trade Journal* 95(1 September 1933):5; *Southern Lumberman* 146(15 June 1933):8.

4. Henry B. Steer, comp., *Lumber Production in the United States, 1799–1946*, USDA Misc. Pub. No. 669(Washington, D.C., 1948), 11–16; Jerome to W. .M. Leuthold, 2 March 1933; to Francis P. Sears, 3 March 1933; to M. N. Brady, 5 April 1933, Merrill & Ring Lumber Co. Records; *West Coast Lumberman* 60(March 1933):28.

5. F. S. Bell to J. P. Weyerhaeuser, Jr., 13 February 1933, J. P. Weyerhaeuser, Jr. Papers, Weyerhaeuser Co. Archives; J. Weyerhaeuser to George F. Jewett, 11 January 1933, George F. Jewett Papers, University of Idaho Library; Carl M. Stevens to R. D. Merrill, 4 February 1933; Merrill to Stevens, 6 February 1933, Merrill & Ring Lumber Co. Records.

6. Stevens to Merrill, 4, 19, and 25 January 1933, Merrill & Ring Lumber Co. Records; Laird Bell to J. Weyerhaeuser, 8 and 20 February and 19 December 1933, Weyerhaeuser Papers; Mark E. Reed to William C. Butler, 28 January 1933, Mark E. Reed Papers, Manuscript Collection, University of Washington Library.

7. The sales figures were somewhat misleading. Featuring a huge increase in shipments to China, Northwest lumber exports were up by thirty-seven percent in the first half of 1933. In contrast, domestic water shipments actually declined by a slight margin. J. Weyerhaeuser to F. Bell, 28 March 1933, Weyerhaeuser Papers; Merrill to Sears, 23 May 1933; to R. W. Wetmore, 6 June 1933, Merrill & Ring Lumber Co. Records; Reed to C. H. Kreienbaum, 5 June and 5 July 1933, Reed Mill Co. Records, Simpson Timber Co., Shelton; *West Coast Lumberman* 60(August 1933):23.

8. *Timberman* 35(November 1933):8; W. B. Greeley, "The Outlook for Timber Management by Private Owners," *Journal of Forestry* 31(February 1933):213; Harold K. Steen, *The U.S. Forest Service: A History* (Seattle, 1976), 199–204.

9. Rodney C. Loehr, ed., *Forests for the Future: The Story of Sustained Yield as Told in the Diaries and Papers of David T. Mason, 1907–1950* (St. Paul, 1952), 110; Jerome to L. Howard Place, 19 April 1933; to E. E. Barthell, 31 March 1933, Merrill

& Ring Lumber Co. Records; *West Coast Lumberman* 60(April 1933):5; (June 1933):19; *Timberman* 34(March 1933):7.

10. J. Weyerhaeuser to F. K. Weyerhaeuser, 22 April 1933, Weyerhaeuser Papers; *American Lumberman* (29 April 1933), 12; (13 May 1933), 14; *Timberman* 34(April 1933):5; *West Coast Lumberman* 60(May 1933):5.

11. W. B. Greeley, "The West Coast Problem of Stabilizing Lumber Production," *Journal of Forestry* 28(February 1930):196–98; Ward Shepard, "Cooperative Control: A Proposed Solution of the Forest Problem," ibid., 116–88; Robert Marshall, "A Proposed Remedy for Our Forest Illness," *Journal of Forestry* 28(March 1930):276–9.

12. *American Lumberman* (27 May 1933), 10; Loehr, ed., *Forests for the Future,* 115; Merrill to Frank H. Brownell, 13 and 17 June 1933; to R. Wetmore, 13 June 1933, Merrill & Ring Lumber Co. Records; Minutes, General Loggers Meeting, 2 June 1933, Pacific Northwest Loggers Association Records; Minutes, Committee on Production, 23 June 1933, West Coast Lumbermen's Association Records, Oregon Historical Society.

13. Minutes, General Loggers Meetings, 2 and 20 June 1933; Minutes, Meeting on British Columbia Imports, 18 July 1933, Pacific Northwest Loggers Association Records; Merrill to R. Wetmore, 21 June 1933; to Sears, 19 June 1933, Merrill & Ring Lumber Co. Records; Minutes, Board of Directors Meeting, 25 July 1933, Council of Forest Industries Records, University of British Columbia Library.

14. F. E. Weyerhaeuser to J. Weyerhaeuser, 12 July 1933, Weyerhaeuser Family Collection, Minnesota Historical Society; J. Weyerhaeuser to F. K. Weyerhaeuser, 16 June 1933, Weyerhaeuser Papers; Minutes, Emergency National Committee, 4 July 1933, Reed Mill Co. Records; Loehr, ed., *Forests for the Future,* 116–17; Franklin D. Roosevelt to Henry A. Wallace, 16 June 1933, in Edgar B. Nixon, ed., *Franklin D. Roosevelt and Conservation, 1911–1945,* 2 vols. (Hyde Park, N.Y., 1957), 1:182; E. T. Allen to G. B. McLeod, 20 June 1933, Jewett Papers.

15. Merrill to P. V. Eames, 14 and 25 July 1933; Eames to Merrill, 20 July 1933, Merrill & Ring Lumber Co. Records; Minutes, Committee on Production, 20 July 1933, West Coast Lumberman's Association Records; J. Weyerhaeuser to F. E. Weyerhaeuser, 18 July 1933, Weyerhaeuser Papers.

16. E. G. Griggs II to E. G. Griggs, 5 July 1933, St. Paul & Tacoma Lumber Co. Records, Manuscripts Collection, University of Washington Library; E. H. Meiklejohn to Jerome, 31 July 1933; Merrill to Sears, 29 July 1933; to Brownell, 4 August 1933; to Eames, 7 August 1933, Merrill & Ring Lumber Co. Records; J. Weyerhaeuser to F. Bell, 5 August 1933, Weyerhaeuser Papers.

17. F. E. Weyerhaeuser to J. Weyerhaeuser, 31 July and 15, 17 and 21 August 1933, Weyerhaeuser Family Collection; A. W. Clapp to J. Weyerhaeuser, 4 and 23 August 1933; J. Weyerhaeuser to Clapp, 17 August 1933; F. E. Weyerhaeuser to Clapp, 10 August 1933; to F. Bell, 12 August 1933, Weyerhaeuser Papers; Loehr, ed., *Forests for the Future,* 118–19.

18. Memorandum of Conference with Recovery Administration, 8 July 1933, Reed Mill Co. Records; Code of Fair Competition, 19 August 1933, Pacific Northwest Loggers Association Records; *West Coast Lumberman* 60(August 1933):3–5; C. W. Bahr, "A Brief Survey of the Provisions of the Lumber and Timber Products Code," *Journal of Forestry* 33(March 1935):218–19.

19. William B. Greeley to Mrs. W. B. Greeley, 9 August 1933, William B. Greeley Papers, University of Oregon Library; Loehr, ed., *Forests for the Future,* 117–19; Allen to Wilson Compton, 21 August 1933, Jewett Papers.

20. F. E. Weyerhaeuser to J. Weyerhaeuser, 21 August 1933, Weyerhaeuser Family Collection; F. E. Weyerhaeuser to Clapp, 10 August 1933; to F. Bell, 12 August 1933, Weyerhaeuser Papers; Jerome to Louis H. Moore, 20 July 1933; Merrill to Eames, 14 July 1933; to C. S. Battle, 28 August 1933, Merrill & Ring Lumber Co. Records.

21. R. Wetmore to Merrill, 23 August 1933; Merrill to R. Wetmore, 8 and 25 September and 17 and 27 October 1933; to Battle, 28 August 1933; Jerome to R. W. Hibberson, 18 November 1933, Merrill & Ring Lumber Co. Records; *West Coast Lumberman* 60(September 1933):3.

22. The appointment of David Mason as executive officer in June 1934 gave to the Authority a distinct Northwest orientation. William C. Ruegnitz to Russell Early, 20 October 1933, William C. Ruegnitz Papers, Manuscripts Collection, University of Washington Library; R. Wetmore to Merrill, 7 September 1933; Merrill to Jerome, 28 August 1933; to Battle, 5 September 1933, Merrill & Ring Lumber Co. Records.

23. Merrill to Eames, 12 September 1933; Eames to Merrill, 18 September 1933; R. Wetmore to Merrill, 29 September 1933; Jerome to H. L. Plumb, 25 September 1933, Merrill & Ring Lumber Co. Records; W. H. Dole to Pacific Northwest Loggers Association, 5 October 1933, Aloha Lumber Co. Records, Manuscripts Collection, University of Washington Library; F. E. Weyerhaeuser to Clapp, 28 November 1933; to F. Bell, 20 December 1933, Weyerhaeuser Papers.

24. F. E. Weyerhaeuser to J. Weyerhaeuser, 9 October 1933, Weyerhaeuser Papers; F. R. Titcomb to W. H. Peabody, 4 November 1933, F. R. Titcomb Papers, Weyerhaeuser Co. Archives; *West Coast Lumberman* 60(November 1933):20, 32; Julian S. Duncan, "The Effect of the N.R.A. Lumber Code on Forest Policy," *Journal of Political Economy* 49(February 1941):93–94; Peter A. Stone, et al., *Economic Problems of the Lumber and Timber Products Industry*, NRA Work Materials No. 79, pp. 94–102.

25. J. Weyerhaeuser to F. Bell, 11 October 1933, Weyerhaeuser Papers; David T. Mason to Eames, 15 November 1933, Jewett Papers; *West Coast Lumberman* 61(January 1934):16; (July 1934), 29; Duncan, "Effect of the N.R.A. Lumber Code," 94, 96–97.

26. Minutes, Production Committee, 23 October 1933, Pacific Northwest Loggers Association Records; Titcomb to Stevens, 5 April 1934, Titcomb Papers, Jerome to Petrus Pearson, 4 April 1934; to Stiles W. Burr, 3 April 1934, Merrill & Ring Lumber Co. Records; Loehr, ed., *Forests for the Future*, 122.

27. *Division of Forestry, Annual Reports, 1928 to 1932* (Olympia, 1933), 39–42; *Seventh Biennial Report of the Department of Conservation and Development, 1934* (Olympia, 1935), 37; Wade DeVries and E. H. McDaniels, "Forest Taxation in Oregon and Washington," *West Coast Lumberman* 69(April 1942):32.

28. *Journal of Forestry* 31(December 1933):883–84; Loehr, ed., *Forests for the Future*, 108; *West Coast Lumberman* 60(April 1933):20; 62(July 1935):7, 33; Mason to Austin Cary, 18 February 1936, David T. Mason Papers, Oregon Historical Society; Alex Polson to Jerome, 3 February 1934, Merrill & Ring Lumber Co. Records.

29. Burt P. Kirkland and Axel J. Brandstrom, *Selective Timber Management in the Douglas Fir Region* (Seattle, 1936), 117; Leo A. Isaac, *Douglas Fir Research in the Pacific Northwest*, Oral history interview by Amelia R. Fry, Regional Oral History Office, University of California, Berkeley, 92–93; L. T. Murray, Sr. oral history interview, Forest History Society, 57–58; L. T. Murray to Harold L. Ickes, 23 April 1940, Harold L. Ickes Papers, Library of Congress.

30. Isaac, *Douglas Fir Research*, 86–101; Walter Lund, *Timber Management in the Pacific Northwest Region, 1927–1965*, Oral history interview by Amelia R. Fry, Regional Oral History Office, 48–49; Thornton T. Munger, *Forest Research in the Northwest*, Oral history interview by Amelia R. Fry, Regional Oral History Office, 123–32; John B. Woods to Phillips A. Haywood, 24 October 1938, National Forest Products Association Records, Forest History Society.

31. C. S. Chapman to Jewett, 11 October 1933, Jewett Papers; Code of Fair Competition, 19 August 1933, Pacific Northwest Loggers Association Records.

32. *West Coast Lumberman* 61(February 1934):11; (April 1934):5, 20; Statement on Industrial Forest Conservation, 23 February 1934, in Nixon, ed., *Roosevelt and Con-*

servation, 1:259–60; William B. Greeley, *Forests and Men* (Garden City, N.Y., 1951), 134–37; Allen to Jewett, 25 November 1933, Jewett Papers; J. Weyerhaeuser to Titcomb, 13 September 1934, Titcomb Papers.

33. Jewett to Allen, 30 March 1934; Allen to Jewett, 2 November 1934, Jewett Papers; Compton to Roosevelt, 16 May 1934; Ovid Butler to Roosevelt, 5 July 1935, in Nixon, ed., *Roosevelt and Conservation*, 1:275–76, 390–91.

34. *West Coast Lumberman* 60(October 1933):4; J. Weyerhaeuser to F. Bell, 19 March 1934, Weyerhaeuser Papers; Ruegnitz to F. F. Duell, 9 January 1934, Ruegnitz Papers; Robert W. Vinnedge to J. B. Fitzgerald, 1 September 1933, Pacific Northwest Loggers Association Records.

35. Ruegnitz to J. W. Lewis, 29 September and 2 November 1933; to Early, 20 October 1933; Early to Ruegnitz, 4 November 1933; M. T. Dunten to Ruegnitz, 14 June 1934, Ruegnitz Papers; T. J. Linton to James A. Taylor, 6 October 1933, Washington State Federation of Labor Records, Manuscripts Collection, University of Washington Library; Vernon H. Jensen, *Lumber and Labor* (New York, 1945), 154–55.

36. Taylor to Roger A. Jones, 9 April 1934, Washington State Federation of Labor Records; Weyerhaeuser to F. K. Weyerhaeuser, 16 March 1934, Weyerhaeuser Papers; Jensen, *Lumber and Labor,* 159–60.

37. H. M. Grimes to John P. Burke, 1 September 1933; George Clausen to Burke, 17 and 24 September, 8 October and 12 November 1933, International Brotherhood of Pulp, Sulphite & Paper Mill Workers Records (microfilm), State Historical Society of Wisconsin; Roger Randall, *Labor Relations in the Pulp and Paper Industry of the Pacific Northwest* (Portland, 1942), 29–30, 52–53.

38. Burke to Grimes, 11 September 1933; to Ronald Macdonald, 2 and 26 October 1933; to C. S. Homer, 24 January and 7 February 1934; to Clausen, 27 August 1933; to H. I. Hansen, 2 and 24 January 1935; to H. J. Peterson, 5 and 29 August 1935, International Brotherhood of Pulp, Sulphite & Paper Mill Workers Records; Randall, *Labor Relations,* 21, 57, 68–99.

39. J. Weyerhaeuser to Executive Committee, 15 May 1934; to F. E. Weyerhaeuser, 14 and 22 June, 9 July and 1 and 18 August 1934, Weyerhaeuser Papers; Merrill to Eames, 15 June 1934; to R. Wetmore, 20 July 1934; to E. D. Wetmore, 2 August 1934; Jerome to E. Wetmore, 21 May 1934, Merrill & Ring Lumber Co. Records; *West Coast Lumberman* 61(June 1934):42; (July 1934):40; *Timberman* 35(July 1934):74.

40. Jones to Taylor, 27 March 1934; Ruegnitz to Hugh S. Johnson, 15 March 1935, Washington State Federation of Labor Records; Ruegnitz to C. L. Billings, 13 March 1934, Ruegnitz Papers; J. Weyerhaeuser to F. Bell, 19 March 1934, Weyerhaeuser Papers.

41. Steer, comp., *Lumber Production,* 15–16, 197; *West Coast Lumberman* 62(February 1935):20; Loehr, ed., *Forests for the Future*:137–38, 141.

42. Although Washington lumber production did not increase in 1934 over 1933 levels, the number of operating sawmills increased by nineteen percent, forcing reductions in individual quotas. Kreienbaum to C. H. Watzek, 7 August 1934; J. Weyerhaeuser to F. E. Weyerhaeuser, 1 August 1934, Weyerhaeuser Papers; J. Weyerhaeuser to Titcomb, 13 September 1934, Titcomb Papers; Steer, comp., *Lumber Production,* 16.

43. Loehr, ed., *Forests for the Future,* 144, 146; J. Weyerhaeuser to F. E. Weyerhaeuser, 27 and 31 August 1934, Weyerhaeuser Papers; Titcomb to H. N. Anderson, 17 and 19 September 1934; to Greeley, 1 September 1934; to H. B. Hewes, 31 October 1934; to R. W. Lea, 28 September 1934, Titcomb Papers; *Timberman* 35(September 1934):57; (October 1934):10, 22, 24–26; 36(November 1934):10–11; *West Coast Lumberman* 61(October 1934):20; (November 1934):5.

44. Jewett to H. Johnson, 9 March 1934; to Will Simons, 9 March 1934, Jewett Papers; Titcomb to R. R. Macartney, 15 September 1934, Titcomb Papers; C. D. John-

son to Greeley, 4 September 1934, Weyerhaeuser Papers; *Timberman* 35(August 1934):26; 36(November 1934), 81; *West Coast Lumberman* 61(October 1934):20; (November 1934):5.

45. Titcomb to Greeley, 1 September 1934, Titcomb Papers; Jerome to Pearson, 25 and 29 September, 15 and 20 October and 2 November 1934, Merrill & Ring Lumber Co. Records.

46. Loehr, ed., *Forests for the Future*, 146–48, 164; J. D. Tennant to Titcomb, 31 December 1934, Titcomb Papers; Corydon Wagner to Greeley, 14 November 1934, Merrill & Ring Lumber Co. Records; *American Lumberman* (22 December 1934), 30; (5 January 1935), 29–30.

47. Titcomb to F. K. Weyerhaeuser, 27 March 1935; Tennant to Mason, 28 March 1935; Mason to Tennant, 4 April 1935, Tennnat Papers; Loehr, ed., *Forests for the Future*, 170–76; *Southern Lumberman* 150(1 April 1935):18; (15 April 1935):20; (1 May 1935), 18; Minutes, Trustees Meeting, 1 April 1935, Pacific Northwest Loggers Association Records; *West Coast Lumberman* 62(April 1935):28; *Timberman* 36(April 1935):9.

48. Sears to Merrill, 29 May 1935; to Jerome, 3 June 1935; Merrill & Ring Lumber Co. Records; *Timberman* 36(June 1935):9; *Southern Lumberman* 150(1 June 1935):18.

49. *Timberman* 35(September 1934):9–10.

50. Merrill to E. Wetmore, 24 February 1934 and 7 May 1935; Jerome to R. P. Minton, 15 January 1935; to E. Wetmore, 21 May 1934; to Battle, 8 November 1934; to S. A. Stamm, 25 April 1934, Merrill & Ring Lumber Co. Records; Jewett to Duncan U. Fletcher, 11 July 1935; to Allen, 11 February 1935, Jewett Papers; F. E. Weyerhaeuser to J. Weyerhaeuser, 2 July 1934, Weyerhaeuser Papers.

16: Like a Moving Film

1. W. B. Greeley, "Lumber Problems and Opportunities," *Timberman* 39(October 1938):12.

2. Timothy Jerome to A. C. Stubb, 2 June 1937, Merrill & Ring Lumber Co. Records, Manuscripts Collection, University of Washington Library.

3. J. A. Ziemer to W. C. Ruegnitz, 25 June 1935, William C. Ruegnitz Papers, Manuscripts Collection, University of Washington Library.

4. J. P. Weyerhaeuser, Jr. to F. E. Weyerhaeuser, 11 December 1934; 12 April and 4 May 1935, J. P. Weyerhaeuser, Jr. Papers, Weyerhaeuser Co. Archives; F. R. Titcomb to C. G. Kinney, 15 April 1935; to Edmund Hayes, 2 May 1935, F. R. Titcomb Papers, Weyerhaeuser Co. Archives; Vernon H. Jensen, *Lumber and Labor* (New York, 1945), 160–61, 166.

5. J. Weyerhaeuser to F. E. Weyerhaeuser, 12 April 1935, Weyerhaeuser Papers; *Timberman* 36(April 1935):66; Robert A. Christie, *Empire in Wood: A History of the Carpenter's Union* (Ithaca, N.Y., 1956), 288–91, 293.

6. Titcomb to Kinney, 15 April 1935; J. Weyerhaeuser to F. E. Weyerhaeuser, 3 July 1935, Titcomb Papers; J. Weyerhaeuser to F. E. Weyerhaeuser, 30 April and 4 May 1935, Weyerhaeuser Papers; Stewart H. Holbrook to M. T. Dunten, 25 May and 18 July 1935, Stewart H. Holbrook Papers, Manuscripts Collection, University of Washington Library.

7. J. Weyerhaeuser to F. E. Weyerhaeuser, 2, 4, 6, 8, and 11 May 1935; F. E. Weyerhaeuser to J. Weyerhaeuser, 22 May and 6 June 1935, Weyerhaeuser Papers; *Timberworker*, 8, 15, 22, and 29 May and 5 June 1936; R. D. Merrill to F. P. Sears,

21 May 1935, Merrill & Ring Lumber Co. Records; Jensen, *Lumber and Labor*, 167–70, 172–74.

8. Titcomb to Clarence D. Martin, 27 June 1935, Titcomb Papers; Jerome to Martin, 7 June 1935; Robert Polson to Jerome, 24 June 1935, Merrill & Ring Lumber Co. Records; *Timberworker*, 12, 19, and 26 June 1936; Jensen, *Lumber and Labor*, 174–76; 178–79. Martin quoted in *American Lumberman* (6 July 1935), 19.

9. J. Weyerhaeuser to F. E. Weyerhaeuser, 3 July 1935, Titcomb Papers; J. Weyerhaeuser to F. E. Weyerhaeuser, 26 June 1935, Weyerhaeuser Papers; Ziemer to Ruegnitz, 25 June 1935, Ruegnitz Papers; *Timberworker*, 26 June 1936; *American Lumberman* (6 July 1935), 19, 45; (20 July 1935), 36.

10. Ziemer to Ruegnitz, 25 June 1935, Ruegnitz Papers, J. Weyerhaeuser to F. E. Weyerhaeuser, 26 June and 9 August 1935, Weyerhaeuser Papers; *Timberworker*, 26 June, 24 July, and 7 August 1936; *American Lumberman* (20 July 1935), 11, 36; (3 August 1935), 40; (17 August 1935), 17; Jensen, *Lumber and Labor*, 178–79, 182–83.

11. *Timberworker*, 3, 10, 17, 24, and 31 July; 7, 14, and 21 August; 4, 11 and 25 September; 2 October and 6 November 1936; Holbrook to Dunten, 18 July 1935, Holbrook Papers; Frank Hobi Reminiscences, Manuscripts Collection, University of Washington Library, 89–92; Jensen, *Lumber and Labor*, 180–84.

12. J. Weyerhaeuser to F. E. Weyerhaeuser, 3 July 1935, Titcomb Papers; J. Weyerhaeuser to F. E. Weyerhaeuser, 16 July 1935, Weyerhaeuser Papers; Merrill to P. V. Eames, 5 July 1935; Jerome to Stiles W. Burr, 10 July 1935, Merrill & Ring Lumber Co. Records.

13. *Fifth Report of the Department of Labor and Industries, 1932–1936* (Olympia, 1937), 18, 21–23; *Sixth Report of the Department of Labor and Industries, 1937* (Olympia, 1938), 14–15; *Seventh Report of the Department of Labor and Industries, 1938–1939* (Olympia, 1940), 16–17; James G. Eddy to William G. Reed, 20 March 1939, Simpson Timber Co. Records, Simpson Timber Co. Archives; Titcomb to C. H. Kreienbaum, 13 May 1936, Titcomb Papers; Jerome to Sears, 21 September 1935; to William M. Fischer, 14 July 1937, Merrill & Ring Lumber Co. Records.

14. *Timberworker*, 11, 18, and 25 October, 1 November and 6 December 1935; 3 January 1936; Jensen, *Lumber and Labor*, 203–4; Christie, *Empire in Wood*, 296.

15. J. Weyerhaeuser to F. E. Weyerhaeuser, 4 November 1935; 24 January and 9 December 1936; Laird Bell to J. Weyerhaeuser, 5 June 1936 and 21 October 1937, Weyerhaeuser Papers; A. H. Landram to W. M. Barry, 3 September 1937, Corydon Wagner Papers, Manuscripts Collection, University of Washington Library.

16. *Timberworker*, 10, 17, and 24 January, 7 February, 17 April, 24 July, 21 August, and 25 September 1936; 2 and 16 April, 7 and 21 May, 4 June, and 2, 23 and 30 July 1937; Jensen, *Lumber and Labor*, 203–12; Jerry L. Lembcke, "The International Woodworkers of America: An Internal Comparative Study of Two Regions," (Ph.D. diss., University of Oregon, 1978), 62–3, 179–82.

17. Minutes, Executive Board Meetings, 8 and 17 September 1938, Harold Pritchett Papers, University of British Columbia Library; *Timberworker*, 12 February and 4 December 1937; Minutes, Policy Committee Meeting, 22 June 1940, Adolph Germer Papers, State Historical Society of Wisconsin.

18. *Timberworker*, 13 and 20 August 1937; Jerome to Merrill, 11 August 1937, Merrill & Ring Lumber Co. Records; Landram to Barry, 3 September 1937, Wagner Papers; A. H. Adams to ?, 9 April 1938, Francis J. Murnane Papers, Oregon Historical Society; Minutes, Executive Board Meetings, 8 September 1938 and 26 February 1939, Pritchett Papers; J. Weyerhaeuser to F. E. Weyerhaeuser, 11 January and 3 and 17 April 1939; 10 September and 10 and 16 December 1940, Weyerhaeuser Papers; Jensen, *Lumber and Labor*, 212–31; 235–43.

19. Minutes, Executive Board Meeting, 12 May 1940, Pritchett Papers; Germer to

Allen S. Haywood, 11 July, 30 September and 10 and 13 October 1940; 25 March and 10 April 1941, Germer Papers; Jensen, *Lumber and Labor,* 229–43, 268–71.

20. Christie, *Empire in Wood,* 299; Bell to J. Weyerhaeuser, 31 December 1938; J. Weyerhaeuser to F. E. Weyerhaeuser, 22 and 30 December 1938; 6, 9, 11, and 20 January 1939; 10 and 13 September 1940, Weyerhaeuser Papers.

21. *West Coast Lumberman* 63(February 1936):13; 64(February 1937):28; J. D. Tennant to William B. Greeley, 24 December 1935, William B. Greeley Papers, University of Oregon Library; Henry B. Steer, comp., *Lumber Production in the United States, 1799–1946,* USDA Misc. Pub. No. 669(Washington D.C., 1948), 14–17, 197; *Timberman* 40(July 1939):10.

22. *Timberman* 38(January 1937):86; 39(May 1938):9; 41(February 1940):9; Greeley to C. F. Truitt, 2 September 1939, Greeley Papers; Greeley annual report, 27 January 1938, West Coast Lumbermen's Association Records, Oregon Historical Society; Edwin T. Coman, Jr. and Helen M. Gibbs, *Time, Tide and Timber: A Century of Pope & Talbot* (Stanford, Cal., 1949), 300, 304.

23. *Timberman* 39(January 1938):9; 40(February 1939):59; Jerome to M. N. Brady, 1 October 1937 and 15 March 1938; to Merrill, 21 October 1938, Merrill & Ring Lumber Co. Records; F. K. Weyerhaeuser to J. Weyerhaeuser, 14 April 1939; Bell to J. Weyerhaeuser, 1 August 1940, Weyerhaeuser Papers; Wagner to A. L. Jenny, 7 October 1938, Wagner Papers.

24. J. Weyerhaeuser to F. E. Weyerhaeuser, 26 February 1935; F. E. Weyerhaeuser to J. Weyerhaeuser, 12 January 1940, Weyerhaeuser Papers; Wilson Compton to George F. Jewett, 21 November 1935, George F. Jewett Papers, University of Idaho Library; Greeley to C. E. .Dant, 9 February 1937; to C. C. Crow, 9 February 1937, Greeley Papers.

25. J. Weyerhaeuser to Bell, 23 November and 2 December 1937; W. H. Peabody to J. Weyerhaeuser, 21 November 1938, Weyerhaeuser Papers; F. K. Weyerhaeuser to J. Weyerhaeuser, 2 November 1937, Weyerhaeuser Family Collection, Minnesota Historical Society; Wagner to Jenny, 15 March 1938, Wagner Papers; Jerome to Petrus Pearson, 30 September and 1 October 1937; to E. C. Weeks, 29 November 1937 and 30 April 1938; to Mrs. E. B. Smith, 28 April 1938, Merrill & Ring Lumber Co. Records; *Timberman* 39(November 1937):85; (February 1938):94; (May 1938):82; 40(February 1939), 59.

26. Annual Report, 1938, Rayonier Incorporated Records, ITT Rayonier, Hoquiam; *Pacific Pulp and Paper Industry* 11(September 1937):10; (October 1937):13; 12(July 1938), 37–39; 13(January 1939), 10; *Business Week* (9 October 1937), 43–44.

27. Jenny to Landram, 16 April 1938, Wagner Papers; E. T. Allen to Jewett, 10 March 1937; Jewett to Allen, 3 March 1937, Jewett Papers; Merrill to Jerome, 24 March 1938; Jerome to Weeks, 30 April 1938; to Stubb, 17 March 1938, Merrill & Ring Lumber Co. Records; F. E. Weyerhaeuser to J. Weyerhaeuser, 14 April 1936, Weyerhaeuser Papers; F. K. Weyerhaeuser to J. Weyerhaeuser, 2 November 1937, Weyerhaeuser Family Collection.

28. Jerome to Merrill, 4 November 1936; Merrill to Stubb, 22 November 1936, Merrill & Ring Lumber Co. Records.

29. David T. Mason to Julian E. Rothery, 23 July 1935, David T. Mason Papers, Oregon Historical Society; Rodney C. Loehr, ed., *Forests for the Future: The Story of Sustained Yield as Told in the Diaries and Papers of David T. Mason, 1907–1950* (St. Paul, 1952), 178–81, 189–90; J. Weyerhaeuser to F. E. Weyerhaeuser, 16 February and 7 March 1935; Carl M. Stevens to J. Weyerhaeuser, 1 April 1935, Weyerhaeuser Papers; *Timberman* 39(December 1937):77; *West Coast Lumberman* 62(July 1935):38; William B. Greeley, *Forests and Men* (Garden City, N.Y., 1951), 137–38.

30. F. A. Silcox, "Foresters Must Choose," *Journal of Forestry* 33(March 1935):198–

204; "Forestry—A Public and Private Responsibility," *Journal of Forestry* (May 1935), 460–68, and "Our Adventure in Conservation," *Atlantic Monthly* 160(December 1937):714–22; *Timberman* 36(April 1935):9–10; *West Coast Lumberman* 63(January 1936):24–32; 66(August 1939):15; F. A. Silcox to Roderic Olzendam, 28 October 1937, Weyerhaeuser Papers; Silcox to Jewett, n.d., but ca. February 1938, National Forest Products Association Records, Forest History Society; Jewett to Compton, 27 November 1935; to A. B. Recknagel, 19 December 1935; C. S. Chapman to Jewett, 30 November 1935 and 19 December 1935, Jewett Papers.

31. Emanual Fritz to Allen, 15 September 1938, Western Forestry & Conservation Association Records, Oregon Historical Society; Chapman to John B. Woods, 15 March 1938; to Bell, 24 July 1939; Bell to J. Weyerhaeuser, 17 November 1938 and 10 March 1939; F. K. Weyerhaeuser to J. Weyerhaeuser, 19 November 1938, Weyerhaeuser Papers; Compton to Woods, 18 November 1937, National Forest Products Association Records; *West Coast Lumberman* 64(June 1937):32, 70–71; 65(September 1938)7.

32. Gifford Pinchot, "Old Evils in New Clothes," *Journal of Forestry*, 35(May 1937):438; *American Forests* 42(April 1936):177; 43(August 1937):397; (October 1937):495; 44(January 1938):23; (March 1938), 121–22; *Journal of Forestry:* 32(December 1934), 927–29.

33. Loehr, ed., *Forests for the Future,* 202–3; *Timberman* 38(June 1937):9–10; (July 1937):11; 39(December 1937):10; C. S. Martin to Fritz, 22 November 1937, Western Forestry & Conservation Association Records; Chapman to Jewett, 11 November 1937, Jewett Papers; *American Lumberman* (20 July 1935), 10; Richard Polenberg, "Conservation and Reorganization: The Forest Service Lobby, 1937–1938," *Agricultural History,* 39(October 1965):233–34.

34. Hugo Winkenwerder to H. H. Chapman, 18 September 1933; to Franklin Reed, 31 October 1933, Hugo Winkenwerder Papers, University of Washington Archives; Greeley to George E. Lammers, 16 December 1939, Greeley Papers; Jewett to Allen, 27 March 1936, Jewett Papers.

35. Irving Brant to Harry Slattery, 26 January 1935, Irving Brant Papers (microfilm), Manuscripts Collection, University of Washington Library; Alfred J. Runte, *National Parks: The American Experience* (Lincoln, Neb., 1979), Chap. 3.

36. In 1933, Franklin Roosevelt transferred administration of the monument from the Forest Service to the Interior Department. John Ise, *Our National Park Policy: A Critical History* (Baltimore, 1961), 383–84; Clifford E. Roloff, "The Mount Olympus National Monument," *Washington Historical Quarterly* 25(July 1934):224–27.

37. R. L. Fromme to Thomas T. Aldwell, 13 October 1916 and 13 December 1919, Thomas T. Aldwell Papers, Manuscripts Collection, University of Washington Library; Aldwell and Chris Morgenroth to Franklin D. Roosevelt, 4 August 1939, Rayonier Incorporated Records, ITT Rayonier, Port Angeles; Asahel Curtis to Russell V. Mack, 1 February 1936, Asahel Curtis Papers, Manuscripts Collection, University of Washington Library.

38. Merrill to Martin F. Smith, 12 April 1935, Merrill & Ring Lumber Co. Records; Curtis to Federation of Women's Clubs, 6 April 1936; to Rene L. DeRouen, 11 April 1936; to Grafton S. Wilcox, 9 May 1936; to B. H. Kizer, 14 May 1936; to Sam C. Massingale, 18 May 1936; to F. W. Mathis, 21 May 1936, Curtis Papers; Hugo Winkenwerder Report to State Planning Council, 17 November 1934, College of Forest Resources Records, University of Washington Archives.

39. Lilian Gustin McEwan to Mrs. Robert C. Wright, 21 February 1938, Washington State Conservation Society Records, Manuscripts Collection, University of Washington Library; Elmo R. Richardson, "Olympic National Park: Twenty Years of Controversy," *Forest History* 12(April 1968):8–9.

40. Willard G. Van Name to Brant, 11 April and 19 October 1936; William C.

Schulz to Brant, 10 January and 13 October 1937; Brant to Schulz, 25 September 1937; to Rosalie Edge, 4 October 1937; Edge to Brant, 13 October 1937; to Schulz, 30 September 1937, Brant Papers.

41. C. S. Cowan, "The Proposed Mount Olympus National Park," *Journal of Forestry* 34(August 1936):747–79; Curtis to Marion A. Zionchek, 14 April 1936; to Compton I. White, 28 May 1936; to Charles C. Marshall, 5 September 1936; to L. A. Nelson, 29 September 1936; to L. G. McClellan, 29 December 1936, Curtis Papers; Greeley to E. H. Meiklejohn, 25 June 1937, Pacific Northwest Loggers Association Records, Manuscripts Collection, University of Washington Library; Greeley to McEwan, 13 March 1936, Washington State Conservation Society Records.

42. Supporters of the large park persisted in the belief that the opposition consisted of "the lumber interests," not the pulp and paper industry. Curtis to David Whitcomb, 31 March 1937, Curtis Papers; Brant to Henry A. Wallace, 7 March 1937; Edge to Brant, 17 February 1937; to Schulz, 27 September 1937; Schulz to Brant, 3 September 1937, Brant Papers; Winkenwerder to Elsie G. Cambridge, 25 May 1937, College of Forest Resources Records; Richardson, "Olympic National Park," 9–10.

43. Walter H. Lund, *Timber Management in the Pacific Northwest Region, 1927–1965*, Oral history interview by Amelia R. Fry, Regional Oral History Office, University of California, Berkeley, 52–53; Diary, 13, 18, 20, 22, and 30 September 1937, H. L. Plumb Papers, Manuscripts Collection, University of Washington Library; Schulz to Brant, 27 October 1937, Brant Papers; Brant to Roosevelt, 22 September 1937, in Edgar B. Nixon, ed., *Franklin D. Roosevelt and Conservation, 1911–1945* 2 vols., (Hyde Park, N.Y., 1957), 2:130–31.

44. There is no documentary evidence for Roosevelt's much-quoted reference to logged-off land: "I hope the lumberman who is responsible for this is roasting in hell." It would be ironic if his feelings about the Quinault influenced the transfer of land from the Forest Service to the Interior Department, the very department responsible for the mess on the reservation. Diary, 30 September and 1 October 1937, Plumb Papers; Jerome to Pearson, 30 September and 1 October 1937, Merrill & Ring Lumber Co. Records; Roosevelt to Harold L. Ickes, 10 December 1938, in Nixon, ed., *Roosevelt and Conservation*, 2:274–75; Brant to Schulz, 1 October 1937, Brant Papers; Richardson, "Olympic National Park," 10.

45. Schulz to Brant, 23, 27, and 28 October and 4 and 22 November 1937; 12 February 1938; Brant to Schulz, 25 October and 4 and 27 November 1937; 12 February 1938; to Edge, 28 October and 1 and 12 November 1937; to Roosevelt, 19 November 1937; O. A. Tomlinson to Brant, 30 October and 23 November 1937, Brant Papers; Curtis to Mathias, 26 October 1937; to George Welch, 2 November 1937, Curtis Papers; Winkenwerder to H. Chapman, 30 November 1937, Winkenwerder Papers; John L. Boettiger to Ickes, 30 April 1938, Harold L. Ickes Papers, Library of Congress.

46. Schulz to Brant, 15 January and 22 February 1938; Brant to Schulz, 12 February and 1 March 1938; to Roosevelt, 24 February 1938; to Edge, 1 March 1938; to Van Name, 4 March 1938, Brant Papers; Mathias to Curtis, 5 March 1938; Curtis to Martin, 7 March 1938, Curtis Papers; Richardson, "Olympic National Park," 10.

47. Martin to Reed, 28 April 1938, Simpson Timber Co. Records; Curtis to W. H. Coles, 26 April 1938; to Mathias, 5 May 1938, Curtis Papers; McEwan to Wright, 25 April 1938; to Mrs. O. B. Thorgrimson, 4 May 1938, Washington State Conservation Society Records; Martin to Roosevelt, 31 March 1938, in Nixon, ed., *Roosevelt and Conservation*, 2:207; Richardson, "Olympic National Park," 10–11.

48. Brant to Van Name, 13 and 28 May and 17 June 1938; to Edge, 16 May 1938; Edge to Brant, 10 June 1938; Schulz to Brant, 9, 13, and 25 June 1938, Brant Papers; Ickes to Boettiger, 30 April 1938, Ickes Papers; Richardson, "Olympic National Park," 11–12.

49. Schulz to Brant, 6 July, 28 August and 28 December 1938; 7 January 1939;

Brant to Edge, 31 December 1938 and 27 November 1939; to Schulz, 31 December 1938; to Van Name, 20 December 1939, Brant Papers; Brant to Ickes, 6 December 1939; Roosevelt to Wallace, 12 December 1939; to Martin, 21 December 1939; E. K. Burlew to Roosevelt, 12 December 1939; Wallace to Roosevelt, 18 December 1939; Roosevelt Press Conference, 2 January 1940, in Nixon, ed., *Roosevelt and Conservation*, 2:389–91, 396–99, 402–4, 408–9; Richardson, "Olympic National Park," 12–14.

50. *Timberman* 40(October 1939):9–10; 41(January 1940):10.

Epilogue

1. William B. Greeley to Barrington Moore, 17 March 1943, William B. Greeley Papers, University of Oregon Library; *Timberman* 43(January 1942):102; (April 1942):85; Fremont E. Kast, "Major Manufacturing Industries in Washington State: Changes in Their Relative Importance and Causes of Changes" (D.B.A. diss., University of Washington, 1956), 100.

2. Kast, "Major Manufacturing Industries," 33–36, 63.

3. Ibid., 36–39, 63; Edwin J. Cohn, Jr., *Industry in the Pacific Northwest and the Location Theory* (New York, 1954), 29–41, 46; Otis W. Freeman and H. F. Raup, "Industrial Trends in the Pacific Northwest," *Journal of Geography* 43(May 1944):175–76.

4. John A. Guthrie and George R. Armstrong, *Western Forest Industry: An Economic Outlook* (Baltimore, 1961), 84; Walter J. Mead, "The Forest Products Economy of the Pacific Northwest," *Land Economics* 32(May 1956):127–29, and "Changing Pattern of Cycles in Lumber Production," *Journal of Forestry* 59(November 1961):808–13; Neal Potter and Francis T. Christy, Jr., *Trends in Natural Resource Commodities* (Baltimore, 1962), 282, 285.

5. W. D. Hagenstein, "Trees Grow: The Forest Economy of the Douglas Fir Region," *American Forests* 60(April 1954):34; *Statistical Handbook of Washington's Forest Industry* (Olympia, 1964), 30–31; National Resources Planning Board, *Puget Sound Region: War and Post-War Development, May 1943* (Washington, D.C., 1943), 85–88, 139.

6. Guthrie and Armstrong, *Western Forest Industry*, 59; Florence K. Ruderman, *Production, Prices, Employment, and Trade in Northwest Forest Industry* (Portland, 1977), 11; William B. Greeley, "It Pay to Grow Trees," *Pacific Northwest Quarterly* 44(October 1953):153.

7. Greeley to Moore, 17 March 1943; to Warren G. Tilton, 22 July 1943; to John E. Gribble, 6 August 1946, Greeley Papers; Oscar R. Levin, "The South Olympic Tree Farm," *Journal of Forestry* 52(April 1954):243–9; Robert Cantwell, *The Hidden Northwest* (Philadelphia, 1972), 199–211.

8. C. M. Granger, "The National Forests at War," *American Forests* 49(March 1943):112–15; Dwight Hair, *Historical Forestry Statistics of the United States*, USFS Stat. Bull. No. 228(Washington, D.C., 1958), 2; Greeley to C. H. Kreienbaum, 11 December 1947, Greeley Papers.

9. Roy O. Hoover, "Public Law 273 Comes to Shelton: Implementing the Sustained-Yield Forest Management Act of 1944," *Journal of Forest History*, 22(April 1978):87–101; W. B. Greeley, "Cooperative Forest Management in the Olympics," *Journal of Forestry* 49(September 1951):627–29.

10. Daniel R. Barney, *The Last Stand: Ralph Nader's Study Group Report on the*

National Forests (New York, 1974), xiii; Paul Brooks, *The Pursuit of Wilderness* (Boston, 1971), 33; John G. Mitchell, "Best of the S.O.B.s," *Audubon* 76(September 1974):49–62; John J. Putnam, "Timber: How Much Is Enough?" *National Geographic* 145(April 1974):486.

11. Bruce Catton, *Waiting for the Morning Train* (Garden City, N.Y., 1972), 37.

Bibliography

Manuscript Sources

Anne Abernethy Reminiscences. Bancroft Library, University of California, Berkeley.
George Abernethy Papers. Oregon Historical Society.
William Penn Abrams Diary. Bancroft Library, University of California, Berkeley.
Albion Documents and Letters. Western Americana Collection, Yale University.
Thomas T. Aldwell Papers. University of Washington Library.
E. T. Allen Papers. Oregon Historical Society.
Aloha Lumber Company Records. University of Washington Library.
Edwin G. Ames Papers. Ames Collection, University of Washington Library.
W. A. Anderson Reminiscences. Bancroft Library, University of California, Berkeley.
Thorpe Babcock Papers. Forest History Society.
Daniel Bagley Papers. University of Washington Library.
Richard A. Ballinger Papers. University of Washington Library.
William T. Ballou Reminiscences. Bancroft Library, University of California, Berkeley.
H. H. Bancroft Papers. Bancroft Library, University of California, Berkeley.
Broussais C. Beck Papers. University of Washington Library.
William N. Bell Reminiscences. Bancroft Library, University of California, Berkeley.
Samuel Benn Reminiscences. Bancroft Library, University of California, Berkeley.
James Birnie Papers. Oregon Historical Society.
David E. and Kate P. Blaine Letters. University of Washington Library.
Erastus Brainerd Papers. University of Washington Library.
Irving Brant Papers. University of Washington Library (microfilm).
Albert Briggs Reminiscences. Bancroft Library, University of California, Berkeley.
Bureau of Land Management Records. Federal Records Center, Seattle.
Howard J. Burnham, "Hudson Bay Co. Sawmill." Oregon Historical Society (typescript).
Nicholas Murray Butler Papers. Columbia University Library.
John Campbell Diary. University of Washington Library.
Thomas Chambers Papers. Washington State Historical Society.
Irving M. Clark Papers. University of Washington Library.
W. W. Clark Oral History Interview. Forest History Society.
Coast Lumber Company Records. Minnesota Historical Society.
College of Forest Resources Records. University of Washington Archives.
Wilson Compton Oral History Interview. Forest History Society.
Council of Forest Industries Records. University of British Columbia Library.
Charles S. Cowan Oral History Interview. Forest History Society.
S. J. Crawford Scrapbook. University of Washington Library.
Asahel Curtis Papers. University of Washington Library.
Samuel T. Dana Papers. Bentley Historical Library, University of Michigan.
Dant & Russell, Inc. typescript history. Oregon Historical Society.
William H. Davis Letters. British Columbia Provincial Archives.
Brice P. Disque Papers. University of Oregon Library.

Brice P. Disque Papers. University of Washington Library.
David Douglas Letters. British Columbia Provincial Archives.
James Douglas Journal. Bancroft Library, University of California, Berkeley.
Douglas Fir Export Company Records. Oregon Historical Society.
David Dring Papers. California Historical Society.
R. G. Dun & Company Credit Reports. Harvard Business School Library.
Albert A. Durham Papers. Oregon Historical Society.
Edward Eldridge Reminiscences. Bancroft Library, University of California, Berkeley.
William Ellis Letters. Yale University Library.
George H. Emerson Letterbooks. University of Washington Library.
O. D. Fisher Oral History Interview. Forest History Society.
Floyd-Jones Family Papers. Bancroft Library, University of California, Berkeley.
Fort Nisqually Journal of Occurences. University of Washington Library.
Francis Frink Oral History Interview. Forest History Society.
Rudolph L. Fromme Oral History Interview. Forest History Society.
General Land Office Records, National Archives.
Adolph Germer Papers. State Historical Society of Wisconsin.
J. J. Gilbert Reminiscences. Bancroft Library, University of California, Berkeley.
Grays Harbor Commercial Company Records. Ames Collection, University of Washington Library.
William B. Greeley Papers. University of Oregon Library.
William D. Hagenstein Oral History Interview. Forest History Society.
Alfred Hall Correspondence. Bancroft Library, University of California, Berkeley.
Mrs. Ben Hartsuck Reminiscences. University of Washington Library.
Hastings Saw Mill Company Records. University of British Columbia Library.
M. E. Hay Papers. Eastern Washington State Historical Society.
Harry C. Heermans Papers. University of Washington Library.
Frank Hobi Reminiscences. University of Washington Library.
Stewart H. Holbrook Papers. University of Washington Library.
Hudson's Bay Company Misc. Records. Oregon Historical Society.
Edward Huggins Papers. Washington State Historial Society.
Humbird Family Papers. University of British Columbia Library.
Harold L. Ickes Papers. Library of Congress.
O. H. Ingram Papers. State Historical Society of Wisconsin.
International Brotherhood of Pulp, Sulphite & Paper Mill Workers Records. State Historical Society of Wisconsin (microfilm).
Leo A. Isaac Oral History Interview. Regional Oral History Office, University of California, Berkeley.
George F. Jewett Papers. University of Idaho Library.
Wesley L. Jones Papers. University of Washington Library.
Josiah P. Keller Papers. Western Americana Collection, Yale University.
J. P. Kinney Papers. Cornell University Library.
James D. Lacey Company records. Forest History Society.
Laird Norton Company Records. Minnesota Historical Society.
Joseph Lane Papers. Oregon Historical Society.
James S. Lawson Reminiscences. Bancroft Library, University of California, Berkeley.
W. G. Leonard Reminiscences. Washington State University Library.
Log of the Mayflower. New York Public Library.
Loyal Legion of Loggers and Lumbermen Convention Proceedings, 1919. Oregon Historical Society.
Michael F. Luark Diary. University of Washington Library.
Walter Lund Oral History Interview. Regional Oral History Office, University of California, Berkeley.

Richard E. McArdle Oral History Interview. Forest History Society.
John J. McGilvra Papers. University of Washington Library.
Eli B. Mapel Reminiscences. Bancroft Library, University of California, Berkeley.
David T. Mason Papers. Oregon Historical Society.
David T. Mason, "Putting the Brakes on Lumber Production." Washington State University Library (typescript).
Meany Pioneer File. University of Washington Library.
Merrill & Ring Lumber Company Records. University of Washington Library.
J. C. Moe Reminiscences. Oregon Historical Society.
Stuart Moir Papers. Oregon Historical Society.
A. W. Moltke Oral History Interview. Forest History Society.
Eldridge Morse Notebooks. Bancroft Library, University of California, Berkeley.
Mumby Lumber & Shingle Company Records. Simpson Timber company Archives, Seattle.
Thornton T. Munger Oral History Interview. Regional Oral History Office, University of California, Berkeley.
Francis J. Murnane Papers. Oregon Historical Society.
L. T. Murray, Sr., Oral History Interview. Forest History Society.
National Forest Products Association Records. Forest History Society.
Northern Pacific Railroad Records. Minnesota Historical Society.
Notes Copied from the Hudson's Bay Company Account Books at Fort Nisqually. University of Washington Library.
Pacific Northwest Loggers Association Records. University of Washington Library.
F. W. Pettygrove Reminiscences. Bancroft Library, University of California, Berkeley.
Phoenix Logging Company records. Simpson Timber Company Archives, Seattle.
Gifford Pinchot Papers. Library of Congress.
Herbert L. Plumb Papers. University of Washington Library.
George Plummer Correspondence. University of Washington Library.
Andrew J. Pope Papers. Pope & Talbot Archives, Port Gamble.
Pope & Talbot Records. Ames Collection, University of Washington Library.
Pope & Talbot Records. Pope & Talbot Archives, Port Gamble.
Port Blakely Mill Company Records. University of Washington Library.
Port of Honolulu Records. Hawaii State Archives.
Nathan S. Porter Diary. Washington State Historical Society.
Power Family Papers. Montana Historical Society.
August Priester Diary. Washington State Historical Society.
Harold Pritchett Papers. University of British Columbia Library.
Puget Mill Company Records. Ames Collection, University of Washington Library.
Puget Sound Customs District Records. Federal Records Center, Seattle.
Antonio B. Rabbeson Reminiscences. Bancroft Library, University of California, Berkeley.
F. C. Ransom Papers. Oregon Historical Society.
Rayonier Incorporated Records. ITT Rayonier, Hoquiam.
Rayonier Incorporated Records. ITT Rayonier, Shelton.
Rayonier Incorporated Records. ITT Rayonier, Port Angeles.
Mark E. Reed Papers. University of Washington Library.
Mark E. Reed Personal Papers. Simpson Timber Company Archives, Seattle.
William G. Reed Papers. Simpson Timber Company Archives, Seattle.
Reed Mill Company Records. Simpson Timber Company, Shelton.
George B. Roberts Reminiscences. Bancroft Library, University of California, Berkeley.
Henry Roder Reminiscences. Bancroft Library, University of California, Berkeley.
Lottie Roeder Roth Reminiscences. University of Washington Library.

William C. Ruegnitz Papers. University of Washington Library.
Edward Rutledge Timber Company Records. University of Idaho Library.
St. Paul & Tacoma Lumber Company Records. University of Washington Library.
George Savage Reminiscences. University of Washington Library.
W. T. Sayward Reminiscences. Bancroft Library, University of California, Berkeley.
Seattle Chamber of Commerce Annual Report, 1891. University of Washington Library.
Eugene Semple Papers. University of Washington Library.
Simpson Logging Company Records. Simpson Timber Company, Shelton.
Simpson Logging/Timber Company Records. Simpson Timber Company Archives, Seattle.
Simpson Timber Company Records. Simpson Timber Company Archives, Seattle.
Harold Slater Papers. Washington State Historical Society.
Sound Timber Company Records. Minnesota Historical Society.
Soundview Pulp Company Minute Books. Washington State University Library.
Hazard Stevens Papers. University of Oregon Library.
Isaac I. Stevens Papers. University of Washington Library.
Thomas D. Stimson, "A Record of the Family, Life and Activities of Charles Douglas Stimson." University of Washington Library (typescript).
Stimson Mill Company Records. University of Washington Library.
Henry Suzzallo Papers. University of Washington Archives.
James G. Swan Letters. University of Washington Library.
Charles Sweet Diary. Washington State Historical Society.
Tacoma Mill Company Records. Washington State Historical Society.
Taholah Indian Agency Records. Federal Records Center, Seattle.
William C. Talbot Papers. Pope & Talbot Archives, Port Gamble.
F. R. Titcomb Papers. Weyerhaeuser Company Archives, Federal Way.
William F. Tolmie Journal. Bancroft Library, University of California, Berkeley.
William F. Tolmie Papers. University of Washington Library.
William F. Tolmie Reminiscences. Bancroft Library, University of California, Berkeley.
William Petit Trowbridge Journal. Washington State Historical Society.
William F. Vilas Papers. State Historical Society of Wisconsin.
Corydon Wagner Papers. University of Washington Library.
Cyrus Walker Letterbooks. Ames Collection, University of Washington Library.
Washington Mill Company Records. University of Washington Library.
Washington State Biography File. University of Washington Library.
Washington State Conservation Society Records. University of Washington Library.
Washington State Federation of Labor Records. University of Washington Library.
Washington Territorial Papers. Federal Records Center, Seattle (microfilm).
West Coast Lumbermen's Asociation Records. Oregon Historical Society.
Western Forestry & Conservation Association Records. Oregon Historical Society.
J. P. Weyerhaeuser, Jr., Papers. Weyerhaeuser Company Archives, Federal Way.
Weyerhaeuser & Company Records. Minnesota Historical Society.
Weyerhaeuser Family Collection. Minnesota Historical Society.
Weyerhaeuser Timber Company Records. Weyerhaeuser Company Archives, Federal Way.
Arch Whisnant Oral History Interview. Forest History Society.
Hugo Winkenwerder Papers. University of Washington Archives.
E. K. Wood Lumber Company Records. Bancroft Library, University of California, Berkeley.
Henry L. Yesler Papers. University of Washington Library.
Henry L. Yesler Reminiscences. Bancroft Library, University of California, Berkeley.

Theses and Dissertations

Brady, Eugene A. "The Role of Government Land Policy in Shaping the Development of the Lumber Industry in the State of Washington." M.A. thesis, University of Washington, 1954.

Buchanan, Iva L. "An Economic History of Kitsap County, Washington, to 1889." Ph.D. diss., University of Washington, 1930.

Buskirk, Richard H. "A Description and Critical Analysis of the Marketing of Douglas Fir Plywood." D.B.A. diss., University of Washington, 1955.

Cairney, Daniel W. "The Effect of Some Economic Disturbances on Lumber Trade of Washington and British Columbia." M.S. thesis, University of Washington, 1935.

Calhoun, Charles E. "Financing the Pulp and Paper Industry of the Pacific Coast." M.B.A. thesis, University of Washington, 1930.

Clark, Donald H. "An Analysis of Forest Utilization as a Factor in Colonizing the Pacific Northwest and in Subsequent Population Transitions." Ph.D. diss., University of Washington, 1952.

Cory, Floyd W. "The Influence of Canadian Competition on the Pulpwood Industry of the Pacific Coast." M.F. thesis, University of Washington, 1927.

Cotroneo, Ross R. "The History of the Northern Pacific Land Grant, 1900–1952." Ph.D. diss., University of Idaho, 1966.

Cox, John H. "Organizations of the Lumber Industry in the Pacific Northwest, 1889–1914." Ph.D. diss., University of California, Berkeley, 1937.

Cranston, Robert B. "The Forests and Forest Industries of British Columbia." M.F. thesis, University of Washington, 1952.

Cross, Clark I. "Factors Influencing the Abandonment of Lumber Mill Towns in the Puget Sound Region." M.S. thesis, University of Washington, 1946.

Dick, Roger S. "History of Lumbering in Cowlitz County." M.A. thesis, University of Washington, 1941.

Doig, Ivan C. "John J. McGilvra: The Life and Times of an Urban Frontiersman, 1827–1903." Ph.D. diss., University of Washington, 1969.

Finger, John R. "Henry L. Yesler's Seattle Years, 1852–1892." Ph.D. diss., University of Washington, 1968.

Fricke, Edgar F. "Soviet Lumber Exports and Their Effect upon the Export of Lumber from the United States." M.B.A. thesis, University of Washington, 1931.

Gedosch, Thomas F. "Seabeck, 1857–1886: The History of a Company Town." M.A. thesis, University of Washington, 1967.

Hazeltine, Jean. "Timber Conservation in Western Washington." M.A. thesis, Ohio State University, 1952.

Hoover, Roy O. "The Public Land Policy of Washington State: The Initial Period, 1889–1912." Ph.D. diss., Washington State University, 1967.

Kast, Fremont E. "Major Manufacturing Industries in Washington State: Changes in Their Relative Importance and Causes of Changes." D.B.A. diss., University of Washington, 1956.

Lembcke, Jerry L. "The International Woodworkers of America: An Internal Comparative Study of Two Regions." Ph.D. diss., University of Oregon, 1978.

MacDonald, Alexander N. "Seattle's Economic Development, 1880–1910." Ph.D. diss., University of Washington, 1959.

Meany, Edmond S., Jr. "The History of the Lumber Industry in the Pacific Northwest to 1917." Ph.D. diss., Harvard University, 1935.

Melendy, Howard B. "One Hundred Years of the Redwood Lumber Industry, 1850–1950." Ph.D. diss., Stanford University, 1952.

Nichols, Claude W., Jr. "Brotherhood in the Woods: The Loyal Legion of Loggers and Lumbermen, A Twenty Year Attempt at 'Industrial Cooperation.'" Ph.D. diss., University of Oregon, 1959.

Pohl, Thomas W. "Seattle, 1851–1861: A Frontier Community." Ph.D. diss., University of Washington, 1970.

Prouty, Andrew W. "Logging with Steam in the Pacific Northwest: The Men, the Camps, and the Accidents, 1885–1918." M.A. thesis, University of Washington, 1973.

Rakestraw, Lawrence. "A History of Forest Conservation in the Pacific Northwest, 1891–1913." Ph.D. diss., University of Washington, 1955.

Salo, Sarah Jenkins. "Timber Concentration in the Pacific Northwest: with Special Reference to the Timber Holdings of the Northern Pacific Railroad and the Weyerhaeuser Timber Company." Ph.D. diss., Columbia University, 1945.

Smythe, Limen T. "The Lumber and Sawmill Workers' Union in British Columbia." M.A. thesis, University of Washington, 1937.

Steen, Harold K. "Forestry in Washington to 1925." Ph.D. diss., University of Washington, 1969.

Tattersall, James N. "The Economic Development of the Pacific Northwest to 1920." Ph.D. diss., University of Washington, 1960.

Tripp, Joseph F. "Progressive Labor Laws in Washington State (1900–1925)." Ph.D. diss., University of Washington, 1973.

Yonce, Frederick J. "Public Land Disposal in Washington." Ph.D. diss., University of Washington, 1969.

Books and Pamphlets

Adams, Harriet L. *A Woman's Journeyings in the New Northwest*. Cleveland: B-P Printing Co., 1892.

Adams, Kramer. *Logging Railroads of the West*. Seattle: Superior Publishing Company, 1961.

Adams, W. Claude. *History of Papermaking in the Pacific Northwest*. Portland: Binfords & Mort, 1951.

Ahern, George P. *Deforested America*. Washington , D.C.: n.p., 1928.

———. *Forest Bankruptcy in America*. 2nd ed. Washington, D.C.: Shenandoah Publishing House, Inc., 1934.

Aldwell, Thomas T. *Conquering the Last Frontier*. Seattle: Superior Publishing Company, 1950.

Allen, Miss A. J., comp. *Ten Years in Oregon: Travels and Adventures of Doctor E. White and Lady, West of the Rocky Mountains*. Ithaca: Andrus, Gauntlet & Co., 1850.

Allen, E. T. *Practical Forestry in the Pacific Northwest*. Portland: Western Forestry & Conservation Association, 1911.

Allen, James B. *The Company Town in the American West*. Norman: University of Oklahoma Press, 1966.

American Tree Association. *Forestry Almanac, 1933*. Washington, D.C.: American Tree Association, 1933.

Andrews, Ralph W. *Glory Days of Logging*. Seattle: Superior Publishing Company, 1956.

———. *This Was Sawmilling*. Seattle: Superior Publishing Company, 1957.

———. *Timber*. Seattle: Superior Publishing Company, 1968.

Badé, William Frederic. *The Life and Letters of John Muir*. 2 vols. Boston: Houghton Mifflin Company, 1924.

Bagley, Clarence B. *History of Seattle, From the Earliest Settlement to the Present Time*. 3 vols. Chicago: The S. J. Clarke Publishing Company, 1916.

———. *History of King County*. 3 vols. Chicago: The S. J. Clarke Publishing Company, 1929.

Baker, Ray Stannard and William E. Dodd, eds. *The Public Papers of Woodrow Wilson*. 6 vols. New York: Harper & Brothers, 1925–1927.

Bancroft, Frederic, ed. *Speeches, Correspondence and Political Papers of Carl Schurz*. 6 vols. New York: G. P. Putnam's Sons, 1913.

Bancroft, H. H. *History of Washington, Idaho, and Montana, 1845–1889*. San Francisco: The History Company, 1890.

———. *Chronicles of the Builders of the Commonwealth*. 7 vols. San Francisco: The History Company, 1892.

Barker, Burt Brown, ed. *Letters of Dr. John McLoughlin, Written at Fort Vancouver, 1829–1832*. Portland: Binfords & Mort, 1948.

Barney, Daniel R. *The Last Stand: Ralph Nader's Study Group Report on the National Forests*. New York: Grossman Publishers, 1974.

Barrett, John W., ed. *Regional Silviculture of the United States*. New York: Ronald Press, 1962.

Beaglehole, J. C., ed. *The Journals of Captain James Cook on His Voyages of Discovery*. Vol. 3, *The Voyage of the Resolution and Discovery, 1776–1780*. Cambridge, Eng.: Hakluyt Society, 1967.

Beard, James Franklin, ed. *The Letters and Journals of James Fenimore Cooper*. 6 vols. Cambridge, Mass.: Harvard University Press, 1960–1968.

Benson, Henry Kreitzer. *The Pulp and Paper Industry of the Pacific Northwest*. Seattle: University of Washington Engineering School, 1929.

Bernstein, Irving. *Turbulent Years: A History of the American Worker, 1933–1941*. Boston: Houghton Mifflin Company, 1970.

Binns, Archie. *The Laurels Are Cut Down*. New York: Reynal & Hitchcock, 1937.

———. *Mighty Mountain*. Portland: Binfords & Mort, 1940.

———. *The Roaring Land*. New York: Robert M. McBride & Company, 1942.

———. *The Timber Beast*. New York: Charles Scribner's Sons, 1944.

Birkeland, Torger. *Echoes of Puget Sound: Fifty Years of Logging and Steamboating*. Caldwell, Id.: The Caxton Printers, Ltd., 1960.

Boddam-Whetham, J. W. *Western Wanderings*. London: Richard Bentley and Son, 1874.

Brandstrom, Axel J. F. *Analysis of Logging Costs and Operating Methods in the Douglas Fir Region*. Seattle: Charles Lathrop Pack Forestry Foundation, 1933.

Brooks, Paul. *The Pursuit of Wilderness*. Boston: Houghton Mifflin Company, 1971.

Brown, Nelson. *The American Lumber Industry*. New York: John Wiley and Sons, 1923.

Budlong, Caroline Gale. *Memoirs of Pioneer Days in Oregon and Washington Territory*. Eugene: Picture Press Printers, 1949.

Cameron, Jenks. *The Development of Governmental Forest Control in the United States*. Baltimore: The Johns Hopkins Press, 1928.

Campbell, Archibald. *A Voyage Round the World, From 1806 to 1812*. Edinburgh: Archibald Constable and Company, 1816.

Cantwell, Robert. *The Land of Plenty*. New York: Farrar & Rinehart, Inc., 1934.

———. *The Hidden Northwest*. Philadelphia: J. B. Lippincott, 1972.

Catton, Bruce. *Waiting for the Morning Train*. Garden City, N.Y.: Doubleday & Company, Inc., 1972.

Christie, Robert A. *Empire in Wood: A History of the Carpenters' Union*. Ithaca: Cornell University Press, 1956.

Churchill, Sam. *Don't Call Me Ma*. Garden City, N.Y.: Doubleday & Company, Inc., 1977.

Chaplin, Ralph W. *Wobbly: The Rough-and-Tumble Story of an American Radical*. Chicago: University of Chicago Press, 1948.

Clark, Donald H. *18 Men and a Horse*. Seattle: The Metropolitan Press, 1949.

Clark, Norman H. *Mill Town: A Social History of Everett, Washington, from Its Earliest Beginnings on the Shores of Puget Sound to the Tragic and Infamous Event Known as the Everett Massacre*. Seattle: University of Washington Press, 1970.

Clayson, Edward, Sr. *Historical Narratives of Puget Sound: Hoods Canal, 1865–1885*. Seattle: R. L. Davis Printing Co., 1911.

Cohn, Edwin J., Jr. *Industry in the Pacific Northwest and the Location Theory*. New York: Columbia University Press, 1954.

Collins, Septima M. *A Woman's Trip to Alaska*. New York: Cassell Publishing Company, 1890.

Colton, Walter. *Three Years in California*. New York: A. S. Barnes & Co., 1850.

Colvocoresses, George M. *Four Years in a Government Exploring Expedition*. New York: Cornish, Lamport & Co., 1852.

Coman, Edwin T., Jr. and Helen M. Gibbs. *Time, Tide and Timber: A Century of Pope & Talbot*. Stanford: Stanford University Press, 1949.

Compton, Wilson. *The Organization of the Lumber Industry*. Chicago: American Lumberman, 1916.

Corney, Peter. *Early Voyages in the North Pacific, 1813–1818*. 1896, reprint. Fairfield, Wa.: Ye Galleon Press ed., 1965.

Cour, Robert M. *The Plywood Age*. Portland: Binfords & Mort, 1955.

Cowan, Charles S. *The Enemy is Fire!* Seattle: Superior Publishing Company, 1961.

Cox, Thomas R. *Mills and Markets: A History of the Pacific Coast Lumber Industry to 1900*. Seattle: University of Washington Press, 1974.

Cuff, Robert D. *The War Industries Board: Business-Government Relations during World War I*. Baltimore: The Johns Hopkins University Press, 1973.

Dana, Richard Henry, Jr. *Two Years Before the Mast*. New York: Harper & Brothers, 1842.

Dana, Samuel Trask. *Forest and Range Policy: Its Development in the United States*. New York: McGraw-Hill Book Company, Inc., 1956.

Decker, Peter R. *Fortunes and Failures: White-Collar Mobility in Nineteenth-Century San Francisco*. Cambridge, Mass.: Harvard University Press, 1978.

Defebaugh, James E. *History of the Lumber Industry in America*. 2 vols. Chicago: American Lumberman, 1906–1907.

Denny, Arthur A. *Pioneer Days on Puget Sound*. Seattle: C. B. Bagley, Printer, 1888.

Diamond, Sigmund. *The Reputation of the American Businessman*. Cambridge, Mass.: Harvard University Press, 1955.

Douglas, David. *Journal Kept By David Douglas During His Travels in North America, 1823–1827*. London: William Wesley & Son, 1914.

Drake, George L. with Elwood R. Maunder. *A Forester's Log: Fifty Years in the Pacific Northwest*. Santa Cruz: Forest History Society, 1975.

Dubofsky, Melvyn. *We Shall Be All: A History of the Industrial Workers of the World*. Chicago: Quadrangle Books, 1969.

Eckardt, H. W. *Accounting in the Lumber Industry*. New York: Harper & Brothers, 1929.

Ekirch, Arthur A., Jr. *Man and Nature in America*. New York: Columbia University Press, 1963.

Elchibegoff, Ivan M. *United States International Timber Trade in the Pacific Area.* Stanford: Stanford University Press, 1949.

Engstrom, Emil. *The Vanishing Logger.* New York: Vantage Press, 1956.

Erickson, Russell F. *The Story of Rayonier Incorporated.* New York: The Newcomen Society, 1963.

Farnham, Thomas. *Travels in the Great Western Prairies, the Anahuac and Rocky Mountains, and in the Oregon Territory.* New York: Greeley & McElrath, 1843.

Ficken, Robert E. *Lumber and Politics: The Career of Mark E. Reed.* Seattle: University of Washington Press, 1979.

Fickle, James E. *The New South and the 'New Competition': Trade Association Development in the Southern Pine Industry.* Urbana: University of Illinois Press, 1980.

Frémont, John Charles. *Memoirs of My Life.* Chicago: Belford, Clarke & Company, 1887.

Friedheim, Robert L. *The Seattle General Strike.* Seattle: University of Washington Press, 1964.

Fries, Robert F. *Empire in Pine: The Story of Lumbering in Wisconsin, 1830–1900.* Madison: The State Historical Society of Wisconsin, 1951.

Gass, Patrick. *Gass's Journal of the Lewis and Clark Expedition.* Chicago: A. C. McClurg & Co., 1904.

Gates, Charles M., ed. *Messages of the Governors of the Territory of Washington to the Legislative Assembly, 1854–1889.* Seattle: University of Washington Press, 1940.

George, Henry, Jr., ed. *The Complete Works of Henry George.* 10 vols. New York: Doubleday, Page & Company, 1904.

Goodman, David Michael. *A Western Panorama, 1849–1875: The Travels, Writings and Influence of J. Ross Browne.* Glendale, Cal.: The Arthur H. Clarke Company, 1966.

Greeley, William B. *Forests and Men.* Garden City, N.Y.: Doubleday and Company, Inc., 1951.

Gunther, Erna. *Ethnobotany of Western Washington: The Knowledge and Use of Indigenous Plants by Native Americans.* Seattle: University of Washington Press ed., 1973.

Guthrie, John A. *The Newsprint Industry: An Economic Analysis.* Cambridge, Mass.: Harvard University Press, 1941.

———. *The Economics of Pulp and Paper.* Pullman: The State College of Washington Press, 1950.

——— and George R. Armstrong. *Western Forest Industry: An Economic Outlook.* Baltimore: Johns Hopkins University Press, 1961.

Hancock, Samuel. *The Narrative of Samuel Hancock.* New York: Robert M. McBride & Company, 1927.

Hanford, Cornelius H. *Seattle and Environs, 1852–1924.* 3 vols. Chicago: Pioneer Historical Publishing Co., 1924.

Hardwick, Walter G. *Geography of the Forest Industry of Coastal British Columbia.* Vancouver, B.C.: Canadian Association of Geographers, 1963.

Harrison, J. M. *Harrison's Guide and Resources of the Pacific Slope.* 2nd ed. San Francisco: C. A. Murdock & Co., 1876.

Harvey, Athelston George. *Douglas of the Fir.* Cambridge, Mass.: Harvard University Press, 1947.

Hawaiian Almanac and Annual for 1876. Honolulu: J. H. Black, 1876.

——— *1878.* Honolulu: H. L. Sheldon, 1878.

——— *1882.* Honolulu: Thos. G. Thrum, 1882.

——— *1885.* Honolulu: Thos. G. Thrum, 1885.

——— *1887.* Honolulu: Press Publishing Company, 1886.

———— *1890*. Honolulu: Press Publishing Company, 1889.

———— *1892*. Honolulu: Press Publishing Company, 1891.

———— *1901*. Honolulu: Hawaiian Gazette Co., 1900.

———— *1902*. Honolulu: Thos. G. Thrum, 1901.

Hawley, Ellis W. *The New Deal and the Problem of Monopoly*. Princeton: Princeton University Press, 1966.

Hawthorne, Julian, ed. *History of Washington, the Evergreen State*. 2 vols. New York: American Historical Publishing Co., 1893.

Hays, Samuel P. *Conservation and the Gospel of Efficiency: The Progressive Conservation Movement, 1890–1920*. Cambridge, Mass.: Harvard University Press, 1959.

Hazeltine, Jean. *The Historical and Regional Geography of the Willapa Bay Area*. South Bend, Wa.: South Bend Journal, 1956.

Heckman, Hazel. *Island in the Sound*. Seattle: University of Washington Press, 1967.

Hibbard, Benjamin H. *A History of the Public Land Policies*. New York: The Macmillan Company, 1924.

Hidy, Ralph W., Frank Ernest Hill and Allan Nevins. *Timber and Men: The Weyerhaeuser Story*. New York: The Macmillan Company, 1963.

Hill, William Bancroft. *Frederick Weyerhaeuser, Pioneer Lumber Manufacturer*. Minneapolis: The McGill Lithographic Company, 1940.

Himmelberg, Robert F. *The Origins of the National Recovery Administration: Business, Government, and the Trade Association Issue, 1921–1933*. New York: Fordham University Press, 1976.

Hines, Gustavus. *Oregon: Its History, Condition and Prospects*. Buffalo: Geo. H. Derby & Co., 1851.

Hittell, John S. *The Commerce and Industries of the Pacific Coast*. 2nd ed. San Francisco: A. L. Bancroft & Company, 1882.

Holbrook, Stewart H. *Holy Old Mackinaw*. New York: The Macmillan Company, 1938.

————. *Burning an Empire: The Story of American Forest Fires*. New York: The Macmillan Company, 1944.

————. *The Columbia*. New York: Rinehart and Company, 1956.

Horn, Stanley F. *This Fascinating Lumber Business*. Indianapolis: The Bobbs-Merrill Company, 1943.

Horowitz, Morris A. *The Structure and Government of the Carpenters' Union*. New York: John Wiley and Sons, Inc., 1962.

Hotchkiss, George W. *History of the Lumber and Forest Industry of the Northwest*. Chicago: George W. Hotchkiss & Co., 1898.

Howay, F. W., ed. *Voyages of the 'Columbia' to the Northwest Coast, 1787–1790 and 1790–1793*. Boston: Massachusetts Historical Society, 1941.

————, W. N. Sage, and H. F. Angus. *British Columbia and the United States*. Toronto: The Ryerson Press, 1942.

Hunt, Herbert. *Tacoma, Its History and Its Builders*. 3 vols. Chicago: The S. J. Clarke Publishing Company, 1916.

———— and Floyd C. Kaylor. *Washington West of the Cascades*. Chicago: The S. J. Clarke Publishing Company, 2 vols., 1917.

Hunt, Reed O. *Pulp, Paper and Pioneers: The Story of Crown Zellerbach*. New York: The Newcomen Society, 1961.

Hussey, John A. *The History of Fort Vancouver and its Physical Surroundings*. Portland: Washington State Historical Society, 1957.

Hyman, Harold M. *Soldiers and Spruce: Origins of the Loyal Legion of Loggers and Lumbermen*. Los Angeles: Institute of Industrial Relations, University of California, Los Angeles, 1963.

Hynding, Alan. *The Public Life of Eugene Semple: Promoter and Politician of the Pacific Northwest*. Seattle: University of Washington Press, 1973.

Ise, John. *The United States Forest Policy.* New Haven: Yale University Press, 1920.
————. *Our National Park Policy: A Critical History.* Baltimore: Johns Hopkins University Press, 1961.
Ickes, Harold L. *The Secret Diary of Harold L. Ickes,* Vol. 2, *The Inside Struggle, 1936–1939.* New York: Simon and Schuster, 1954.
Jackson, Donald, ed. *Letters of the Lewis and Clark Expedition.* 2nd ed. 2 vols. Urbana: University of Illinois Press, 1978.
Jensen, Vernon H. *Lumber and Labor.* New York: Farrar & Rinehart, Inc., 1945.
Johannsen, Robert L. *Frontier Politics on the Eve of the Civil War.* Seattle: University of Washington Press, 1955.
Jones, Edward G. *The Oregonian's Handbook of the Pacific Northwest.* Portland: The Oregonian Publishing Co., 1894.
Judson, Phoebe. *A Pioneer's Search for an Ideal Home.* Bellingham: Union Printing, Binding and Stationery Company, 1905.
Kinney, J. P. *Indian Forest and Range: A History of the Administration and Conservation of the Redman's Heritage.* Washington, D.C.: Forestry Enterprises, 1950.
Kipling, Rudyard. *From Sea to Sea and Other Sketches.* 2 vols. Garden City, N.Y.: Doubleday, Page & Company, 1925.
Kirkland, Burt P. and Axel J. F. Brandstrom. *Selective Timber Management in the Douglas Fir Region.* Seattle: Charles Lathrop Pack Forestry Foundation, 1936.
Kreienbaum, C. H. with Elwood R. Maunder. *The Development of a Sustained-Yield Industry: The Simpson-Reed Lumber Interests in the Pacific Northwest, 1920s to 1960s.* Santa Cruz: Forest History Society, 1972.
Larson, Agnes M. *History of the White Pine Industry in Minnesota.* Minneapolis: University of Minnesota Press, 1949.
Leighton, Caroline C. *Life at Puget Sound.* Boston: Lee and Shepard, 1884.
LeWarne, Charles P. *Utopias on Puget Sound, 1885–1915.* Seattle: University of Washington Press, 1975.
Lewis, Howard T. and Stephen I. Miller, eds. *The Economic Resources of the Pacific Northwest.* Seattle: Lowman & Hanford Company, 1923.
Lewis, Sinclair. *Free Air.* New York: Harcourt, Brace and Howe, 1919.
Lillard, Richard G. *The Great American Forest.* New York: Alfred A. Knopf, 1947.
Loehr, Rodney C., ed. *Forests for the Future: The Story of Sustained Yield as Told in the Diaries and Papers of David T. Mason, 1907–1950.* St. Paul: Forest Products History Foundation, 1952.
Lombard, B., Jr. and W. A. Lombard. *Report on Oregon and Washington and Idaho Territories.* New York: Lombard Investment Company, 1888.
Lotchin, Roger W. *San Francisco, 1846–1856: From Hamlet to City.* New York: Oxford University Press, 1974.
Lower, A. R. M. *The North American Assault on the Canadian Forest.* Toronto: The Ryerson Press, 1938.
Lucia, Ellis. *Head Rig.* Portland: Overland West Press, 1965.
————. *The Big Woods.* Garden City, N.Y.: Doubleday & Company, Inc., 1975.
Ludington, Townsend, ed. *The Fourteenth Chronicle: Letters and Diaries of John Dos Passos.* Boston: Gambit Incorporated, 1973.
Lyon, Leverett S., et al. *The National Recovery Administration: An Analysis and Appraisal.* Washington, D.C.: The Brookings Institution, 1935.
Mark Twain's Speeches. New York: Harper & Brothers, 1923.
March, George P. *The Earth as Modified by Human Action.* New York: Charles Scribner's Sons, 1907.
Marriott, Elsie Frankland. *Bainbridge through Bifocals.* Seattle: Gateway Publishing Company, 1941.

McKinley, Charles. *Uncle Sam in the Pacific Northwest*. Berkeley: University of California Press, 1952.

McClelland, John M., Jr. *Longview: The Remarkable Beginnings of a Modern Western City*. Portland: Binfords & Mort, 1949.

McCormick's Almanac, 1870. Portland: S. J. McCormick, 1873.

McGeary, M. Nelson. *Gifford Pinchot, Forester-Politician*. Princeton: Princeton University Press, 1960.

Meares, John. *Voyages Made in the Years 1788 and 1789, from China to the North West Coast of America*. 2 vols. London: Logographic Press, 1791.

Meeker, Ezra. *Washington Territory West of the Cascade Mountains*. Olympia: Transcript Office, 1870.

―――. *Pioneer Reminiscences of Puget Sound*. Seattle: Lowman & Hanford, 1905.

Mercer, A. S. *Washington Territory*. Utica, N.Y.: L. C. Childs, 1865.

Merk, Frederick, ed. *Fur Trade and Empire: George Simpson's Journal*. rev. ed. Cambridge, Mass.: Harvard University Press, 1968.

Miles, Charles and O. B. Sperlin, eds. *Building a State: Washington, 1889–1939*. Tacoma: Washington State Historical Society, 1940.

Miller, Wallace J. *South-western Washington*. Olympia: Pacific Publishing Company, 1890.

Morgan, George T., Jr. *William B. Greeley: A Practical Forester, 1879–1955*. St. Paul: Forest History Society, 1961.

Morgan, Murray. *Skid Road*. New York: The Viking Press, 1951.

―――. *The Northwest Corner*. New York: The Viking Press, 1962.

―――. *The Last Wilderness*. Seattle: University of Washington Press ed., 1976.

―――. *The Mill on the Boot: The Story of the St. Paul & Tacoma Lumber Company*. Seattle: University of Washington Press, 1982.

―――. *Puget's Sound: A Narrative of Early Tacoma and the Southern Sound*. Seattle: University of Washington Press, 1979.

Morison, Elting E., ed. *The Letters of Theodore Roosevelt*. 8 vols. Cambridge, Mass.: Harvard University Press, 1951–1954.

Muir, John. *Travels in Alaska*. Boston: Houghton Mifflin Company, 1915.

Nesbit, Robert C. *"He Built Seattle": A Biography of Judge Thomas Burke*. Seattle: University of Washington Press, 1961.

Newcombe, C. F., ed. *Menzies' Journal of Vancouver's Voyage, April to October 1792*. Victoria: William H. Cullin, 1923.

Newell, Gordon, ed. *The H. W. McCurdy Marine History of the Pacific Northwest*. Seattle: Superior Publishing Company, 1966.

Parker, Carleton H. *The Casual Laborer and Other Essays*. Seattle: University of Washington Press ed., 1972.

Parker, Samuel. *Journal of an Exploring Tour Beyond the Rocky Mountains*. 2nd ed. Ithaca: Mack, Andrus, & Woodruff, Printers, 1840.

Pearce, Charles A. *NRA Trade Programs*. New York: Columbia University Press, 1939.

Peattie, Donald Culross. *A Natural History of Western Trees*. Boston: Houghton Mifflin Company, 1953.

Pinchot, Gifford. *The Fight for Conservation*. Seattle: University of Washington Press ed., 1967.

―――. *Breaking New Ground*. Seattle: University of Washington Press ed., 1972.

Pinkett, Harold T. *Gifford Pinchot: Private and Public Forester*. Urbana: University of Illinois Press, 1970.

Potter, Neal and Francis T. Christy, Jr. *Trends in Natural Resource Commodities*. Baltimore: Johns Hopkins University Press, 1962.

Prosch, Charles. *Reminiscences of Washington Territory*. Seattle: n.p., 1904.

Prosser, William F. *A History of the Puget Sound Country.* 2 vols. New York: The Lewis Publishing Company, 1903.

Puter, S. A. D. and Horace Stevens. *Looters of the Public Domain.* Portland: The Portland Printing House, 1908.

Quigg, Lemuel Ely. *New Empires in the Northwest.* New York: The Tribune Association, 1889.

Randall, Roger. *Labor Relations in the Pulp and Paper Industry of the Pacific Northwest.* Portland: Northwest Regional Council, 1942.

Redfield, Edith Sanderson. *Seattle Memories.* Boston: Lothrop, Lee & Shepard Co., 1930.

Reed, William G. with Elwood R. Maunder. *Four Generations of Management: The Simpson-Reed Story.* Santa Cruz: Forest History Society, 1977.

Rich, E. E., ed. *The Letters of John McLoughlin, from Fort Vancouver to the Governor and Committee, First Series, 1825–38.* Toronto: Champlain Society, 1941.

————. *The Letters of John McLouglin, from Fort Vancouver to the Governor and Committee, Second Series, 1839–44.* Toronto: Champlain Society, 1943.

————. *The Letters of John McLoughlin, from Fort Vancouver to the Governor and Committee, Third Series, 1844–46.* Toronto: Champlain Society, 1944.

————. *Part of a Dispatch From George Simpson . . . to the Governor and Committee of the Hudson's Bay Company, London, March 1, 1829.* Toronto: Champlain Society, 1947.

Richardson, Albert D. *Beyond the Mississippi: From the Great River to the Great Ocean.* Hartford, Conn.: American Publishing Company, 1867.

Richardson, Elmo R. *The Politics of Conservation: Crusades and Controversies, 1897–1913.* Berkeley: University of California Press, 1962.

————. *Dams, Parks and Politics: Resource Development and Preservation in the Truman-Eisenhower Era.* Lexington: University Press of Kentucky, 1973.

Ripley, Thomas Emerson. *Green Timber: On the Flood Tide to Fortune in the Great Northwest.* New York: Ballantine Books ed., 1972.

Rowan, James. *The I.W.W. in the Lumber Industry.* Seattle: Lumber Workers Industrial Union No. 500, n.d.

Robbins, Roy. *Our Landed Heritage.* Princeton: Princeton University Press, 1942.

Ruffner, W. H. *A Report on Washington Territory.* New York: Seattle, Lake Shore and Eastern Railway, 1889.

Runte, Alfred J. *National Parks: The American Experience.* Lincoln: University of Nebraska Press, 1979.

Russell, Charles Lord of Killowen. *Diary of a Visit to the United States of America in the Year 1883.* New York: The United States Catholic Historical Society, 1910.

Schmoe, Floyd. *A Year in Paradise.* New York: Harper & Brothers, 1959.

Schwantes, Carlos A. *Radical Heritage: Labor, Socialism, and Reform in Washington and British Columbia.* Seattle: University of Washington Press, 1979.

Seattle Chamber of Commerce. *Seattle, the Puget Sound Country and Western Washington, 1901: Their Resources and Opportunities.* Seattle: The Trade Register, 1901.

Simpson, Sir George. *Narrative of a Journey Round the World, During the Years 1841 and 1842.* 2 vols. London: Henry Colburn, 1847.

Smith, David C. *History of Papermaking in the United States, 1691–1969.* New York: Lockwood Publishing Co., 1970.

Snowden, Clinton A. *History of Washington: The Rise and Progress of an American State.* 4 vols. New York: The Century History Company, 1909.

Spaulding, Kenneth A., ed. *On the Oregon Trail: Robert Stuart's Journey of Discovery.* Norman: University of Oklahoma Press, 1953.

Spencer, Lloyd and Lancaster Pollard. *A History of the State of Washington.* 4 vols. New York: The American Historical Society, 1937.

Steen, Harold K. *The U.S. Forest Service: A History*. Seattle: University of Washington Press, 1976.

Stevens, James. *Paul Bunyan*. Garden City, N.Y.: Garden City Publishing Co., Inc., 1925.

Strahorn, Carrie Adell. *Fifteen Thousand Miles by Stage*. New York: G. P. Putnam's Sons, 1911.

Stuart, Mrs. A. H. H. *Washington Territory: Its Soil, Climate, Productions and General Resources*. Olympia: Washington Standard, 1875.

Swain, Donald C. *Federal Conservation Policy, 1921–1933*. Berkely: University of California Press, 1963.

Swan, James G. *The Northwest Coast, or, Three Years' Residence in Washington Territory*. Seattle: University of Washington Press ed., 1972.

Teale, Edwin Way, ed. *The Wilderness World of John Muir*. Boston: Houghton Mifflin Company, 1954.

Thayer, William M. *Marvels of the New West*. Norwich, Conn.: The Henry Bill Publishing Company, 1887.

Throckmorton, Arthur L. *Oregon Argonauts: Merchant Adventurers on the Western Frontier*. Portland: Oregon Historical Society, 1961.

Thwaites, Reuben Gold, ed. *Ross's Adventures of the First Settlers on the Oregon or Columbia River, 1810–1813*. Cleveland: The Arthur H. Clark Company, 1904.

———. *Original Journals of the Lewis and Clark Expedition, 1804–1806*. 8 vols. New York: Dodd, Mead & Company, 1905.

Tocqueville, Alexis de. *Democracy in America*. New York: George Dearborn & Co., 1838.

Turner, Frederick Jackson. *The Frontier in American History*. New York: Henry Holt and Company, 1950.

———. *The Significance of Sections in American History*. Gloucester, Mass.: Peter Smith, 1959.

Twining, Charles E. *Downriver: Orrin H. Ingram and the Empire Lumber Company*. Madison: The State Historical Society of Wisconsin, 1975.

———. *Phil Weyerhaeuser: Lumberman*. Seattle: University of Washington Press, 1985.

Tyler, Robert L. *Rebels of the Woods: The I.W.W. in the Pacific Northwest*. Eugene: University of Oregon Books, 1967.

Tyrrell, J. B., ed. *David Thompson's Narrative of His Explorations in Western America, 1784–1812*. Toronto: The Champlain Society, 1916.

Vancouver, George. *A Voyage of Discovery to the North Pacific Ocean, and Round the World*. 3 vols. London: G. G. and J. Robinson; and J. Edwards, 1798.

Van Hise, Charles R. *The Conservation of Natural Resources*. New York: The Macmillan Company, 1913.

Victor, Frances Fuller. *All over Oregon and Washington*. San Francisco: John H. Carmany & Co., 1872.

Vinnedge, Robert W. *The Pacific Northwest Lumber Industry and Its Development*. New Haven: Yale University School of Forestry, 1923.

Watt, Roberta Frye. *The Story of Seattle*. Seattle: Lowman & Hanford Company, 1932.

Weeks, Lyman Horace. *A History of Paper-Manufacturing in the United States, 1690–1916*. New York: The Lockwood Trade Journal Company, 1916.

Welsh, William D. *A Brief History of Port Angeles*. Port Angeles: Port Angeles Chamber of Commerce, 1940.

White, Richard. *Land Use, Environment, and Social Change: The Shaping of Island County, Washington*. Seattle: University of Washington Press, 1980.

Williams, Glyndwar, ed. *London Correspondence Inward from Sir George Simpson, 1841–42*. London: The Hudson's Bay Record Society, 1973.

Williams, Guy W. *Logger-Talk: Some Notes on the Jargon of the Pacific Northwest Woods*. Seattle: University of Washington Book Store, 1930.
Winkenwerder, Hugo. *Forestry in the Pacific Northwest*. Washington, D.C.: The American Tree Association, 1928.
Winser, Henry J. *The Pacific Northwest*. New York: n.p., 1882.
Woods, S. D. *Lights and Shadows of Life on the Pacific Coast*. New York: Funk & Wagnalls Company, 1910.
Wright, E. W., ed. *Lewis & Dryden's Marine History of the Pacific Northwest*. New York: Antiquarian Press, Ltd. ed., 1961.

Articles

"A British Report on Washington Territory, 1885." *Pacific Northwest Quarterly* 35(April 1944):147–56.
Adams, Bristow. "The War and the Lumber Industry." *American Forestry* 20(September 1914), 617–25.
———. "Wood on the Wing." *American Forestry* 23(October 1917):583–89.
Adams, Kramer A. "Blue Water Rafting: The Evolution of Ocean Going Log Rafts." *Forest History* 15(July 1971):16–27.
Albason, William. "History of Willapa Harbor Lumbering." *Timberman* 33(June 1932):20, 32.
Allen, E. T. "What Protective Co-operation Did," *American Forestry* 16(November 1910):641–43.
———. "Handling the Fire Peril." *American Forestry* 17(June 1911):329–32.
———. "America's Transition from Old Forests to New." *American Forestry* 29(February 1923):67–71, 106; (March 1923):163–68; (April 1923):235–40; (May 1923):307–11.
Ames, E. G. "Port Gamble, Washington." *Washington Historical Quarterly* 16(January 1925):17–19.
Anderson, Bern, ed. "The Vancouver Expedition: Peter Puget's Journal of the Exploration of Puget Sound, May 7–June 11, 1792." *Pacific Northwest Quarterly* 30(April 1939):177–217.
Appleman, Roy E. "Timber Empire from the Public Domain." *Mississippi Valley Historical Review* 26(September 1939):193–208.
Bahr, C. W. "A Brief Survey of the Provisions of the Lumber and Timber Products Code." *Journal of Forestry* 33(March 1935):217–24.
Beckham, Stephen Dow. "Asa Mead Simpson, Lumberman and Shipbuilder." *Oregon Historical Quarterly* 68(June 1967):259–73.
Bek, William G., trans. "From Bethel, Missouri, to Aurora, Oregon: Letters of William Keil, 1855–1870." *Missouri Historical Review* 48(January 1954):141–53.
Benson, H. K. "Industrial Resources of Washington." *Journal of Geography* 14(May 1916):353–56.
Bergman, H. J. "The Reluctant Dissenter: Governor Hay of Washington and the Conservation Problem." *Pacific Northwest Quarterly* 62(January 1971):27–33.
Berner, Richard C. "The Port Blakely Mill Company, 1876–89." *Pacific Northwest quarterly* 57(October 1966):158–71.
Bien, Morris. "The Public Lands of the United States." *North American Review* 192(September 1910):387–402.

Billington, Ray Allen. "The Origin of the Land Speculator as a Frontier Type." *Agricultural History* 19(October 1945):204–12.

"Bloedel Donovan Lumber Company Mills." *American Lumberman* (19 July 1913), 43–57.

Bloedel, J. H. "Relation of the Panama Canal to Pacific Coast Lumbering." *Pacific Lumber Trade Journal* 17(June 1911):27–29.

Botting, David C., Jr. "Bloody Sunday." *Pacific Northwest Quarterly* 49(October 1958):162–72.

Boyle, L. C. "Lumber Production and Prices." *Lumber World Review* 41(10 September 1921):27–33.

Brown, Nelson C. "Recent Developments in Lumber Distribution." *Journal of Forestry* 22(January 1924):62–64.

Bryant, R. C. "Lumbermen and Our National Development." *American Forestry* 19(December 1913):946–51.

———. "The Lumber Industry." *American Forestry* 19(October 1913):671–76.

———. "Social Welfare and the Lumber Industry." *American Forestry* 19(August 1913):551–55.

———. "The European War and the Lumber Trade." *American Forestry* 20(December 1914):881–86.

———. "The War and the Lumber Industry." *Journal of Forestry* 17(February 1919):125–34.

Buchanan, Iva L. "Lumbering and Logging in the Puget Sound Region in Territorial Days." *Pacific Northwest Quarterly* 27(January 1936):34–53.

"Bunyan in Broadcloth—The House of Weyerhaeuser." *Fortune* 9(April 1934):63–65, 170–82.

Burgess, Sherwood. "Lumbering in Hispanic California." *California Historical Quarterly* 41(September 1962):237–48.

"Business Broadside of 1853." *Washington Historical Quarterly* 20(July 1929):228–32.

Cary, Austin. "How Lumbermen in Following Their Own Interest Have Served the Public." *Journal of Forestry* 15(March 1917):272–73.

Chapman, H. H. "The Profession of Forestry and Professional Ethics." *Journal of Forestry* 21(May 1923):452–57.

———. "Second Conference on the Lumber Code." *Journal of Forestry* 32(March 1934):272–74.

———. "Some Important Trends in Forestry in the United States." *Journal of Forestry* 36(July 1938):653–58.

Clark, Donald H. "The Yacolt Burn, Forest Graveyard." *American Forests* 60(September 1954):18–20.

Clepper, Henry. "The Ten Most Influential Men in American Forestry." *American Forests* 56(May 1950):10–11, 30, 37–39.

———. "Gifford Pinchot and the SAF." *Journal of Forestry* 63(August 1965):590–92.

———. "The Forest Service Backlashed." *Forest History* 11(January 1968):6–15.

Coffman, N. B. "When I Came to Washington Territory." *Washington Historical Quarterly* 26(April 1935):94–106.

Cole, Douglas and Maria Tippett. "Pleasing Diversity and Sublime Desolation: The 18th-Century British Perception of the Northwest Coast." *Pacific Northwest Quarterly* 65(January 1974):1–7.

Compton, Wilson. "Recent Tendencies in the Reform of Forest Taxation." *Journal of Political Economy* 23(December 1915):971–79.

———. "The Present Conditions in the Lumber Industry." *Journal of Forestry* 15(April 1917):387–93.

————. "The Price Problem in the Lumber Industry." *American Economic Review* 7(September 1917):582–97.

————. "Our Future Forest Needs." *Journal of Forestry* 28(February 1930):138–46.

————. "Lumber—an Old Industry—and the New Competition." *Harvard Business Review* 10(January 1932):161–69.

Conlin, Joseph R. "'Old Boy, Did You Get Enough of Pie?': A Social History of Food in Logging Camps." *Journal of Forest History* 23(October 1979):164–85.

Cornwall, George F. "Fifty Years: History of St. Paul & Tacoma Lumber Company." *Timberman* 39(19 May 1938):16–30.

————. "Dealers in Billions: Pacific Lumber Inspection Bureau." *Timberman*, 43(May 1942):18–32.

Cornwall, George M. "The Men in Our Industry." *Timberman* 34(November 1932):8, 41–42, 44.

Cotroneo, Ross R. "Western Land Marketing by the Northern Pacific Railway." *Pacific Historical Review* 27(August 1968):299–320.

Cowan, C. S. "The Proposed Mount Olympus National Park." *Journal of Forestry* 34(August 1936):747–49.

Cox, John H. "Trade Associations in the Lumber Industry of the Pacific Northwest, 1899–1914." *Pacific Northwest Quarterly* 41(October 1950):285–311.

Cox, Thomas R. "Lumber and Ships: The Business Empire of Asa Mead Simpson." *Forest History* 14(July 1970):16–26.

————. "Pacific Log Rafts in Economic Perspective." *Forest History* 15(July 1971):18–19.

Cox, William T. "Recent Forest Fires in Oregon and Washington." *Forestry & Irrigation* 8(November 1902):462–70.

Davis, Minot. "George S. Long, Pioneer Logger and Lumberman." *Timberman* 35(November 1933):14–15.

DeVoto, Bernard. "The West: A Plundered Province." *Harpers Magazine* 169(August 1934):355–63.

DeVries, Wade and E. H. McDaniels. "Forest Taxation in Oregon and Washington." *West Coast Lumberman* 69(April 1942):32–33.

"Diary of an Emigrant of 1845." *Washington Historical Quarterly* 1(April 1907):138–58.

Dickinson, Fred E. "Development of Southern and Western Freights on Lumber." *Southern Lumberman* 185(15 December 1952):229–45.

Doig, Ivan C. "John J. McGilvra and Timber Trespassing: Seeking a Puget Sound Timber Policy, 1861–1865." *Forest History* 13(January 1970):6–17.

Dollar, Robert. "Lumber Trade and the Canal." *American Forestry* 20(July 1914):499–500.

Donovan, J. J. "The Problem of our Logged Off Lands." *American Forestry* 18(July 1912):467–68.

Drake, George L. "The Douglas Fir Logger Looks at Selective Logging." *Journal of Forestry* 34(July 1936):705–7.

————. "The U.S. Forest Service, 1905–1955: An Industry Viewpoint." *Journal of Forestry* 53(February 1955):116–20.

Dumke, Glenn S. "The Real Estate Boom of 1887 in Southern California," *Pacific Historical Review*. 11(December 1942):425–38.

Duncan, Julian S. "The Effect of the N.R.A. Lumber Code on Forest Policy." *Journal of Political Economy* 49(February 1941):91–102.

Dunham, Harold H. "Some Crucial Years of the General Land Office, 1875–1890." *Agricultural History* 11(April 1937):117–41.

Durham, George. "Canoes from Cedar Logs: A Study of Early Types and Designs." *Pacific Northwest Quarterly* 46(April 1955):38–39.

"Edmund Sylvester's Narrative of the Founding of Olympia." *Pacific Northwest Quarterly* 36(October 1945):331–39.

Emerson, George H. "Logging on Grays Harbor." *Timberman* 8(September 1907), 20–21.

Evenson, W. T. "Ocean Log Rafts." *Timberman* 27(July 1926):37–38.

Fairchild, Fred Rogers. "The Taxation of Timber Lands." *Forestry Quarterly* 6(December 1908):383–86.

Fernow, B. E. "Taxation of Woodlands." *Forestry Quarterly* 5(December 1907):373–84.

———. "Why Do Lumber Men Not Apply Forestry?" *American Forestry* 18(October 1912):613–15.

———. "Forestry and the War." *Journal of Forestry* 16(February 1918):149–54.

Ficken, Robert E. "Mark E. Reed: Portrait of a Businessman in Politics." *Journal of Forest History* 20(January 1976):4–19.

———. "The Port Blakely Mill Company, 1888–1903." *Journal of Forest History* 21(October 1977):202–17.

———. "Weyerhaeuser and the Pacific Northwest Timber Industry, 1899–1903." *Pacific Northwest Quarterly* 70(October 1979):146–54.

Finger, John R. "Seattle's First Sawmill, 1853–1869: A Study of Frontier Enterprise." *Forest History* 15(January 1972):24–31.

"Forest Devastation: A National Danger and a Plant to Meet It," *Journal of Forestry,* 17(December 1919), 911–45.

"Fort Langley Correspondence." *British Columbia Historical Quarterly* 1(July 1937):187–94.

Freeman, Otis W. and H. F. Raup. "Industrial Trends in the Pacific Northwest." *Journal of Geography* 43(May 1944):175–84.

Gates, Charles M. "Daniel Bagley and the University of Washington Land Grant, 1861–1868." *Pacific Northwest Quarterly* 52(April 1961):56–67.

Gates, Paul W. 'The Homestead Law in an Incongruous Land System." *American Historical Review* 41(July 1936):652–81.

Gibbs, Helen M. "Pope & Talbot's Tugboat Fleet." *Pacific Northwest Quarterly* 42(October 1951):302–23.

Goodman, R. B. "The Lumber Industry and the Income Tax." *Lumber World Review* 37(25 July 1919):25–27.

Granger, C. M. "The National Forests at War." *American Forests* 49(March 1943):112–15.

Graves, Henry S. "The Protection of Forests from Fire." *American Forestry* 16(September 1910):509–18.

———. "Federal Forestry." *American Forestry* 19(December 1913):909–18.

———. "A National Lumber and Forest Policy." *Journal of Forestry* 17(April 1919):351–63.

———. "Who Is Practicing Forestry?" *Lumber World Review* 36(1 August 1919):123–25.

Greeley, William B. "The Problem of Forest Conservation." *American Forestry* 21(September 1915):928.

———. "Self-Government in Forestry." *Journal of Forestry* 18(February 1920): 103–5.

———. "Economic Aspects of Our Timber Supply." *Journal of Forestry* 20(December 1922):837–47.

———. "The Westward Ho of the Sawmill." *Sunset* 50(June 1923):56–58, 96–99.

———. "Forest Management on Federal Lands." *Journal of Forestry* 23(March 1925):223–35.

————. "How the Pacific Northwest Lumber Industry Can Be Made Prosperous." *West Coast Lumberman* 54(15 June 1928):15–16.

————. "The West Coast Problem of Stabilizing Lumber Production." *Journal of Forestry* 28(February 1930):191–98.

————. "The Outlook for Timber Management by Private Owners." *Journal of Forestry* 31(February 1933):208–14.

————. "Lumber Problems and Opportunities." *Timberman* 39(October 1938):12–13.

————. "Can We Accept the New Rules?" *Timberman* 40(January 1939):10.

————. "Cooperative Forest Management in the Olympics." *Journal of Forestry* 49(September 1951):627–29.

————. "It Pays to Grow Trees." *Pacific Northwest Quarterly* 44(October 1953):152–56.

Griggs, Everett G. "Self-Government in the Lumber Business." *West Coast Lumberman* 50(1 June 1926):35, 50.

Gross, James A. "The Making and Shaping of Unionism in the Pulp and Paper Industry." *Labor History* 5(Spring 1964):183–208.

Hagenstein, W. D. "Trees Grow: The Forest Economy of the Douglas Fir Region." *American Forests* 60(April 1954):31–37.

Hall, R. C. "The Federal Income Tax and Forestry." *Journal of Forestry* 21(October 1923):553–62.

Hamilton, Lawrence S. "The Federal Forest Regulation Issue: A Recapitulation." *Forest History* 9(April 1965):2–11.

Harrison, J. D. B. "American Forestry in World Perspective." *Journal of Forestry* 49(March 1951):172–76.

Highsmith, R. M., Jr. "Resources and the Regional Economy." *Pacific Northwest Quarterly* 46(January 1955):25–29.

"History of the Robertson Log Raft." *Timberman* 30(January 1929):37–40.

Holbrook, Stewart H. "Ghost Towns Still Walk." *American Forests* 43(May 1937):216–17, 242, 258.

————. "Timber Ships." *American Forests* 43(November 1937):529–31, 562.

————. "Sawdust on the Wind." *American Heritage* 4(Summer 1953):48–53.

————. "Greeley Went West." *American Forests* 64(March 1958):16–23, 52–62.

————. "Daylight in the Swamp." *American Heritage* 9(October 1958):10–19, 77–80.

————. "The First Fifty Years." *Loggers Handbook* 19(1959):7, 11–26.

Hoover, Roy O. "Public Law 273 Comes to Shelton: Implementing the Sustained-Yield Forest Management Act of 1944." *Journal of Forest History* 22(April 1978):86–101.

Hosmer, Ralph S. "Some Recollections of Gifford Pinchot, 1898–1904." *Journal of Forestry* 43(August 1945):558–62.

Hough, Emerson. "The Slaughter of the Trees." *Everybody's Magazine* 18(May 1908):579–92.

Howay, F. W. "Early Days of the Maritime Fur-Trade on the Northwest Coast." *Canadian Historical Review* 4(March 1923):26–44.

————. "Brig Owhyee in the Columbia, 1827." *Oregon Historical Quarterly* 34(December 1933):324–29.

————. "Early Shipping in Burrard Inlet, 1863–1870," *British Columbia Historical Quarterly*. 1(January 1937):1–20.

Howd, Cloice R. "Development of Lumber Industry of West Coast." *Timberman* 25(August 1924):194–98.

Ingersoll, Ernest. "From the Fraser to the Columbia—Part II," *Harper's New Monthly Magazine* 68(May 1884):869–82.

Karlin, Jules Alexander. "The Anti-Chinese Outbreak in Tacoma, 1885." *Pacific Historical Review* 23(August 1954):271–83.

Kellogg, R. S. "The Rise in Lumber Prices." *Forestry & Irrigation* 12(February 1906):68–70.

——. "What Forest Conservation Means." *Conservation* 15(May 1909):283–86.

Kinney, J. P. "Forestry on Indian Reservations." *Forestry Quarterly* 10(September 1912):471-72.

——. "Forestry Administration on Indian Reservations." *Journal of Forestry* 19(December 1921):836–43.

——. "The Administration of Indian Forests." *Journal of Forestry* 28(December 1930):1041–52.

Kirkland, Burt P. "Continuous Forest Production of Privately Owned Timberlands as a Solution of the Economic Difficulties of the Lumber Industry." *Journal of Forestry* 15(January 1917):15–64.

——. "Continuous Forest Production in the Pacific Northwest." *Commonwealth Review* 3(July 1918):63–78.

——. "Nation-Wide Solution of Forest Production Problems of the United States." *Journal of Forestry* 28(April 1930):430–35.

Knapp, F. C. "Development of Lumbering in the Pacific Northwest." *American Lumberman* (2 February 1924), 47.

Kreienbaum, C. H. with Elwood R. Maunder. "Forest Management and Community Stability: The Simpson Experience." *Journal of Forest History* 12(July 1968):6–19.

Lamb, W. Kaye. "Early Lumbering on Vancouver Island, 1844–1866." *British Columbia Historical Quarterly* 2(January 1938):31–53; (April 1938):95–121.

——. "The Founding of Fort Victoria." *British Columbia Historical Quarterly* 7(April 1943):71-92.

Langille, H. D. "Canadian Lumber Competition." *American Forestry* 21(February 1915):130–39.

Leader, Herman, ed. "A Voyage from the Columbia to California in 1840: From the Journal of Sir James Douglas." *California Historical Society Quarterly* 8(June 1929):97–115.

Levin, Oscar R. "The South Olympic Tree Farm." *Journal of Forestry* 52(April 1954):243–49.

Loehr, Rodney C. "Saving the Kerf: The Introduction of the Band Saw Mill." *Agricultural History* 23(July 1949):168–72.

Lovejoy, P. S. "Review of Lumber Industry Affairs." *Journal of Forestry* 17(March 1919):245–59.

McCabe, James O. "Arbitration and the Oregon Question." *Canadian Historical Review* 41(December 1960):308–27.

McClelland, John M., Jr. "Terror on Tower Avenue." *Pacific Northwest Quarterly* 57(April 1966):65–72.

McKean, Gertrude L. "Tacoma, Lumber Metropolis." *Economic Geography* 17(July 1941):311–20.

M.A.R. "An Autumn Ramble in Washington Territory." *Overland Monthly* 7(January 1886):41–45.

Marshall, Robert. "A Proposed Remedy for Our Forestry Illness." *Journal of Forestry* 28(March 1930):273–80.

Martinson, Arthur D. "Mount Rainier National Park: First Years." *Forest History* 10(October 1966):26–33.

"Marvelous Growth and Development of City of Smokestacks." *American Lumberman* (5 March 1910):56–57.

Maunder, Elwood R., ed. "Memoirs of a Forester: An Excerpt from Oral History Interviews with David T. Mason." *Forest History* 10(January 1967):6–12, 29–35.

Mead, Walter J. "The Forest Products Economy of the Pacific Northwest." *Land Economics* 32(May 1956):127–29.

————. "Changing Pattern of cycles in Lumber Production." *Journal of Forestry* 59(November 1961):808–13.

Meany, Edmond S., ed. "New Log of the Columbia by John Boit." *Washington Historical Quarterly* 12(January 1921):3–50.

Melendy, H. Brett. "Two Men and a Mill: John Dolbeer, William Carson, and the Redwood Lumber Industry in California." *California Historical Society Quarterly* 38(March 1959):59–71.

Merz, Charles. "Tying Up Western Lumber." *New Republic* 12(29 September 1917):242–44.

Mitchell, John G. "Best of the S.O.B.s." *Audubon* 76(September 1974):49–62.

Mittelman, Edward B. "The Loyal Legion of Loggers and Lumbermen: An Experiment in Industrial Relations." *Journal of Political Economy* 31(June 1923):313–41.

Morgan, George T., Jr. "Conflagration as Catalyst: Western Lumbermen and American Forest Policy," *Pacific Historical Review*, 47(May 1978), 167–87.

Muir, John. "The American Forests." *Atlantic Monthly* 80(August 1897):145–57.

————. "The Wild Parks and Forest Reservations of the West." *Atlantic Monthly* 81(January 1898):15–28.

Munger, Thornton T. "The Pulpwood Forests of Oregon and Washington." *Pacific Pulp and Paper Industry* 8(September 1934):20–23.

————. "A Look at Selective Cutting in Douglas Fir." *Journal of Forestry* 48(February 1950):97–99.

Murphy, J. "Summer Ramblings in Washington Territory." *Appleton's Journal* 3(November 1877):385–95.

Nash, Roderick. "The American Wilderness in Historical Perspective." *Forest History* 6(Winter 1963):2–13.

Nelson, Milton O. "The Lumber Industry of America." *Review of Reviews* 36(November 1907):561–75.

Neuberger, Richard L. "How Much Conservation?" *Saturday Evening Post* 212(15 June 1940):12–13, 89–96.

Norcross, Charles P. "Weyerhaeuser—Richer than John D. Rockefeller." *Cosmopolitan* 42(January 1907):252–59.

Olmsted, Frederick E. "The Year's Accomplishments." *Journal of Forestry* 18(February 1920):93–97.

————. "Professional Ethics." *Journal of Forestry* 20(February 1922):106–12.

Peters, William Harrison. "In the Forests of the Olympics." *American Forests* 42(April 1936):170–76.

Peterson, Charles E. "Pioneer Prefabs in Honolulu." *Hawaiian Journal of History* 5(1971):24–38.

Pinchot, Gifford. "The Immediate Future in Forest Work." *Forestry & Irrigation* 8(January 1902):18–21.

————. "The Forester and the Lumberman." *Forestry & Irrigation* 9(April 1903):176–78.

————. "What the Forest Service Stands For." *Forestry & Irrigation* 13(January 1907):26–29.

————. "Where We Stand." *Journal of Forestry* 18(May 1920):441–47.

————. "The Economic Significance of Forestry." *North American Review* 213(February 1921):157–67.

————. "How Conservation Began in the United States." *Agricultural History* 11(October 1937):255–65.

Polenberg, Richard. "Conservation and Reorganization: The Forest Service Lobby, 1937–1938." *Agricultural History* 39(October 1965):230–39.

————. "The Great Conservation Contest." *Forest History* 10(January 1967):13–23.

Rakestraw, Lawrence. "Uncle Sam's Forest Reserves." *Pacific Northwest Quarterly* 44(October 1953):145–51.

———. "The West, State's Rights, and Conservation: A Study of Six Public Land Conferences." *Pacific Northwest Quarterly* 48(July 1957):89-99.

———. "Before McNary: The Northwestern Conservationist, 1889–1913." *Pacific Northwest Quarterly* 51(April 1960):49–56.

Recknagel, A. B. "Sustained Yield for a Permanent and Sufficient Supply of Forest Products." *Journal of Forestry* 28(December 1930):1053–56.

Reed, Franklin W. "The United States Timber Conservation Board: Its Origin and Organization; Its Purpose and Progress." *Journal of Forestry* 29(December 1931): 1202–5.

———. "Conference on Article X." *Journal of Forestry* 31(December 1933):891-96.

Rhodes, J. E. "Lumbermen and Forestry." *American Forestry* 19(May 1913):318–20.

Richardson, Elmo R. "Olympic National Park: Twenty Years of Controversy." *Forest History* 12(April 1968):6–15.

———. "Was There Politics in the Civilian Conservation Corps?" *Forest History* 16(July 1972):12–21.

Robbins, Roy M. "The Federal Land System in an Embryo State." *Pacific Historical Review* 4(December 1935):356–75.

———. "The Public Domain in the Era of Exploitation, 1862–1901." *Agricultural History* 13(April 1939):97–108.

Rollins, Montgomery, "Pitfalls of Timber Bond Issues." *Forestry Quarterly* 12(December 1914):548–58.

Roloff, Clifford E. "The Mount Olympus National Monument." *Washington Histrocial Quarterly* 25(July 1934):214–28.

Russell, Charles Edward. "The Mysterious Octopus: Story of the Strange and Powerful Organization That Controls the American Lumber Trade." *World Today* 21(February 1912):1735–50.

Sargent, Charles S. "The Protection of the Forests." *North American Review* 135(October 1882):386–401.

Scammon, C. M. "Lumbering in Washington Territory." *Overland Monthly* 5(July 1870):55–60.

Seeman, Albert L. "Economic Adjustments on the Olympic Peninsula." *Economic Geography* 8(July 1932):299–310.

Seymore, W. B. "Port Orchard Fifty Years Ago." *Washington Historical Quarterly* 8(October 1917):257–60.

Sharp, Paul F. "The War of the Substitutes: The Reaction of the Forest Industries to the Competition of Wood Substitutes." *Agricultural History* 23(October 1949):274–49.

Shaw, S. B. "Modifications in Forests of the Pacific Slope Due to Human Agencies." *Journal of Forestry* 24(May 1926):500–6.

Shepard, Ward. "Cooperative Control: A Proposed Solution of the Forest Problem." *Journal of Forestry* 28(February 1930):116–88.

Silcox, F. A. "Foresters Must Choose." *Journal of Forestry* 33(March 1935):198–204.

———. "Forestry—A Public and Private Responsibility." *Journal of Forestry* 33(May 1935):460–68.

———. "Our Adventure in Conservation." *Atlantic Monthly* 160(December 1937):714–22.

Smalley, Brian H. "Some Aspects of the Maine to San Francisco Trade, 1849–1850." *Journal of the West* 5(October 1967):593–603.

Smith, Herbert A. "The Early Forestry Movement in the United States." *Agricultural History* 12(October 1938):326–46.

Stevens, Carl M. "The Forest Industries and the Income Tax." *Journal of Forestry* 18(April 1920):329–37.

Stevens, James. "Sawmills I Have Loved." *West Coast Lumberman* 48(1 May 1925):122–26.

Stuart, R. Y. "National Forest Timber and the West Coast Lumber Industry." *Journal of Forestry* 31(January 1933):45–50.

Swain, Donald C. "The National Park Service and the New Deal, 1933–1940." *Pacific Historical Review* 41(August 1972):312–32.

Thorne, Marco G., ed. "Bound for the Land of Canaan Ho!: The Diary of Levi Stowell." *California Historical Society Quarterly* 27(March 1948):33–50.

Throckmorton, Arthur L. "George Abernethy, Pioneer Merchant." *Pacific Northwest Quarterly* 48(July 1957):76–88.

Toumey, J. W. "Recent Programs and Trends in Forestry in the United States." *Journal of Forestry* 23(January 1925):1–9.

Trimble, William J. "The Influence of the Passing of the Public Lands." *Atlantic Monthly* 113(June 1914):755–67.

Tugwell, Rexford G. "The Casual of the Woods." *Survey* 44(3 July 1920):472-74.

"Two Railroad Reports on Northwest Resources." *Pacific Northwest Quarterly* 37(July 1946):175–91.

Tyler, Robert L. "The Everett Free Speech Fight." *Pacific Historical Review* 23(February 1954):19–30.

———. "I.W.W. in the Pacific N.W.: Rebels of the Woods." *Oregon Historical Quarterly* 55(March 1954):3–44.

———. "The Rise and Fall of an American Radicalism: The I.W.W." *Historian* 19(November 1956):48–65.

Vassault, F. I. "Limbering in Washington." *Overland Monthly* 20(July 1882):23–32.

Vastokas, Joan M. "Architecture and Environment: The Importance of the Forest to the Northwest Coast Indians." *Forest History* 13(Octoer 1969):12–21.

Weiss, Howard F. "Forest Products Investigations." *Journal of Forestry* 23(July–August 1925):565–73.

Wentworth, Lloyd J. "Where Pacific Coast Export Business Has Gone." *West Coast Lumberman* 63(May 1936):38, 48.

White, Richard. "Indian Land Use and Environmental Change: Island County, Washington, A Case Study." *Arizona and the West* 17(Winter 1975):327–38.

———. "Poor Men on Poor Lands: The Back-to-the Land Movement in the Early Twentieth Century—A Case Study." *Pacific Historical Review* 49(February 1980):105–31.

Williams, Asa. "Logging By Steam." *Forestry Quarterly* 6(March 1908):1–31.

Williams, William J. "Bloody Sunday Revisited." *Pacific Northwest Quarterly* 71(April 1980):50–62.

Winkenwerder, Hugo. "The Forests of Washington." *Journal of Geography* 14(May 1916):332–36.

Wollenberg, R. P. and E. N. Cooper. "Labor in the Pacific Coast Paper Industry: A Case in Collective Bargaining." *Harvard Business Review* 16(Spring 1938):366–72.

Woods, John B. "The Lumber Code Situation." *Journal of Forestry* 33(June 1935):579–81.

———. "The Post-Code Status of Conservation." *Journal of Forestry* 33(August 1935):710–12.

———. "Status of the Article X Joint Conservation Program." *Journal of Forestry* 33(December 1935):958–63.

———. and Franklin Reed. "The National Problem of Forest Ownership." *Journal of Forestry* 33(September 1935):789–91.

Bibliography

X. "An Answer to Dr. Compton's Fourteen Points." *Journal of Forestry* 17(December 1919):946–54.
Yonce, Frederick J. "The Public Land Surveys in Washington." *Pacific Northwest Quarterly* 63(October 1972):129–41.
———. "Lumbering and the Public Timberlands in Washington: The Era of Disposal." *Journal of Forests History* 22(January 1978):4–17.
Zieger, Robert H. "The Limits of Militancy: Organizing Paperworkers, 1933–1935." *Journal of American History* 63(December 1976):638–57.

Government Publications

Andrews, H. J. and Robert W. Cowlin. *Forest Resources of the Douglas Fir Region*. PNFES Research Note No. 13. Portland: Pacific Northwest Forest Experiment Station, 1934.
Annual Report of the Commissioner of the General Land Office, 1875. Washington, D.C.: General Printing Office, 1875.
Bolsinger, Charles L. *Changes in Commercial Forest Area in Oregon and Washington, 1945–1970*. USFS Resource Bull. PNW-46. Portland: Pacific Northwest Forest and Range Experiment Station, 1973.
Bureau of Corporations. *The Lumber Industry*. 3 vols. Washington, D.C.: General Printing Office, 1913–14.
Bureau of the Census. *Manufactures, 1905, Part 2, States and Territories*. Washington, D.C.: General Printing Office, 1907.
———. *Manufactures, 1905, Part 3, Special Reports on Selected Industries*. Washington, D.C.: General Printing Office, 1908.
Bureau of Labor Eleventh Biennial Report, 1917–1918. Olympia: State Printer, 1918.
Division of Forestry Annual Reports, 1928 to 1932. Olympia: State Printer, 1933.
Eells, Myron. "The Twana, Chemakum, and Klallam Indians, of Washington Territory." *Annual Report of the Board of Regents of the Smithsonian Institution*. Washington, D.C.: General Printing Office, 1889.
Eighth Biennial Report of the Bureau of Labor Statistics and Factory Inspection, 1911–1912. Olympia: State Printer, 1912.
Eleventh Biennial Report of the Commissioner of Public Lands, 1911. Olympia: State Printer, 1911.
Eleventh Census of the United States, 1890. Washington, D.C.: General Printing Office, 1895.
Fairchild, Fred Rogers. *Forest Taxation in the United States*. USDA Misc. Pub. No. 218. Washington, D.C.: General Printing Office, 1935.
Fifteenth Census of the United States, Manufactures: 1929, Vol. 2. *Reports by Industries*. Washington, D.C.: General Printing Office, 1933.
Fifth Biennial Report of the Bureau of Labor Statistics and Factory Inspection, 1905–1906. Olympia: State Printer, 1907.
Fifth Report of the Department of Labor and Industries, 1932–1936. Olympia: State Printer, 1937.
First Annual Report of the Industrial Insurance Department, 1912. Olympia: State Printer, 1912.
First Biennial Report of the Commissioner of Public Lands, 1890. Olympia: State Printer, 1890.
First Report of the Secretary of State, 1890. Olympia: State Printer, 1891.

Fourteenth Census of the United States, 1920, Vol. 9, *Manufactures, 1919.* Washington, D.C.: General Printing Office, 1923.

Fourth Report of the Department of Labor and Industries, 1932. Olympia: State Printer, 1933.

Gates, Paul W. *History of Public Land Law Development.* Washington, D.C.: General Printing Office, 1968.

Garrett, Paul Willard. *Government Control Over Prices.* Washington, D.C.: General Printing Office, 1920.

Gibbs, George. *Tribes of Western Washington and Northern Oregon.* Washington, D.C.: General Printing Office, 1877.

Hair, Dwight. *Historical Forestry Statistics of the United States.* USFS Stat. Bull. No. 228. Washington, D.C.: General Printing Office, 1958.

Howd, Cloice R. *Industrial Relations in the West Coast Lumber Industry.* BLS Bull. No. 349. Washington, D.C.: General Printing Office, 1924.

Johnson, Herman M., comp. *Production of Lumber, Lath, and Shingles in Washington and Oregon, 1869–1936.* Forest Research Notes No. 24. Portland: Pacific Northwest Forest and Range Experiment Station, 1938.

Mattoon, Wilbur R. *Forest Trees and Forest Regions of the United States.* USDA Misc. Pub. 217. Washington, D.C.: General Printing Office, 1936.

Measurement of the Social Performance of Business. TNEC Monograph No. 7. Washington, D.C.: General Printing Office, 1940.

Message of Chas. E. Laughton, Acting Governor, to the Legislature of 1891. Olympia: State Printer, 1891.

Message of the Governor of Washington Territory. Olympia: Public Printer, 1857.

Mitchell, Wesley C. *History of Prices During the War.* Washington, D.C.: General Printing Office, 1919.

National Resources Planning Board. *Puget Sound Region: War and Post-War Development, May 1943.* Washington, D.C.: General Printing Office, 1943.

Nelson, Saul. *Minimum Price Regulation under Codes of Fair Competition,* NRA Work Materials No. 56, 1936.

Ninth Biennial Report of the Bureau of Labor Statistics and Factory Inspection, 1913–1914. Olympia: State Printer, 1914.

Ninth Census. Washington, D.C.: General Printing Office, 1872.

Nixon, Edgar B., ed. *Franklin D. Roosevelt and Conservation, 1911-1945.* 2 vols. Hyde Park, N.Y.: Franklin D. Roosevelt Library, 1957.

Pratt, Edward E. *The Export Lumber Trade of the United States.* USDC Misc. Series No. 67. Washington, D.C.: General Printing Office, 1918.

Proceedings of a Conference of Governors, May 13–15, 1908. Washington, D.C.: General Printing Office, 1909.

Report of the Commissioner of the General Land Office, 1867. Washington, D.C.: General Printing Office, 1867.

Report of the Commissioner of the General Land Office, 1868. Washington, D.C.: General Printing Office, 1868.

Report of the National Conservation Commission, February 1909. Washington, D.C.: General Printing Office, 3 vols., 1909.

Report of the State Forester to the State Board of Forest Commissioners, 1914. Olympia: State Printer, 1915.

Ruderman, Florence K. *Production, Prices, Employment, and Trade in Northwest Forest Industry.* Portland: Pacific Northwest Forest and Range Experiment Station, 1977.

Second Biennial Report of the Washington State Planning Council. Olympia: State Printer, 1936.

Seventh Biennial Report of the Department of Conservation and Development, 1934. Olympia: State Printer, 1935.

Seventh Report of the Department of Labor and Industries, 1938–1939. Olympia: State Printer, 1940.

Sixth Report of the Department of Labor and Industries, 1937. Olympia: State Printer, 1938.

Sixteenth Census of the United States: 1940, Manufactures, 1939. Washington, D.C.: General Printing Office, 1942.

Sixth Biennial Report of the Bureau of Labor Statistics and Factory Inspection, 1907–1908. Olympia: State Printer, 1908.

Statistical Handbook of Washington's Forest Industry. Olympia: Department of Commerce and Economic Development, 1964.

Steer, Henry B., comp. *Lumber Production in the United States, 1799–1946.* USDA Misc. Pub. No. 669. Washington, D.C.: General Printing Office, 1948.

Stone, Peter A., et al. *Minimum Price Regulation under Codes of Fair Competition,* NRA Work Materials No. 56, 1936.

Swan, James G. "The Indians of Cape Flattery." *Smithsonian Contributions to Knowledge.* Washington, D.C.: Smithsonian Institution, 1870.

Tenth Biennial Report of the Bureau of Labor Statistics and Factory Inspection, 1915–1916. Olympia: State Printer, 1916.

Tenth Census of the United States. Vol. 9. *Report on the Forests of North America.* Washington, D.C.: General Printing Office, 1884.

Thirteenth Census of the United States, 1910, Abstract of the Census with Supplement for Washington. Washington, D.C.: General Printing Office, 1913.

Twelfth Census of the United States, 1900. Vol. 9, Part 3. Washington, D.C.: General Printing Office, 1902.

U.S. Bureau of Statistics. *Commercial Australia in 1900.* Washington, D.C.: General Printing Office, 1901.

Washington Pioneer Project. *Told By the Pioneers: Reminiscences of Pioneer Life in Washington.* 3 vols. Olympia: W.P.A., 1937–39.

Washington State Division of Forestry. *Forest Resources of Washington.* Olympia: State Printer, 1904.

Wilcox, Clair. *Competition and Monopoly in American Industry.* TNEC Monograph No. 21. Washington, D.C.: General Printing Office, 1941.

Index

Aberdeen, Wash.: growth as lumbering center, 58, 105; lumber strike at, 212
Abernethy, George, 19, 20, 25
Adams, William J., 36
Adams, Blinn & Co., 28. *See also* Washington Mill Co.
Admiralty Logging Co., 143
Alameda, Cal., 53
Alaska, 87–88
Alger, Horatio, 119
Alger, Russell, 100
Alki Point, 19, 28, 30
Allen, E. T.: on lumbermen as conservationists, 118; and practical forestry, 129; on Timber Conservation Board, 185; on FDR administration, 216
American Federation of Labor: early organizing efforts of, 133–34, 137; and IWW, 135; and workmen's compensation, 136; WWI strike of, 138, 141; use of employer concessions by, 139–40; and Disque, 146, 150, 151; WWI organization drive of, 150–51; post-WWI activities of, 155, 162; New Deal organizing efforts of, 197, 203, 204; and 4L, 205; and 1935 lumber strike, 209–10; and craft-industrial union dispute, 210, 213; and rivalry with IWA, 214
American Forestry Association, 112, 126
American Lumberman: on Wall Street crash, 182; on need for change in industry, 192; on National Industrial Recovery Act passage, 195
American River, 21
Ames, Edwin G.: and recollections of Walker, 30, 43; on mill towns, 32; on Puget Mill Co. land purchases, 51; on timber fraud, 52; on undermining of combinations, 66; arrives at Port Gamble, 69; as protege of Walker, 69, 116–17, 242n44; personality of, 69; on Knights of Labor, 74; on David Skinner, 101, 115; on San Francisco earthquake, 104; on pricefixing, 109; and lumber inspection, 110; on Panama Canal tolls controversy, 113; on need for a

"Moses," 114; on eight-hour day, 114; on impact of WWI, 117, 139; on state educational system, 122; on T. Roosevelt and federal regulation, 131; on working conditions, 131–32; on value of improved living conditions, 132; on unmanliness of workers, 133; on IWW, 134, 136–37; and workmen's compensation, 136; on AFL organizers, 140, 150; on Everett Massacre, 140; on draft exemptions, 140; and 1917–18 strike, 141, 142, 143, 144, 146, 148–49; on SPD, 150; on postwar reconstruction, 153; on Seattle General Strike, 154; and Bolshevik Revolution, 154; on postwar IWW, 154–55; on income tax, 157; on Mason, 158; on determination of timberland values, 159; on 1920 election, 160; on 1920s market conditions, 160, 161; on importance of 4L, 162; and effort to abandon eight-hour day, 162–63; on Snell Bill, 165; on Long-Bell Lumber Co., 168, 169, 173; on Longview, 169; and sale of Puget Mill Co., 169; on falling 1924 prices, 174; on troubles of Charles R. McCormick Lumber Co., 175
Anderson, A. H., 83, 250n33
Andrews, W. T., 158–59
Article X, 196–97, 202–3. *See also* Lumber Code
Astor, John Jacob, 11
Astoria, Or.: establishment of, 11; logging at, 11–12; cession of, 12; manufacturing at, 20, 25
Australia: lumber shipments to, 24, 64–65, 88; economic conditions in, 35, 64; duration of voyages to, 35; return cargoes from, 35, 64

Bagley, Daniel: and UW land sales, 41–42, 52
Bainbridge Island, 28
Ballard, Wash., 60, 67, 133
Ballinger, Richard, 130
Baltimore, Md., 161
Bank Holiday, 193–94
Bell, F. S., 188, 194, 270n20